State-of-the-Art Liquid Crystals Research in UK

State-of-the-Art Liquid Crystals Research in UK

Editor

Ingo Dierking

MDPI • Basel • Beijing • Wuhan • Barcelona • Belgrade • Manchester • Tokyo • Cluj • Tianjin

Editor
Ingo Dierking
University of Manchester
UK

Editorial Office
MDPI
St. Alban-Anlage 66
4052 Basel, Switzerland

This is a reprint of articles from the Special Issue published online in the open access journal *Crystals* (ISSN 2073-4352) (available at: https://www.mdpi.com/journal/crystals/special_issues/state-of-the-art-LC).

For citation purposes, cite each article independently as indicated on the article page online and as indicated below:

LastName, A.A.; LastName, B.B.; LastName, C.C. Article Title. *Journal Name* **Year**, *Volume Number*, Page Range.

ISBN 978-3-0365-6111-0 (Hbk)
ISBN 978-3-0365-6112-7 (PDF)

Cover image courtesy of Dr. Ingo Dierking

© 2022 by the authors. Articles in this book are Open Access and distributed under the Creative Commons Attribution (CC BY) license, which allows users to download, copy and build upon published articles, as long as the author and publisher are properly credited, which ensures maximum dissemination and a wider impact of our publications.
The book as a whole is distributed by MDPI under the terms and conditions of the Creative Commons license CC BY-NC-ND.

Contents

About the Editor vii

Preface to "State-of-the-Art Liquid Crystals Research in UK" ix

Ingo Dierking
State-of-the-Art Liquid Crystals Research in UK
Reprinted from: Crystals **2022**, 12, 1725, doi:10.3390/cryst12121725 1

John W. Goodby and Stephen J. Cowling
Conception, Discovery, Invention, Serendipity and Consortia: Cyanobiphenyls and Beyond
Reprinted from: Crystals **2022**, 12, 825, doi:10.3390/cryst12060825 9

David Dunmur
Anatomy of a Discovery: The Twist–Bend Nematic Phase
Reprinted from: Crystals **2022**, 12, 309, doi:10.3390/cryst12030309 41

John E. Lydon
Reappraisal of The Optical Textures of Columnar Phases in Terms of Developable Domain Structures with Relaxed Constraints and a Rationale for The Striated Texture
Reprinted from: Crystals **2022**, 12, 1180, doi:10.3390/cryst12081180 55

Corrie T. Imrie, Rebecca Walker, John M. D. Storey, Ewa Gorecka and Damian Pociecha
Liquid Crystal Dimers and Smectic Phases from the Intercalated to the Twist-Bend
Reprinted from: Crystals **2022**, 12, 1245, doi:10.3390/cryst12091245 71

Bartlomiej Czubak, Nicholas J. Warren and Mamatha Nagaraj
Steroid-Based Liquid Crystalline Polymers: Responsive and Biocompatible Materials of the Future
Reprinted from: Crystals **2022**, 12, 1000, doi:10.3390/cryst12071000 99

Yuan Shen and Ingo Dierking
Recent Progresses on Experimental Investigations of Topological and Dissipative Solitons in Liquid Crystals
Reprinted from: Crystals **2022**, 12, 94, doi:10.3390/cryst12010094 119

Peter John Tipping and Helen Frances Gleeson
Ferroelectric Smectic Liquid Crystals as Electrocaloric Materials
Reprinted from: Crystals **2022**, 12, 809, doi:10.3390/cryst12060809 137

Mark R. Wilson, Gary Yu, Thomas D. Potter, Martin Walker, Sarah J. Gray, Jing Li and Nicola Jane Boyd
Molecular Simulation Approaches to the Study of Thermotropic and Lyotropic Liquid Crystals
Reprinted from: Crystals **2022**, 12, 685, doi:10.3390/cryst12050685 149

J. C. Jones, S. A. Jones, Z. R. Gradwell, F. A. Fernandez and S. E. Day
Defect Dynamics in Anomalous Latching of a Grating Aligned Bistable Nematic Liquid Crystal Device
Reprinted from: Crystals **2022**, 12, 1291, doi:10.3390/cryst12091291 173

Bohan Chen, Zimo Zhao, Camron Nourshargh, Chao He, Patrick S. Salter, Martin J. Booth, Steve J. Elston and Stephen M. Morris
Laser Written Stretchable Diffractive Optic Elements in Liquid Crystal Gels
Reprinted from: Crystals **2022**, 12, 1340, doi:10.3390/cryst12101340 195

About the Editor

Ingo Dierking

Ingo Dierking received his PhD in 1995 from the University of Clausthal in Germany. After a stint at the IBM TJ Watson Research Center in the US, working on electronic paper, he joined Chalmers University in Gothenburg, Sweden, as a Humboldt fellow. There he was appointed as lecturer, before joining the University of Darmstadt in Germany for several years and eventually moving to the Department of Physics & Astronomy at the University of Manchester in 2002. Ingo has published more than 150 scientific papers, as well as several books on topics of liquid crystal research. He is the 2009 awardee of the Hilsum medal of the British Liquid Crystal Society (BLCS) and the 2016 winner of the Samsung Mid-Career Award for Research Excellence of the International Liquid Crystal Society (ILCS). In 2021 he was awarded the internationally prestigious GW Gray Medal, the highest prize of the BLCS. Ingo is the former Secretary and current President of the ILCS and the former Chair of the BLCS. His current research interests are broadly focused on soft matter systems with an emphasis on liquid crystals and LC-based composites with polymers and nanoparticles.

Preface to "State-of-the-Art Liquid Crystals Research in UK"

Besides Germany and France, the UK is certainly one of the countries that must be credited with significant contributions to the research and development of the field of liquid crystals. This is the case not only from a synthetic chemistry point of view, which laid the groundwork for all modern liquid crystal display (LCD) applications, but also in terms of physical theory and experiments, as well as computer simulation work. It is thus fitting to edit a volume that showcases the history and present-day state-of-the-art research in this area of soft matter research. This Topical collection brings together a range of leading UK scientists, discussing the recent developments in this highly interdisciplinary field in all its breadth, from physics to chemistry, and from material aspects to applications.

Ingo Dierking
Editor

Editorial

State-of-the-Art Liquid Crystals Research in UK

Ingo Dierking

Department of Physics and Astronomy, University of Manchester, Oxford Road, Manchester M13 9PL, UK; ingo.dierking@manchester.ac.uk

A number of countries could reasonably produce a collection represented a name such as "State-of-the-Art Research in Liquid Crystals". The history of liquid crystal research [1–3] dates back to approximately 135 years ago in the German speaking countries, when Friedrich Reinitzer (Figure 1), a botanist from Austria, discovered the two supposed melting points of cholesteryl benzoate [4] at the German University in Prague. In his article, he clearly described the liquid crystalline behaviour of this compound, including the phenomenon of selective reflection, but he was unable to precisely explain the observed behaviour. This explanation was provided within a year's time by the German physicist Otto Lehmann (Figure 1), who was the successor of Heinrich Hertz at the University of Karlsruhe. Lehmann first coined the phrase "liquid crystal" [5] and spent the rest of his life in research attempting to foster the acceptance of his concepts by much of the physical chemistry community in Germany, with his greatest opponent being Gustav Tammann. Today, of course, we know that Otto Lehmann's view of the liquid crystal state is correct. This leads us to mention Daniel Vorländer (Figure 1), a synthetic chemist at the University of Halle, who remains among the top ten most productive listed liquid crystal organic chemists as the field expands through new research on new phases of matter [6]. Also well-known in the field of physical chemistry is the name Rudolf Schenk (Figure 1), who studied chemistry in Halle and later moved to the University of Marburg, where he investigated liquid crystals [7], producing results in favour of Lehmann's views rather than those of Tammann.

Figure 1. The German liquid crystal researchers of the early days: Friedrich Reinitzer, Otto Lehmann, Daniel Vorländer, and Rudolf Schenk.

At the same time, interest in liquid crystals was also growing in France, where Charles-Victor Maugin (Figure 2) explored the polarisation properties of twisted liquid crystals and cholesteric phases [8] and the effects of magnetic fields [9]. Francois Grandjean (Figure 2) was the first to describe the importance of surface interactions for the orientation of liquid crystals [10] and developed a method for measuring the cholesteric pitch in wedge cells. It was finally Georges Friedel (Figure 2) who, in 1922, categorised the then-discovered liquid crystals according to their structural properties in the first detailed and extensive review of the subject [11]. He proposed three mesomorphic states, including the *nematic, cholesteric,* and *smectic*. Grandjean had already described the layered structure of smectics based on his observation of the steps of uncovered droplets on a clean plane glass substrate [10]. Today, of course, we know that this classification is not remotely detailed enough to account for all the different liquid crystal phases, nor does it truly make much sense as a method used to distinguish between the nematic and cholesteric states, as the latter is only the chiral version of the former, albeit with a greatly different spontaneous structure and properties.

Figure 2. The French liquid crystal researchers of the early days: Charles Maugin, Francois Grandjean, and Georges Friedel.

Smectic polymorphism and the layered structure of the smectic phase also constituted the focus of one of the famous early meetings and, quite possibly, the first international conference on liquid crystals, the Faraday Discussion Meeting of the Royal Society, held in 1933. The contributions and discussions are documented in the *Transactions of the Faraday Society* 1933, vol 29, issue 140 (https://ur.booksc.me/journal/24896/29/140, last accessed on 14 September 2022).

Following the birth of liquid crystal research and the establishment of this fourth state of matter, research efforts became more international in nature. Arguably, one of the pioneering figures in the field was the Russian physicist Vsevolod Konstantinovich Frederiks (Figure 3), who was the first to describe the magnetic [12] and, later, the electric reorientation of the director by externally applied fields, now known as the Frederiks transition, the basis of most of the display devices that have enabled the significant progression in liquid crystal research since the 1970s. In this category, we must also mention the Swedish scientist Carl Wilhelm Oseen (Figure 3), who was the first to formulate a description of the elastic properties of nematics [13], and Marian Miesowicz (Figure 3) from the Mining Academy in Cracow, Poland, who studied the viscosities of liquid crystals under magnetic fields [14] and later introduced the three viscosity coefficients that now bear his name [15]. Equally worthy of note are Hans Ernst Werner Zocher (Figure 3), a German born researcher who worked at the University of Prague and later emigrated to Brazil, who is known for his work on vanadium pentoxide lyotropic liquid crystals [16] but more so for his theoretical contributions regarding field effects on nematic liquid crystals [17]. The Russian scholar Victor Nikolaevich Tsvetkov (Figure 3) introduced the orientational order parameter [18], whose temperature dependence was predicted in the classic publications of Wilhelm Maier and Alfred Saupe (Figure 3) [19,20], thus producing the first working theory of the nematic liquid crystalline phase. It took more than 70 years from the time of the discovery of liquid crystals to their first functional description. This basic field description was later supplemented by the Landau-type description of the isotropic to nematic transition of Pierre-Gilles de Gennes (Figure 3) [21], who was awarded with the Nobel Prize for Physics in 1991 for his work on liquid crystals and polymers. His book on *The Physics of Liquid Crystals* [22] is still considered as a kind of bible for researchers in this field.

Figure 3. Pioneering international liquid crystal scientists before the display era: Fredericks (Russia), Oseen (Sweden), Miesowicz (Poland), Zocher (Germany, Brazil), Tsvetkov (Russia), Maier and Saupe (Germany), and de Gennes (France).

By this point, the time was ripe for researchers to not only engage in fundamental research on liquid crystals and the synthesis of novel, increasingly complicated mesogens but also to think critically about the applications and structure–property relationships that enable this synthesis. This was also the time when liquid crystal research truly blossomed in the UK. Frederick Charles Frank (Figure 4) did not published a great deal on the topic of liquid crystals, yet his influence was far-reaching, as he revised the theory of elasticity formulated by Oseen to produce the form in which it is known today [23], with the three bulk elastic constants describing splay, twist, and bend deformations. Frank's description also opened the door to the experimental determination of these functionally vital constants.

Figure 4. Sir Charles Frank had a lasting impact on British liquid crystal research, despite having not published many papers on the topic himself.

Having been introduced to liquid crystals by Brynmor Jones, George Gray, Head of Department at Hull University, published his first paper in the field in 1951, entitled "Mesomorphism of some alkoxynaphthoic acids" [24]. During the 1960s, George William Gray (Figure 5) was working as a synthetic chemist at the University of Hull with the aim of producing liquid crystals at room temperature, the ultimate prerequisite for any display device. Being aware of the structure–property relationships of liquid crystals, which were also established during this time by the Halle group in Germany, Gray succeeded in his task, and the family of the cyano-biphenyls that are still employed today was born, comprising single-component liquid crystals with room temperature nematic and smectic phases [25]. The melting and clearing temperatures could still be improved using mixtures with similar rod-like mesogens, and materials which possessed a reasonable nematic temperature range for the operation of displays were obtained. Ben Sturgeon (Figure 5), a senior scientist at BDH (later part of Merck) was instrumental in scaling-up the production of the liquid crystalline cyano-biphenyls, thus enabling the worldwide development of the display industry that is now a global multi-billion dollar business. This would not have been possible without the achievements of George Gray, the risks taken by Ben Sturgeon, and the engineering skills of Cyril Hilsum (Figure 5), an engineer who pioneered the development of modern display technology, and his political persuasion of the Ministry of Defence.

Figure 5. UK liquid crystal scientists who were instrumental in the development of the LCD technology: George Gray, Ben Sturgeon, and Cyril Hilsum.

The displays of Heilmeier and Zanoni (Figure 6) [26] or Schadt and Helfrich (Figure 6) [27] could only function and attain market-competitive value with room-temperature materials such as 5CB (or K15, as it was known in those days).

Figure 6. The pioneers of the invention of liquid crystal displays: Heilmeier and Zanoni (dynamic scattering display), as well as Schadt and Helfrich (twisted nematic display).

After the development of room-temperature materials and the invention of the first displays, naturally, both fundamental and applied liquid crystal research continued in the UK. The mathematical physicist Frank Matthews Leslie (Figure 7) (University of Strathclyde), together with Jerald Ericksen, who is mostly known for his development of a continuous theory of the mechanical behaviour of nematics [28], introduced the viscosity parameters known as Leslie coefficients. Geoffrey Luckhurst (Figure 7), a chemical physicist at the University of Southampton, has made seminal contributions to the field with respect to the synthesis and NMR and ESR characterisation of materials, as well as computer simulations and theory. Peter Raynes (Figure 7), an optoelectronics engineer at Oxford University, has long been involved in research on display design and characterisation. Together with Hulme and Harrison, he invented a method for formulating mixtures of cyano-biphenyls, which was used by Ben Sturgeon at BDH in the commercialization of liquid crystal mixtures, such as E7, first used in LCDs. Raynes has also contributed significantly to our understanding of the electro-optics of liquid crystals. During his time at the Royal Signals and Radar Establishment (RSRE), he invented the super-twisted nematic (STN) display (with Colin Waters), which can be found in most of the early alpha-numeric flat-panel screens. Similarly, Roy Sambles (Figure 7) is an experimental physicist who worked at the University of Exeter, where he studied the interaction between light and matter, with a special emphasis on surface plasmons and microwave photonics. Of course, one cannot fail to mention John Goodby (Figure 7), who worked at the University of Hull and, later, the University of York. His contributions to research on the synthesis of liquid crystal materials in the fields of ferroelectric liquid crystals, low birefringent materials, and chiral liquid crystals, in general, were truly seminal. He later broadened his interests to encompass self-organising materials and complex fluids, and through his work, he brought partially ordered fluids into the realm of materials chemistry and functional materials. The late Mark Warner (Figure 7) from the University of Cambridge was a theoretical physicist, who was one of the founders of the field of liquid crystalline elastomers and actuators.

Figure 7. The next generation of UK liquid crystal researchers who made seminal and pioneering contributions to the field: Frank Leslie, Geoffrey Luckhurst, Peter Raynes, Roy Sambles, John Goodby, and Mark Warner.

In this Special Issue of the journal *Crystals*, "State-of-the-Art Liquid Crystal Research in the UK", we have collected a range of articles and reviews that reflect on this previous work and provide an overview of the variety and diversity of current liquid crystal research in the UK. It includes papers on the history of liquid crystal contributions in the UK, experimental physics and chemistry, simulations, and novel applications. The publications were mainly produced by active members of the research community, including researchers from the universities of Leeds, Aberdeen, Manchester, Durham, Oxford, and York. This range of contributions demonstrates the breadth of liquid crystal research and provides a useful cross-section of the work in progress in the UK today.

Funding: This research received no external funding.

Acknowledgments: I would like to thank John Goodby for carefully reading the first draft of this introductory paper and for pointing out several valuable additions to the UK LC story based on the first draft.

Conflicts of Interest: The authors declare no conflict of interest.

References

1. Sluckin, T.J.; Dunmur, D.A.; Stegemeyer, H. *Crystals That Flow*; Taylor & Francis: London, UK, 2004.
2. Dunmur, D.; Sluckin, T. *Soap, Science and Flat-Screen TVs: A History of Liquid Crystals*; Oxford University Press: Oxford, UK, 2011.
3. Koide, N. (Ed.) *The Liquid Crystal Display Story*; Springer: Tokyo, Japan, 2014.
4. Reinitzer, F. Beiträge zur Kenntiss des Cholesterins. *Mon. Für Chem.* **1888**, *9*, 421. [CrossRef]
5. Lehmann, O. Über fliessende Krystalle. *Z. Für Phys. Chem.* **1889**, *4*, 462. [CrossRef]
6. Vorländer, D. Zeitschrift für Physikalische Chemie A. *Int. J. Res. Phys. Chem. Chem. Phys.* **1906**, *57*, 357.
7. Schenck, R. Über die Natur der Kristallinischen Flüssigkeiten und der Flüssigen Kristalle. *Z. Für Elektrochem.* **1905**, *11*, 951. [CrossRef]
8. Maugin, C.-V. Sur les cristaux liquides de Lehmann. *Bull. De La Soc. Fr. De Mineral.* **1911**, *34*, 71. [CrossRef]
9. Maugin, C.-V. Orientation des cristaux liquids par le champ magnetique. *Comptes Rendus De L'academie Des Sci.* **1911**, *152*, 1680.
10. Grandjean, F. L'orientation des liquides anisotropies sur les cristaux. *Bull. De La Soc. Fr. De Mineral.* **1916**, *39*, 164.
11. Friedel, G. Les etats mesomorphes de la matière. *Ann. De Phys.* **1922**, *18*, 273. [CrossRef]
12. Freedericksz, V.; Zolina, V. On the use of a magnetic field in the measurement of the forces tending to orient an anisotropic liquid in a thin homogeneous layer. *Trans. Am. Electrochem. Soc.* **1929**, *55*, 85.
13. Oseen, C.W. The Theory of Liquid Crystals. *Trans. Faraday Soc.* **1933**, *29*, 883. [CrossRef]
14. Miesowicz, M. Influence of a Magnetic Field on the Viscosity of Para-azoxyanisol. *Nature* **1936**, *136*, 261. [CrossRef]
15. Miesowicz, M. The Three Coefficients of Viscosity of Anisotropic Liquids. *Nature* **1946**, *158*, 27. [CrossRef]
16. Zocher, H. Über freiwillige Strukturbildung in Solen. (Eine neue Art anisotrop flüssiger Medien). *Z. Für Anorg. Und Allg. Chem.* **1925**, *147*, 91. [CrossRef]
17. Zocher, H. The Effect of a Magnetic Field on the Nematic State. *Trans. Faraday Soc.* **1933**, *29*, 945. [CrossRef]
18. Zwetkoff, W. On Molecular Order in the Anisotropic Liquid Phase. *Acta Physicochim. URSS* **1942**, *15*, 132.
19. Maier, W.; Saupe, A. Eine einfache Molekular-statistische Theorie der nematischen kristallinflüssigen Phase. Teil I. *Z. Für Nat.* **1959**, *14A*, 882. [CrossRef]
20. Maier, W.; Saupe, A. Eine einfache Molekular-statistische Theorie der nematischen kristallinflüssigen Phase. Teil II. *Z. Für Nat.* **1960**, *15A*, 287. [CrossRef]
21. de Gennes, P.G. Short Range Order Effects in the Isotropic Phase of Nematics and Cholesterics. *Mol. Cryst. Liq. Cryst.* **1971**, *12*, 193. [CrossRef]
22. de Gennes, P.G. *The Physics of Liquid Crystals*; Oxford University Press: Oxford, UK, 1974.
23. Frank, F.C.I. Liquid Crystals: On the Theory of Liquid Crystals. *Discuss. Faraday Soc.* **1958**, *25*, 19. [CrossRef]
24. Gray, G.W.; Jones, B. Mesomorphism of some alkoxynaphthoic acids. *Nature* **1951**, *167*, 83. [CrossRef] [PubMed]
25. Gray, G.W.; Harrison, K.J.; Nash, J.A. New Family of Nematic Liquid Crystals for Displays. *Electron. Lett.* **1973**, *9*, 130. [CrossRef]
26. Heilmeier, G.H.; Zanoni, L.A.; Barton, L.A. Dynamic Scattering: A New Electrooptic Effect in Certain Classes of Nematic Liquid Crystals. *Proc. IEEE* **1968**, *56*, 1162. [CrossRef]
27. Schadt, M.; Helfrich, W. Voltage-dependent Optical Activity of a Twisted Nematic Liquid Crystal. *Appl. Phys. Lett.* **1971**, *18*, 127. [CrossRef]
28. Leslie, F.M. Distortion of Twisted Orientation Patterns in Liquid Crystals by Magnetic Fields. *Mol. Cryst. Liq. Cryst.* **1970**, *12*, 57. [CrossRef]

Review

Conception, Discovery, Invention, Serendipity and Consortia: Cyanobiphenyls and Beyond

John W. Goodby * and Stephen J. Cowling *

Department of Chemistry, The University of York, York YO10 5DD, UK
* Correspondence: john.goodby@york.ac.uk (J.W.G.); stephen.cowling@york.ac.uk (S.J.C.)

Abstract: In the 1960s, a world-wide change in electronic devices was about to occur with the invention of integrated circuits. The chip was upon us, which instantly created the need for a revolution in visual communication displays. From the watch to the computer monitor, to TVs, to the phone, nearly all everyday applications were affected. A strange connection in technology underpinned these changes; the linkage between silicon semiconductors and organic compounds that did not know if they were solids or liquids. Liquid crystals had been known since 1888 and had seen little usage until they were inserted between conducting glass slides and an applied electric field. Suddenly, the possibility of driving images with low voltage fields became obvious. Many major companies took up the challenge of commercialisation, but in the UK a curious combination of government research facilities, electronic companies and one small university came together in 1970 to form a consortium and within two years the basis for new technologies had been founded. Chemistry is part of this story, with new conceptions, discoveries and inventions, and the luck to be in the right place at the right time.

Keywords: nematic; smectic; ferroelectric; birefringence; dielectrics; chirality

1. Introduction

The Awakening: Our story is one about a technology push where research and development in a new technology drive the creation of new products, rather than market pull, which refers to the need for a solution to a problem that comes from the marketplace. The new development was the giant technology shift caused by the arrival in the 1960s of the integrated circuit—the chip. In the world of liquid crystals, the new technology meant the interfacing of, as yet unknown solid-state electronic devices with as yet unknown display concepts. This meant out of the window would go all of the valves and cathode ray tubes that were used in our TVs, and in would come a myriad of new concepts, discoveries, materials, devices and applications, to the world of communications, see Figure 1. To major international communications laboratories there was a recognition that only two things were important to their businesses—*silicon* and *displays*. Both required unifying inventions—solid-state semiconductor devices and new flexible liquid-like materials. Here, we discuss revolutions made in materials, and in particular liquid crystals.

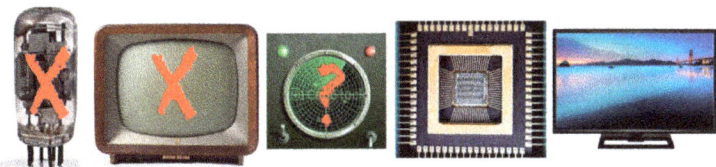

Figure 1. Technology push from the valve to the chip, where X means redundant and ? means possibly continuing in use.

Close to the beginning of the development of silicon-based electronics in the UK, an odd figure stepped onto the stage, no less than John Stonehouse MP, who served in Prime Minister Harold Wilson's government as Minister of Posts and Telecommunications [1]. In March 1967, Stonehouse made his first visit to the Royal Radar Establishment (RRE) to meet with the Director, Dr. George MacFarlane [2,3]. Between them, they discussed monetary returns from inventions, with MacFarlane noting a story that the UK royalties paid to the American company RCA on the shadow-mask colour TV tube were thought to have been more than the development costs for the supersonic jet Concorde. Given the rapid technology changes, and seeing the opportunity, Stonehouse reported back after the meeting that the UK should mount a programme to invent a solid-state alternative to the cathode ray tube. MacFarlane met with his senior staff and discussed with them the possibilities of RRE developing a flat panel colour TV. One might expect the answer to come back as no, but in typical fashion the answer was that a Working Party to study the topic should be set up. Subsequently, the Party had to assess components and materials, unknown at the time, for such a programme. The CVD (Committee for Valve Development) was responsible for their procurement, and contractual and funding arrangements for the Armed Services. Among the topics identified early in 1968 as being of interest were liquid crystals. Things were now starting to move in a direction to meet the wishes of Stonehouse, but what happened to him? Later in his career the national newspapers reported he was a Czech spy, who then walked into the sea at Miami, before resurfacing in Australia, and subsequently, spent time in jail over financial offences [1].

Turning back to the research developments, to quote Hilsum [2,4], there was a recommendation "in December 1969 that the UK Government should fund research on flat-panel electronic displays, with LCs as the first priority". Though formal approval of this recommendation would normally have taken some months and indeed, was never granted, The Royal Radar Establishment (RRE) had anticipated approval and justified their action on the urgent need for displays for the portable radar sets they had invented. They established two consortia, one for materials, involving the Royal Radar Establishment (RRE was formed in 1953), Hull University and BDH Chemicals (originally BDH stood for British Drug Houses, and in 1973 BDH Chemicals was acquired by E. Merck of Germany when Glaxo decided to concentrate on its mainstream activities), and one for devices, involving RRE, the Royal Aircraft Establishment (RAE was formed in 1892), the Services Electronics Research Laboratory (SERL formed in 1945), Marconi (Chelmsford, UK), and Standard Telephone Laboratories Ltd. (STL), Harlow, UK". It appears in reality that one Consortium was created at the beginning and later split into two, one for displays, the other for materials, whereby a UK collaboration between the MoD, industry and universities through consortia was already an established policy [2]. Such consortia were valuable to the fundamental base of science in the UK as they also allowed universities still to have access to Research Council grants as well as government contracts.

Therefore, in the following, we focus on the story of the chemistry of the design of molecular structures of liquid crystals for the purposes of applications, as George Gray commented, " ... *I was happy to see that society in its widest international sense was benefiting by my science, and was not too troubled that the coffers of Electronic Companies did not profit thereby. It did however please me that the UK chemical industry benefited financially from my work - a conveniently forgotten fact. Also, I would like to emphasize some other much wider, and to me equally important advantages and consequences, that stemmed from our simple discovery of the cyanobiphenyls ...* "

2. Materials and Methods

In the following sections, there are a number of differing materials classes under separate subheadings, for which there is a description of results and a short discussion of their applications. Extensive and detailed synthetic procedures, methodologies, equipments, techniques, and modelling and simulations are given in the references section, also with permissions of the funding bodies.

The Consortium on Materials: We turn now to the next steps in the process of the creation of liquid crystal consortia. Between 1968 and 1971, there was a concerted effort to acquire information and in some cases samples of devices and materials on liquid crystal displays (LCDs). Turning inwards to the UK, an effort was put in to see what the academic community could offer, and what technologically advanced companies might become involved with. In typical fashion a "Town Meeting" was subsequently arranged which took place on 1 October 1968. During the meeting the lead speaker, regarded as a UK authority on liquid crystals, became confused in answering questions from the floor, which were put to him at the end of his talk. In searching through a book, he spilled his notes onto the floor. In the ensuing panic a quiet voice from the rear of the audience said, "I wonder if I can help". This was George Gray. In conclusion, at the end of the meeting, Cyril Hilsum commented, "*We must put the man from Hull on a contract*", even though at the time it was thought that "Liquid crystals would make no impact on Black and White or Colour TV". The stories of these events can be found in various publications by Hilsum, Gray and in a Royal Society memoir about the life of George Gray [2,3,5–8].

In April 1970, CVD offered Gray a two-year contract to work as Hull University's contract holder (PI) on "*Substances Exhibiting Liquid Crystal States at Room Temperature*" at a maximum expenditure of GBP 2177 pa. This allowed Gray to appoint a fresh postdoctoral research assistant called John Nash on 1 October 1970. The first meeting the pair attended was in early 1971, with RRE, RAE, SERL and Marconi Ltd. also being present. At any new venture, the start was based on what was happening in the current research environment. Consequently, Hull began by examining potential structures for new liquid crystals for devices based on materials and devices reported by RCA (Radio Corporation of America, New York, NY, USA) and IBM (International Business Machines, Armonk, NY, USA) [1,9–13]. The preferred devices at that time utilised the dynamic scattering (DSM) [13] and cholesteric-nematic phase change modes [14], which often required materials of negative dielectric anisotropy ($-\Delta\varepsilon$), and so low melting materials with rod-like structures and lateral dipoles relative to the molecular long axes were sought after.

3. Results
3.1. Purity of Materials

Common materials used in the USA and Europe in various liquid crystal applications, including displays, were based on the Schiff's bases MBBA and EBBA [9,15], see Figure 2, and on alkyl carbonato-alkoxyphenyl benzoates. Heilmeier at RCA reported privately a three-component formulation at that time, which gave the best transitions for a nematic phase of 24 to 76 °C. In the UK, it was well known that such materials were electrolytically unstable, easily oxidised and could degrade on exposure to UV radiation. This can happen in their preparation, storage and usage. Their use in dynamic scattering mode devices was found to result in displays having differing lifetimes depending on material purity. Hull worked on a variety of synthesized Schiff's bases, stilbenes, carbonates, carboxylic esters, ultra-pure Schiff's bases, etc., but to little effect. There were difficulties in purifying them. In particular, there were problems with low melting variants, which, when it came to recrystallisation at a low temperature, one would need to sit inside a large-size refrigerator to perform vacuum filtrations at -10 °C so that emulsification or the formation of lyotropic phases did not occur.

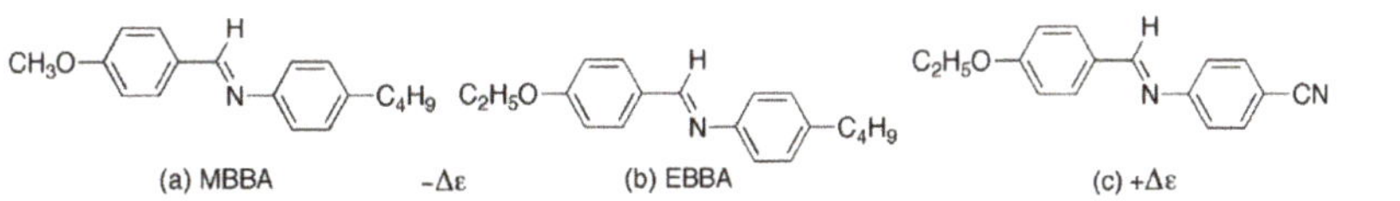

Figure 2. Schiff's bases (**a**,**b**) have negative dielectric anisotropies, whereas compound (**c**) is positive.

Although coloured, Schiff's bases, in addition to having negative dielectric anisotropies, are relatively easy to prepare, thereby providing a variety of materials for use in multicomponent mixtures. However, even in the design and preparation of these materials they still had further problems due to exchange reactions occurring whilst being located in devices. As a consequence, the components of mixtures were required to possess structures where one of the exchangeable moieties is the same. For MBBA and EBBA, the exchangeable unit is the right-hand side of the molecules as shown in Figure 2. The aniline part of the molecule can flip from MBBA to EBBA without changing the mixture composition. This of course limits the selection of material components for mixtures. The fact that electro- and photochemistry is still ongoing in devices is often not realised, particularly where small changes in purity can affect properties. Not only is this apparent in DSM devices, it also can occur in smectic A devices, dye devices, OLEDs, etc.

Nevertheless, MBBA was still particularly popular with the industrial laboratories in the 1970s because it was found to be a room temperature nematogen, however, the feedback on its transition temperatures was somewhat variable. It was not appreciated at the time that the purity of a material for use in an electronic device should be aimed at reaching a purity that was near that of the electronic components of the device. Therefore, it became an objective to provide a supply of ultrapure MBBA of purity greater than expected from organic laboratories.

There were also big problems of materials analysis in those days to evaluate purities as there were no Fourier transform instruments, or high-performance chromatography equipment, thus making pure materials dependent on the skills of the chemist and the use of thin-layer chromatography (TLC). To this day, there are many materials that are commercially available or from academic laboratories that are not pure enough. In the following, the preparation of MBBA is described using the methods of that era in the UK.

MBBA could be prepared via the condensation of equimolar amounts of 4-butylaniline and 4-methoxybenzaldehyde by heating them in ethanolic solution with a trace of glacial acetic acid as a catalyst. However, the products from such a method produced other materials that were not so easy to separate. So, the first point was to use a method that produced less in terms of by-products. Thus, it appeared important to allow the reaction mixture to stand in the dark for 12 h after the reaction had been completed, but longer times were found to yield more highly coloured products. Deep refrigeration of the reaction mixture gave crystals that could be filtered off, which were often difficult to separate from solution because any rise in temperature during filtration would result in the formation of a gel or lyotropic phase. The transition temperatures for a purified product were Cryst to N at 21 °C and N to liquid at 41–43 °C, which was similar to commercial materials. Two crystallisations with light petroleum (bp 40–60 °C) gave constant transition temperatures of Cryst to N at 21 °C and N to liquid at 45 °C, determined by thermal polarised light microscopy [16,17]. Distillation via vacuum sublimation under reduced pressure did not improve the situation. MBBA was also found to be sensitive to moisture, and prolonged evacuation over $CaCl_2$ was necessary. Various storage conditions were then applied, and it was found that being under vacuum in the presence of P_2O_5 at room temperature for several days improved the transition temperatures from Cryst to N at 21 °C and N to liquid at 47 °C. These constants have not been improved since, and the temperatures are considered to be those for pure MBBA. Such a product was found to give a single spot-on neutral TLC plate (N.B. slightly acidic or basic plates produce a cleavage of the Schiff's base). The analysis of the purified material was found to be consistent with IR spectrum and mass spectrometric data for the structure of MBBA. Within the UK, pure samples of MBBA were supplied to various team members, either to be stored in sealed ampoules or when in use in a vacuum over a desiccant such a P_2O_5. Under these conditions, in comparison to commercial samples, the ultrapure material was much paler, and preliminary information indicated that an electro-optic effect was found under normal electrical addressing. For such a pure product, it was estimated that MBBA could be prepared at a cost of around 0.1 pence per gram for an overall yield of ~47%. In contrast, commercial samples of MBBA were found to be considerably impure, wet or both, and as such the physical studies of such samples were doubtful, and applications could be rendered worthless.

Generally, the measurements of dielectric coefficients $\varepsilon_{\parallel}$ and $\varepsilon_{\perp}$ became possible, along with the determination of the resistivities of nematogens. High purity MBBA had a resistivity of 2×10^{11} Ωcm. It is interesting that at such high resistivities, dynamic scattering devices had relatively short lifetimes, lasting approximately 10 h, indicating that these devices required the incorporation of ionic dopants in order to generate more stable displays (the need to incorporate ionic dopants in "DSM-type" devices tended to make them unsuitable for commercial applications). Under these conditions, MBBA has a resistivity of ~10^{10} Ωcm. In comparison, materials used in the twisted nematic construction [18] were required to have resistivities of ~10^9 Ωcm. Consequently, high purity Schiff's bases possessing terminal-substituted cyano-units and resistivities in the region of 10^{10} Ωcm were a factor in obtaining good quality devices, and to chemists, measurements of the purities of their materials became determined by resistivity.

Thus, MBBA was found to be useful only as an experimental material that would never be applied in practice. For practical applications, a material should be as pure as possible, and in a simple experiment, TLC can be used to indicate if the material is a single species (such tests are used today in lateral flow tests for COVID infections). In the cases of electronic devices, purities should be in the region of 99% or have resistivities of ~10^{10} Ωcm or better. A material should also give reproducible physical results. Its transition temperatures should remain constant with time and temperature, and without decomposition [19]. As George Gray would retort: *"To be a reputable synthetic chemist is to supply collaborators with materials checked to be of the highest purities, and for recipients in doubt, TLC can be used as a check for worries over purities"*.

3.2. Materials Designed to Fit

Once purity was recognised as important, two other items came into view in the development of device materials. With the inventions of new property-testing methods, property-structure correlations were amassing and allowing for the specific design of the structures of new molecular entities via feedback mechanisms. The utilisation of materials in various mixture formulations for a variety of devices requires upscaling for the production of materials, which meant at that time, the development of new synthetic methods in order to move from grams to tens of grams to kilograms of high purity compounds. With discovery aspects ongoing at Hull, BDH was contracted to provide a supply of selected materials, thereby increasing the pace of the activities of the UK Consortia. As noted above, MBBA still remained popular because its negative dielectric anisotropy suited applications in DSM devices. However, owing to its stability issues, with lifetimes rarely exceeding 3000 h, it started to become replaced by azo-benzenes and carbonate esters as reported by Heilmeier [12], which also had negative dielectric anisotropies, but of a stronger level in comparison, due to the larger lateral molecular dipole. At Hull, azo-benzenes were not perused due to their poor UV stability, whereas the carbonate esters were more stable, but unfortunately, they had higher melting points. By placing a polar group at the terminus in analogues of MBBA, a dipole pointing along the molecular long axis was achieved, the dielectric anisotropy was positive, and the material was suitable for use in the twisted nematic display (TNLCD) of Schadt and Helfrich [18,20,21]. Other devices also came into play [22], including the phase change device and the electrically controlled birefringence (ECB) display, which now required further tuning to the molecular design.

Materials such as Schiff's bases sufficed for the early work, but with low values of $-\Delta\varepsilon$, they were soon replaced with carbonate esters, which had larger lateral dipoles. Carbonates are not particularly stable and so derivatives of *trans*-stilbene were thought to be suitable alternatives. For such materials, a lateral polar unit (Cl or CN) could be placed in the linking chain between the phenyl moieties. However, the stabilities of stilbenes with respect to *cis–trans* isomerisation were in question, as were the difficulties in their syntheses. At a similar time, heterocyclic systems were also being investigated where a phenyl unit was replaced with a pyridine ring, where the nitrogen atom could be located in different positions. Although of interest, they still had *cis–trans* stability issues and high transition temperatures. Therefore, the family of carbonate esters was also extended via the preparation of 4'-alkoxyphenyl 4-alkoxybenzoates.

Many materials at this particular time were unsuitable to be used in devices; they did not have suitable transition temperatures, were too high melting, unstable in devices, impure, difficult to synthesise, etc. Gray's futuristic targets before the discovery of cyanobiphenyls in 1972, included materials based on the incorporation of bicyclo-octanes and bicyclo-octenes and ring systems such as cyclohexyl moieties [23,24], see Figure 3, which no doubt he may have discussed previously with Dewer, who was his PhD examiner. Indeed, the abstract from Dewar and Goldberg's 1972 paper [25] states that *"Many compounds forming nematic mesophases contain p-phenylene units. It is shown that these perform a dual function, providing rigid linear groupings and contributing to the polarizability of the molecule. These conclusions are based on a comparison with compounds where benzene rings are replaced by cyclohexane or bicyclo[2.2.2.]octane"*. The inclusion of such moieties in molecules possessing rod-like structures (see the top of Figure 3) was probably not envisioned to be for materials with positive dielectric anisotropies, but the lower structures in the figure were probably being considered. It was not obvious at the time, nor was there an immediate need to produce them because of the synthetic complexities involved.

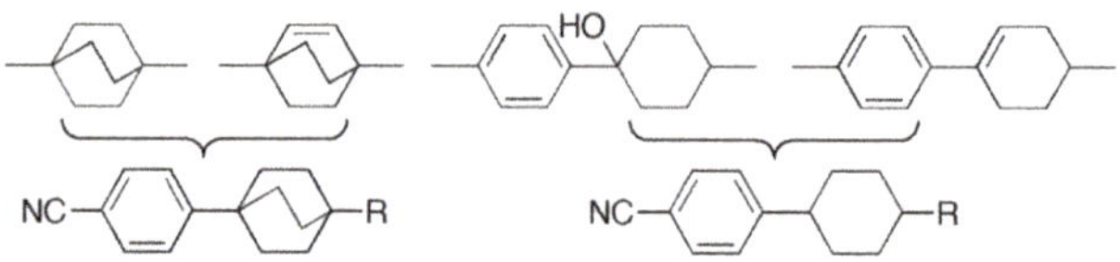

Figure 3. Chemical moieties to be possibly incorporated into rod-like mesomorphic materials. The upper moieties were prospective possibilities, whereas the lower compounds were realised by others.

It should be noted that subsequent work by Gray and Kelly resulted in the production of bicylo-octane mimics of cyanobiphenyls [26–31]. The use of cyclohexyl moieties in molecular architectures was performed by Deutscher et al. [32–34], Osman and Revesz [35] and the well-known device materials by researchers at E. Merck [36–41]. Frustration, however, was growing as most materials were missing the mark. George Gray sought for a common feature in molecular design of all of the materials that he thought might be causing a problem, and he settled on the linking unit between phenyl rings. Its removal, he thought, could provide the change in structure that might make the materials useful liquid crystals.

3.3. A Scientific Revolution

Across research laboratories in the USA, UK and Europe, by the middle of 1972, it appeared that there were still problems with the purities of Schiff's bases, the quantities of materials were still tight, the temperature ranges not suitable, and the dynamic scattering mode was not fully reproducible even though it appeared better than the twisted nematic device; something had to give. Gray's futuristic concepts were set aside, and the simple idea of eliminating the central linkage (-CH=N-, -CH=CH-, -COO-) from the materials previously prepared to give 4,4'-substituted biphenyls took precedence. Gray had already used biphenyls along with lateral fluorination (see later) 15 years earlier [42] in the study of smectic liquid crystals. A second accompanying idea was the use of the nitrile (CN) group in the terminal position to give colourless materials with strong positive dielectric anisotropies. The cyano-equivalent in Schiff's bases were first reported in Castellano in 1968 [43], and later by Boller and Scherrer [44]. The remarkable combination of biphenyls and nitrile resulted in nematic materials that would operate at room temperature in devices that used materials with positive dielectric anisotropy [45–47]. Thus, in 1972, with many still working in the area of light scattering devices, 4-pentyl-4'-cyanobiphenyl (5CB, K15) took to the stage, and when it was incorporated into the Schadt-Helfrich (+Δε) device, the revolutionary twisted nematic display (TNLCD) [18] was born.

As a consequence, Gray would often use the discovery of cyanobiphenyls as an example of the importance of basic research; to quote him from his award of the 1995 Kyoto Prize of the Inamori Foundation of Japan [48], he said, *"I knew what I was doing by using the cyano-group to compensate for loss of molecular length, while at the same time providing the strongly polar molecular structure needed for the electric field to switch on the display. This I stress was not luck . . . the fundamental science was secure . . . we knew what we were doing"*.

To us, and many others, George Gray emphasised "the idea behind the programme was simply the elimination of the central linkage in all of the previous systems to give stable 4′,4-disubstituted biphenyls, and that Ken Harrison undertook their syntheses". However, we knew about the extensive fundamental studies on liquid crystals Gray had performed over the previous 20 years.

In reporting the invention of the cyanobiphenyls to the Consortium, members were reminded of the confidential nature of the work and that the new materials were the invention and property of the group in Hull, and it was the duty of the rest of the group to protect their position. The realisation of the Hull outcome struck home with the need for a more secure source of materials. So, it was decided that BDH would also supply the new materials, leaving the Hull group free to carry on with research, and BDH would become subsequently a full member of the network.

Gray also raised a number of points that he thought were important for the development of cyanobiphenyls [48], of which we identify a few that are relevant here.

"Production of a new class of commercial materials cannot be driven to maximum advantage by an individual or a university group. The success therefore owed much to the partnership we had with the Defence Research Agency at Malvern and the commercial producer of our materials—BDH Chemicals Ltd. in the UK-now Merck (UK) Ltd. (London, UK). Without these alliances, the impact of the materials would not have occurred. The importance of collaboration also lies in the ease with which development problems and new needs can be tackled swiftly. The rapid commercialisation of the new materials owed a great deal to the easy relationship which developed between the University chemists and the large-scale industrial chemists and to the marketing skills of the staff at BDH Ltd. (Poole, UK)"

After the discovery and development of the cyanobiphenyls, the group at Hull became involved with the syntheses of various analogues of the cyanobiphenyls. The family of biphenyls included the alkoxy-cyanobiphenyls, alkyl- and alkoxy-terphenyls, and chiral cyanobiphenyls, di-esters and fluorenes, as shown in Figure 4. It is interesting that many of these motifs for materials construction were discussed in Gray's textbook *"Molecular Structure and the Properties of Liquid Crystals"*, published by the Academic Press Incorporated, London and New York, in 1962 [42]. Figure 4 shows how close Gray was to preparing cyanobiphenyls in 1962 and even to materials of interest today [49]. For example, the nitrobiphenyls and nitroterphenyls shown in Figure 4, are mimics of the cyano analogues as they have strong longitudinal polarities and large positive dielectric anisotropies, the biphenyl carboxylic acids are synthetically only one step away from the cyano-materials, whereas the lateral halogeno-materials are the forerunners of lateral-fluorinated compounds that dominated material design in the 1990s through to today. Surprisingly, Gray also prepared a lateral nitro-substituted biphenyl-carboxylic acid that was later found to exhibit a new phase called the D phase [50], and subsequently characterised as being cubic or a bicontinuous phase [51,52].

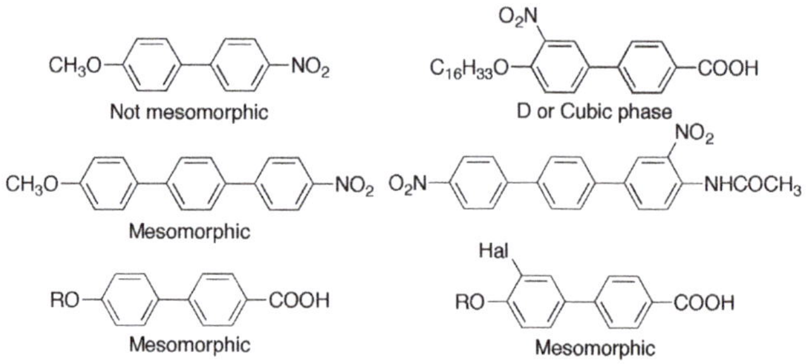

Figure 4. Mimics of cyanobiphenyls, synthetic possibilities in carboxylic acids, and lateral substitutions by nitro- and halogeno-units. The mesomorphic labels refer to Gray's identification of materials that are liquid crystals [42].

In terms of lateral substituents in Schiff's bases, Gray also examined many property–structure correlations for smectics and nematics and found many trends. In combination with Dewer [25], the group efficiency order obtained by Dewer was

$$NO_2 > CH_3 > N(CH_3)_2 > CH_3 > Cl > Br > H.$$

After the initial discovery of 4-pentyl-4'-cyanobiphenyl (5CB), various alkyl and alkoxy cyano-substituted materials, as shown in Figure 5, were prepared in order to provide a range of materials for mixture formulations. The terphenyls were to be used to raise the clearing points, and the chiral analogue, derived from (S)-2-methylbutanol, was used for chiral nematic mixtures and potential applications in phase change devices. The fluorene materials were designed to give the phenyl units in a potentially flat molecular architecture in the hope of improving relative physical properties. However, their melting points were much higher than the biphenyl analogues, and either the materials were non-mesogenic or had very low clearing points, and so they were not pursued further.

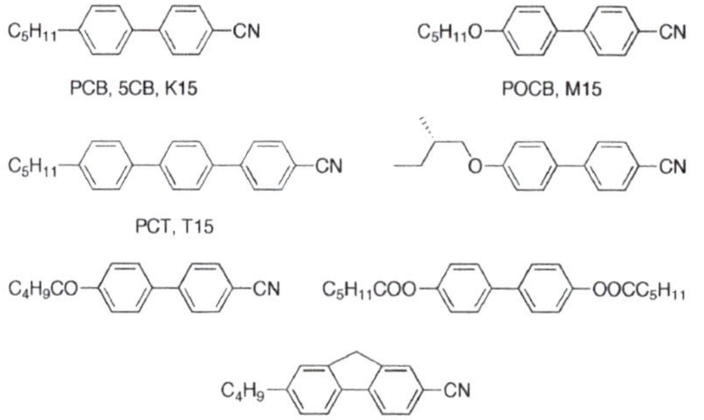

Figure 5. Range of structures of materials subsequently prepared based on the cyano-biphenyls [46,53–55].

In 1962, Gray [42] also reported on 4-methoxy-4'-nitrobiphenyl, as shown in Figure 4; but as it was nonmesogenic, a question was left open—could the nitro terminal group offer a better option than the nitrile as a terminal substituent? Therefore, a separate property–structure investigation was launched to prepare the nitro analogues of the alkoxy cyanobiphenyls as shown in Table 1. The table shows the comparative transition temperatures for the alkoxy cyano- and nitro-biphenyls with the same alkyl chain lengths (C_5 to C_8). It can be seen that the nitro-analogues had comparatively lower clearing (isotropisation) points than the nitrile compounds, and also shorter mesophase temperature ranges. It was concluded for practical purposes that the nitro-compounds were not competitive with the nitrile-analogues. Thus, this was the start of the development of numerous *structure–property correlations* for applications of materials in devices.

Table 1. Transition temperatures of the 4-alkoxy-4'-cyanobiphenyls and the 4-alkoxy-4'-nitrobiphenyls (°C) for property–structure correlations related to the use of cyano- over nitro-terminal groups [56].

X	Y	Cryst—Sm, N or Iso Liq Temp. (°C)	Sm—N Temp. (°C)	N—Iso Liq Temp. (°C)
CN	OC_5H_{11}	48	-	67.5
CN	OC_6H_{13}	58	-	76.5
CN	OC_7H_{15}	53.5	-	75
CN	OC_8H_{17}	54.5	67	90
NO_2	OC_5H_{11}	54.5	-	(<42)
NO_2	OC_6H_{13}	67	-	(32.5)
NO_2	OC_7H_{15}	36.5	(30.5)	38.5
NO_2	OC_8H_{17}	49; 51.5	49.5	51.5

Abbreviations: Cryst—crystal, Sm—smectic, N—nematic, Iso Liq—isotropic liquid.

3.4. Formulation of Mixtures

Property–structure correlations certainly gave the synthesis of materials at least some form of prediction on target design, but for practical applications there is always a need to formulate mixtures with wide temperature ranges, including room temperature. This means there is a requirement to know details about the eutectic points, for example, the relative proportions of the components in the mixture at the eutectic point. Why the eutectic point?

For two-component systems, provided that the equilibrium relies only on temperature, pressure and concentration, the phase rule states that the number of degrees of freedom (F) of the system is related to the number of components (C) and of phases (P) present at equilibrium by the equation: $F = C - P + 2$. At the eutectic point, the solids of the two components are in equilibrium with the liquid phase. There are consequently three phases present, and since the system involves two components, there can be only one degree of freedom according to the phase rule. Since the pressure is arbitrarily fixed at 1 atm, this represents one degree of freedom, and therefore the system has effectively no degree of freedom. This means that the eutectic point is completely defined and there is only one temperature where this equilibrium is possible. The point at which this happens is the lowest temperature at which any liquid mixture can be in equilibrium with the solid phases of the two components and is also the lowest temperature at which any mixture of the two will melt. Thus, the eutectic point similarly appears to be the melting point of a pure compound. With the liquid-crystal-to-isotropic-liquid transition, the nematic and liquid phases are fluids, and the transition temperatures from the nematic phase to the liquid tend to vary linearly with concentration in binary mixtures. Therefore, the broadest temperature range nematic phase appears to be between the lowest melting point (eutectic) and the corresponding clearing point (N to I).

In principle, the eutectic mixtures from two different binary phase diagrams can be used to produce a third eutectic point, thereby reducing the relative melting point of the third eutectic composition of four components. This means that the solidification can be suppressed, whereas the clearing point can be weight-averaged. Ultimately multicomponent mixtures can be developed to further suppress the melting point while at the same time weight-averaging the clearing point. In practice, for binary mixtures the eutectic point can be determined experimentally, whereas for a multicomponent system determining the eutectic point is very time consuming, and a hit or miss process. Rapid evaluations are therefore required by theory and verified by experiment.

Early studies were performed on mixtures of MBBA and EBBA, but with cyanobiphenyls the aim by experiment was simply to obtain the lowest possible melting points consistent with not sacrificing nematic thermal stability too greatly. The objective being to obtain mixtures melting, as distinct from solidifying at <0 °C and giving nematic properties up to about 50 °C.

The results were obtained by optical microscopy using heated or cooled stages, whereby the confirmations of melting temperatures were obtained by differential thermal analysis (DTA) [17]. DTA was used because of the problems of detecting the melting point by microscopy due to the paramorphotic defect textures of the solid state affecting or overpowering the textures of the liquid crystal. The results obtained were reasonable attempts to produce mixtures that could be compared to results found via various theories. However, to yield mixtures with the highest N-I value and the lowest mp at the time would have been serendipity. Better theories started to be deployed in 1973 for the estimations of eutectic points of mixtures, in the shape of the Schröder–van Laar Equation (1)

$$T_i = \Sigma \Delta H_{oi} \div [\Sigma(\Delta H_{oi}/T_{oi}) - \Sigma R \ln x_i] \tag{1}$$

where ΔH_{oi} is the molar heat of fusion for component I, T_{oi} is the melting point of pure component i (K), R is the gas constant and x_i is the mol fraction of component i. The results obtained were not as accurate as desired and so the theory was extended for determining the eutectic point of multicomponent mixtures via a semiempirical form of the Schröder–van Laar equation [57–59]. Melting points of eutectic mixtures were usually obtained to within 5 °C of the experimental results. These methodologies were used to create mixtures such as E7, E8, etc., as shown in Figure 6, with E7 becoming one of the most popular formulations used in research.

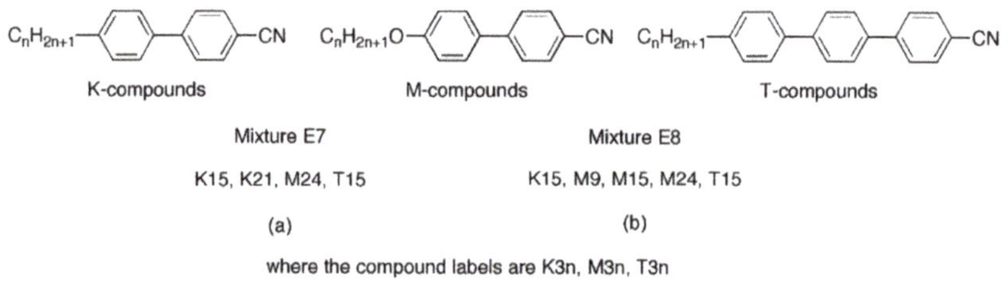

C_nH_{2n+1}—⬡—⬡—CN C_nH_{2n+1}O—⬡—⬡—CN C_nH_{2n+1}—⬡—⬡—⬡—CN

K-compounds   M-compounds   T-compounds

Mixture E7   Mixture E8

K15, K21, M24, T15   K15, M9, M15, M24, T15

(a)   (b)

where the compound labels are K3n, M3n, T3n

Figure 6. The components of the commercial mixtures (a) E7 and (b) E8.

From the beginning of the consortia in 1970, over a period of around 6 to 10 years, the development of the flat-panel industry was being revolutionised by international companies, universities and research establishments. Firstly, in the UK there were searches for information on displays and their developments from external sources, such as with companies as RCA, Bell Laboratories Ltd., TI Instruments and Ilixco, the formation of the Optel Corporation, input on materials from Merck, papers from conferences at the 1970 IEEE conference on Display Devices in NY, the International Conference on Liquid Crystals, articles from leaders (Castlelarno, Helfich and Schadt) in the field, etc. There were also an extremely large variety of inputs from the various members of the consortium, which included a number of government establishments, companies and surprisingly only one university (Hull). Externally, there were other inputs from materials suppliers and skill sets from universities and academics. People also joined the consortium, noticeably for Hull, as the university's short contract was extended with an increase in research assistants to two, when Ken Harrison joined John Nash towards the end of 1971. Even in the first year of existence many new ideas were floating around the materials side of the network. For example, it was realised that for the development of the dynamic light-scattering mode (DSM) and electrically controlled birefringence (ECB) devices, materials with larger negative dielectric anisotropies ($-\Delta \varepsilon$) were required, and in addition, methods for material formulations in mixtures were critical to expanding operational temperature ranges. For devices, new methods were being developed for the homeotropic and homogeneous alignment of the liquid crystal mixtures and for bonding devices.

3.5. Recognition

At this juncture, individual desired materials had been prepared in large quantities, they had been used in formulating various eutectic mixtures, which were finding ways into the marketplace for utilisation in flat, thin displays, and so the commercialisation process was under way. The research successes and transformative applications brought recognition in the form of the *"Queen's Award for Technological Achievement"* to various members of the consortia, including RRE, BDH and Hull University. For Hull, it was the first award of its type to a university in the UK. Figure 7 shows a photograph of Kirton, Hilsum, Raynes (RRE), Gray (Hull), Sturgeon and Pellatt (BDH) at the University of Hull for the Queen's Award, along with the physical appearance of 0.5 kg of 5CB in a one-litre flask at room temperature. Below is the quotation from the document announcing the Queen's award for technological achievement.

Figure 7. Left, John Kirton, Cyril Hilsum, Peter Raynes (RRE), George Gray (University of Hull), Ben Sturgeon and Martin Pellatt (BDH) at Hull for the Queen's Award for Technological Achievement in 1979; centre and right: the chemical structure of 5CB and its physical appearance at room temperature in a one-litre flask.

THE DEPARTMENT OF CHEMISTRY, UNIVERSITY OF HULL

Greetings!

We being cognisant of the outstanding achievement of the said body as manifested in the application of Technology in Our United Kingdom of Great Britain and Northern Ireland, Our Channel Islands and Our Island of Man and being desirous of showing Our Royal Favour do hereby confer upon it

THE QUEEEN'S AWARD FOR TECHNOLOGICAL ACHIEVEMENT

For a period of five years from the twenty-first day of April 1979 until the twentieth day of April 1984 and do hereby give permission for the authorised flag of the said award to be flown during that time by the said body and for the device thereof to be displayed in the manner authorised by Our Warrant of the fifth day of April 1976.

And We do further hereby authorise the said body during the five years of the currency of this Our Award further to use and display in like manner the flags and devices of any current former Awards by it received as prescribed in the eighth Clause of Our said Warrant.

Given at Our Court at St. James's under Our Royal Sign Manual this twenty-first day of April in the year of Our Lord 1979 in the twenty-eighth year of Our Reign.

By the Sovereign's Command

The materials effort in the UK continued to focus on the development of new and improved nematogens for applications in displays such as the TNLCD device. Other displays, such as the dynamic scattering device, lost favour, whereas other new concepts vied for interest. Thus, there was a need for materials for the two-frequency switching mode, the supertwist nematic device (STN) and for multiplexed passive and active TFT addressed displays. These new applications required faster and sharper switching modes, bistable operation, better contrast and brightness, a wider viewing angle and lower operating voltages, and materials were sought with appropriate physical properties to meet these demands. In the meantime, the quantities of liquid crystals required grew into the tonnage scale. Applications other than displays also became of interest, for example in telecommunications, sensors, spatial light modulators, beam steering and switches, polymers for adhesives and alignment agents, etc. But for materials research? Gray commented on the discovery of the cyanobiphenyls, *"Many alternatives that did emerge during the next eight years were in fact cyanobiphenyl mimics or look-a-likes. Once chemists understood the strategy we used, cyclohexyl, pyrimidyl, and dioxanyl analogues appeared. The point is however that to be of greatest effect, an invention has to be timely—again bringing in something of the element of luck or chance."* [48].

In its academia links to the consortia, Hull was free to research into other areas that that were still in their infancy. Liquid crystal research areas of interest included smectic and discotic phases, new synthetic methodologies which would allow access to nematogenic materials that were hitherto inaccessible, and high and low birefringent materials for display and nondisplay applications. Overarching these topics was also the possibility of introducing chirality either at a molecular level or in a structure of a mesophase. As Gray began this new frontier, he was joined by other academic staff in Hull, including Drs Toyne, Scrowston, Biggs and Lacey, and subsequently, near his retirement, he was joined by his selected successor Goodby, brought back from AT&T Bell Laboratories, USA, whom Gray had arranged to be an industrially funded reader in Hull by Thorn EMI and STL. A year later Goodby became professor and the consortium contract holder in Hull, head of the Liquid Crystals Group and Organic Chemistry, and subsequently head of the Department of Chemistry. The chart shown in Figure 8 shows how the materials activities expanded rapidly in the new areas laid down by Gray and Goodby, starting with compounds based on cyano-biphenyl where interests lay now in their optical rather than electrical properties. Both high and low birefringencies were of interest, with high values being investigated for non-display applications and low values for thin display devices. Chirality and ferroelectricity were also topical for bistable fast switching devices. Therefore, the interconnected activities expanded rapidly as shown in the chart in Figure 8, along with an expansion in research applications, reporting, publishing and patenting.

Conversely, the establishments and agencies of the Ministry of Defence (MoD) were reorganised and streamlined in 1991 by creating the Defence Research Agency (DRA), which included RAE and RSRE. In 1995, this metamorphosed to include other agencies in the formation of the Defence Evaluation and Research Agency (DERA). Subsequently in 2001, the MoD split the DERA into two: QinetiQ, which became the sixth largest defence contractor in the UK, and the Defence Science and Technology Laboratory (Dstl). Goodby steered the research activities of the consortium group at Hull through all of these changes until defence needs became redefined.

3.6. Nonlinear Optics

Nonlinear optics was a forthcoming area of interest to those working in liquid crystals in the 1980s, particularly on materials with positive dielectric anisotropies that would have donor–acceptor groups. Second and third order effects were being explored in the examination of the surface organisation of molecules, in beam steering devices and wave guides, light scattering modes, optical processing, optical filters and in various switching effects, e.g., for telecom devices [60,61].

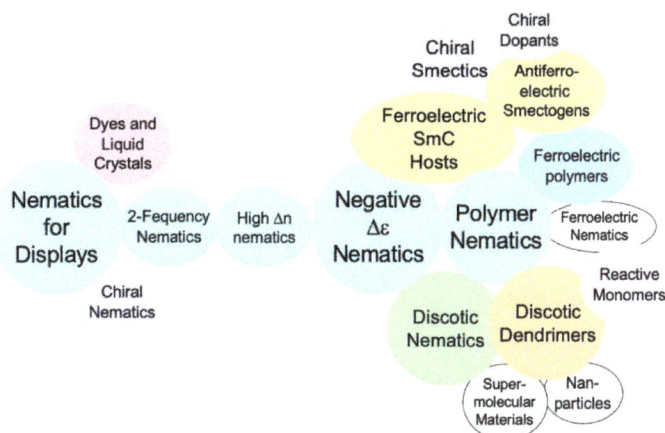

Figure 8. Topics of research on materials starting from ~1970, expanding around 1980, and ending in ~2012. The size of the discs indicated the degree of activity on each topic.

Many organic materials that exhibited NLO effects had similar structures to cyanobiphenyls, and therefore it was easy to take a side-step switch to explore new material designs. Often infrared and microwave light were the target for novel device concepts, particularly in the areas of optical-switching, frequency-doubling and frequency-tripling applications. Often this meant having control over the birefringence of a mesophase; for example, high birefringence was of interest in microwave applications (Δn ~0.3 to 0.5), whereas low birefringence was of more interest in thin displays, such as those found in surface-stabilised ferroelectric devices (SSFLCDs, Δn ~ 0.05 to 0.2).

There were also differences between material systems possessing rod-like shaped molecules and those having disc-like molecules. With rod-like molecules, materials could be designed to have donor and acceptor groups, and hence could exhibit second and third order effects, whereas the symmetry of disc-like molecules meant that such materials favoured third-order properties. Thus, the Universities of Leeds and UEA joined the materials consortia collaborating with Hull on certain aspects of synthetic methods.

Knowing how to control the magnitude of the birefringence became one of developing structure–property correlations. Raising the birefringence (Δn) became a task of basically increasing the number of delocalised π-electrons, and relative polarisability ($\Delta \alpha$), within a molecular architecture without losing mesophase properties or other required materials properties, such as colour in dyed systems. Conversely, lowering the birefringence became a task of replacing π-bonds with σ-bonds. In Figure 9, a chart is shown for pentyl-substituted molecules, with C_5H_{11} denoted as R. Starting at the left-hand column, the structures of the molecules are essentially the same with the only major change to the acceptor groups on the right-hand side, i.e., CN to NCS and the addition of F; the terminal polar groups increase the longitudinal dipole and so the electrons are moving left to right and therefore the birefringence increases, which is mirrored by an increase in the polarisability, whereas the values for the order parameters remain roughly the same. For the centre column, the argument is slightly different as this time, the left-hand sides of the molecules are donating electrons into the central core, but the result is the same: the birefringence increases down the column but not to the extent that it does in the first column. In the right-hand column, the central core of the molecules is increased, extending the extent of the delocalised electrons. Coupled with this are changes to the terminal polar groups, but for the molecules, the longitudinal dipoles increase, thereby the birefringence also increases. For the three columns, the order parameter (S) does not appear to play a major role in determining birefringence, but the linked polarisability and the polarity have a degree of balance in determining the birefringence.

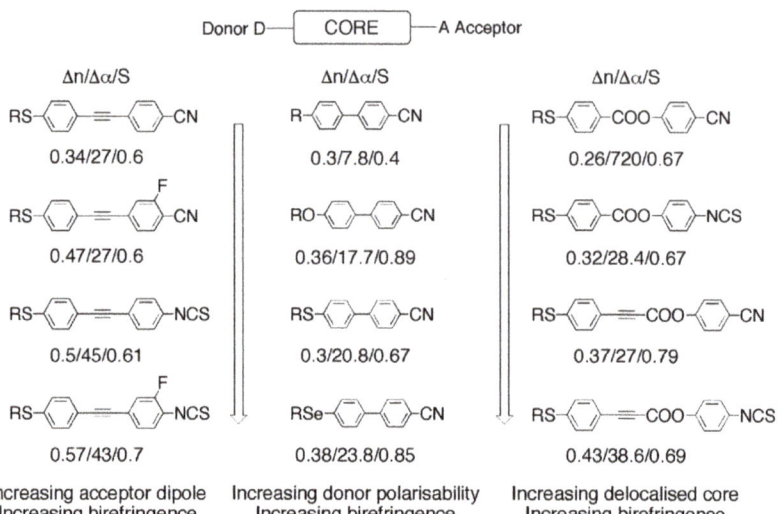

Figure 9. Correlations for birefringence in rod-like materials based primarily on biphenyl motifs.

Such linear correlations shown in Figure 9 can be expanded into a larger number with some crossovers between the correlations, thereby creating arrays of linkages between the physical properties of the materials. For instance, Figure 10 shows how a number of property–structure activities can be linked together to create a two-dimensional landscape for materials that have donor–acceptor molecular structures. The arrays allow for projection beyond those materials that had been synthesised, thereby leading to the prospect of target selection at a distance away from the original development strands and improving the possibility of discovery and invention.

Acceptor groups for materials that might be predicted to exhibit mesomorphism include NO_2, CN and NCS, whereas donor groups include R, OR, SR, NH, etc. In addition, the length for a rod-like material affects the charge separation between the acceptor and donor (A and D) groups, and thereby the polarisability and dipole. The larger the longitudinal dipole, the greater the birefringence. However, the longer the molecule is on average, the greater the effect is on the melting point, with the bigger the molecule, the higher the melting point. Unfortunately, for certain applications a lower melting point was desired. For disc-like molecules, due to their symmetry, usually there are no donor and acceptor groups, and so the strength of any nonlinear effect is dependent on the number of π-electrons and the polarisability. Overall, numerous materials were prepared, and many examples of families of materials, their property–structure correlations, birefringencies ($n_{||}$, $n_\perp$ and Δn), polarisabilities ($\Delta\alpha$) and order parameters (S) were reported in the literature [62–69].

3.7. Nematogens and Smectogens with Negative Dielectric Anisotropies

This particularly important research programme was based on the Nobel Prize awarded to Heck, Suzuki and Miyaura in 2010, for innovations in synthetic chemistry utilising palladium-based coupling reactions [70–72]. In the Suzuki variation, a double-bond-containing molecule is replaced by an organoboron substrate, with palladium acting as a catalyst in a cross-coupling procedure between two different substrates, one featuring the organoboron moiety and the other a good leaving group (Br, I, triflate), thereby it was possible to link two different phenyl units together to give asymmetrically substituted biphenyl and terphenyl products. Studying the publication by Suzuki and Miyaura in the late 1980s, Ken Toyne thought that the synthetic technique could be employed in the synthesis of laterally substituted fluorinated terphenyls, which might be used to make materials with negative dielectric anisotropies. Similar thoughts were ongoing at E. Merck (now owner of BDH), and so the two teams met through the consortium to discuss relative activities. At the beginning [73,74], both groups worked generally on the synthesis of a variety materials, but ultimately E. Merck took the lead in research on nematics for large area displays, whereas Hull tended to focus on smectics for small area microdisplays, and for larger area bistable multiplexed devices.

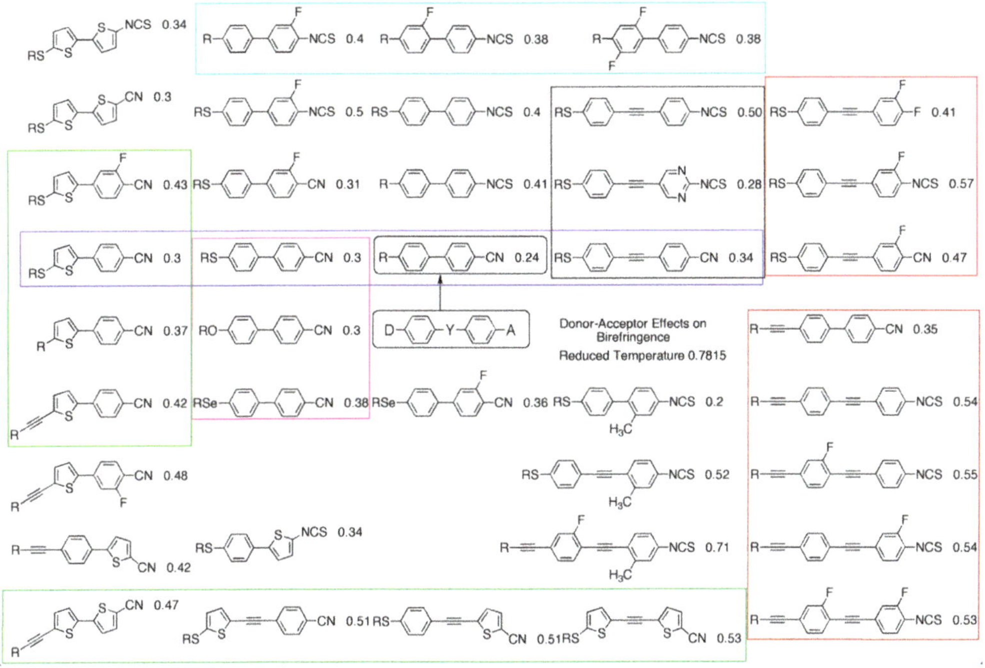

Figure 10. Linked property–structure correlations for birefringencies in donor–acceptor liquid crystals. The coloured boxes are used to identify correlations of materials families, the colour depicting overlapping families.

For nematics with negative dielectric anisotropy, it was already known that it was possible to create electrically controlled birefringence (ECB) displays. In the off state, the molecules are aligned orthogonally with respect to the substrates of a device. Upon applying an electric field, the molecules tilt away from the orthogonal state to give a bright on state. Such devices were being examined within the network in the early 1970s. This so-called vertically aligned configuration was much later called VA, after vertical alignment, and in the initial devices, switching times of around 25 ms [75] were possible. It was also found that VA displays could provide a high brightness with good viewing angles. The reasonable response times and the possibility of creating controlled multidomains that could give symmetrical and wide-angle viewing meant that this mode might be adapted for TV applications.

For VAN/ECB-LCD modes, materials [76] were needed that possessed large dipoles located across the molecular long axes, so that the materials would have a large negative dielectric anisotropy. In addition, they were required to have suitable birefringencies relative to the spacing thickness of the device, and with a relatively low viscosity for fast response times.

In comparison, materials with positive dielectric anisotropies possess longitudinal dipoles, and as such molecular rotation around the long axis, which is fast (10^{11} s^{-1}) and has less effect than for materials with negative dielectric anisotropy, where the lateral dipole is more affected by the slower rotation of 10^6 s^{-1}. In addition, the incorporation of multiple polar groups positioned along the long axis was not easy to achieve synthetically or to have them all point in the same direction across the molecule, unless they were fixed to the same phenyl ring. Furthermore, the lateral moiety that is polar would also have to be relatively small in order to retain mesomerism and at the same time not to increase viscosity. Consequently, fluorine was preferred over moieties such as nitrile, which has the adverse effect of raising viscosity due steric hinderance. So, as with nematogens of positive dielectric anisotropy, fluoro-substitution held the key to developing practical device materials for negative dielectric anisotropy. Achieving all of these goals was taking molecular engineering to a new level, which was only possible at the time via use of the Suzuki–Miyaura coupling methodologies in synthetic pathways.

Deploying fluorine instead of nitrile in the material design did have some drawbacks in terms dielectric anisotropy as shown in Figure 11. Some examples of nematogens that possess two lateral polar groups (F and CN) fixed to the same side of a phenyl ring are shown along with the relative dielectric anisotropies. The materials have much larger negative dielectric anisotropies as one might expect. However, materials designed in this way, with two substituents on the same ring, have additive effects from the polarities of both polar groups. Comparatively, two fluorine atoms attached to adjacent positions on an aromatic ring, when combined, have a polarity a little bit less than a nitrile unit, but at the same time imparting less towards the viscosity, as shown in the figure. Moreover, additional fluoro-substituents can be added to give di-, tri, and tetra-analogues, with little change to the dielectricanisotropy. In addition, the conversion of one of the aromatic rings to cyclohexane can generate important nematogens that are suitable for displays.

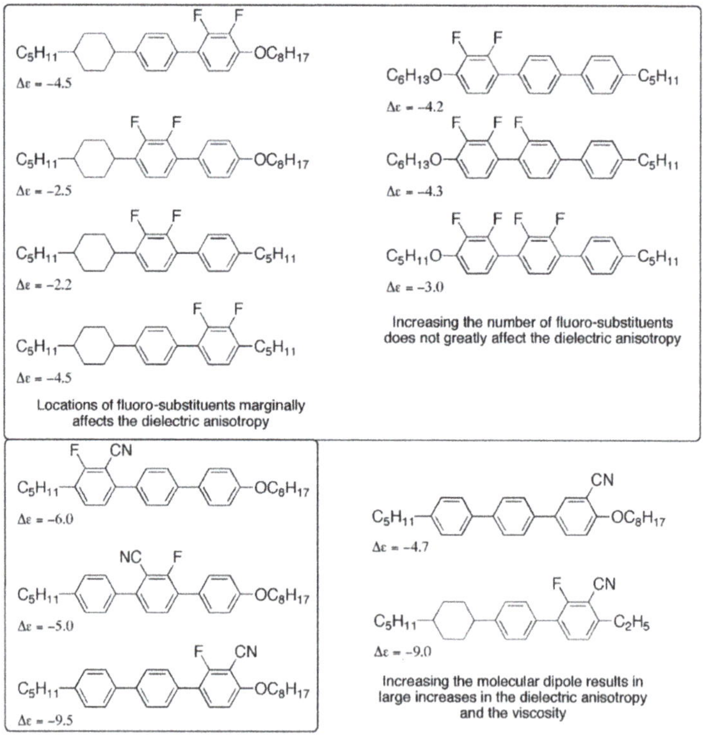

Figure 11. The effect of the incorporation of polar lateral groups in disubstituted terphenyls and cyclohexylbiphenyls [77].

Interestingly, the mesomorphic behaviour is determined by the lengths of the external aliphatic chains, and whether or not they are alkoxy or alkyl, as shown in Table 2. Similarly, the location of the phenyl ring carrying lateral polar group(s) can be used to determine mesomorphic behaviour in terms of transition temperatures, phase sequences, dielectric anisotropies and viscosities, and to what devices the materials are best suited for. Apart from nematic devices, the materials as shown in Table 2 also exhibit smectic C phases, as the aliphatic content is extensive. If the materials shown are substituted with a chiral aliphatic chain instead of a normal chain, the smectic C phase will also become chiral, or alternatively, if they are doped with a chiral material, then the mixture will also exhibit chirality. In both cases if this results in a chiral smectic C phase being formed, then it will exhibit ferroelectric properties. At this point a divergence had occurred in the research paths of Merck and the consortium, and Hull followed a path along research into the syntheses and various properties of ferroelectric materials.

Table 2. Comparison of transition temperatures (°C) as a function of terminal chain length for the dialkyl-2′,3′-difluoroterphenyls.

Structure	Mesomorphic Behaviour
C_5H_{11}—⟨⟩—⟨F,F⟩—⟨⟩—C_5H_{11}	Cryst 60.0 N 120.0 Liq
C_5H_{11}—⟨⟩—⟨F,F⟩—⟨⟩—C_7H_{15}	Cryst 36.5 (SmC 24.0) N 111.5 Liq
C_5H_{11}—⟨⟩—⟨F,F⟩—⟨⟩—C_9H_{19}	Cryst 42.5 SmC 66.0 N 110 Liq

Abbreviations: Cryst—crystal, Sm—smectic, N—nematic, Liq—isotropic liquid.

In the design of smectogens that will either act as hosts or be chiral, it is important to understand the elements that will drive the molecules to tilt in the smectic state, and thereby to form synclinic (ferroelectric) smectic C or anticlinic (antiferroelectric) smectic C_A phases. The generation of the smectic C phase has been thought to be related to what were termed "outboard" terminal dipoles, or the location of a polar atom between a terminal aliphatic chain and a rigid core unit, as shown by the materials in Figure 12. The presence of terminal polar groups was theorised by McMillan [78] to reduce the molecular rotation around the long molecular axis, thereby allowing for a molecular torque to occur in the planes of the layers in the smectic phase, resulting in the generation of a molecular tilt and hence the formation of a potential smectic C phase. This result is also obtained for the fluorinated terphenyls in the figure, whereby oxygen being located at one end or both ends of the aromatic core unit can be used to control phase sequences, transition temperatures and other related properties.

Although we have shown studies with difluoro-substitution on one phenyl ring, as noted, the mesophases properties depend on which ring in terphenyl, for example, is substituted. It is possible to have fluorination on adjacent rings and also have more than two fluorine atoms attached to the aromatic core, see Figure 11. For terphenyls, this gives a plethora of possible substitutions and associated isomers that may be prepared, and which might suit certain mesophase types and applications. Moreover, although configurational isomers can be fixed by synthesis, conformational variants have not been fully explored, in particular in relation to rotations about the phenyl–phenyl bonds in terphenyls which cause interannular twisting. The possibilities of examining conformational interactions have been examined through dielectric studies of both anisotropy and biaxiality and theoretically through computer simulations [79]. However, to this date only a small area of the isomer landscape has been accessed; nevertheless, such materials have found practical and commercial uses in displays for example in VAN-, SSFLCD- and τ_{vmin}LCD-mode devices. In the development of materials created via coupling reactions, Kingston Chemicals Ltd. was created in 2000 to serve small technology companies and universities with advanced organic materials, mostly based on fluorinated compounds.

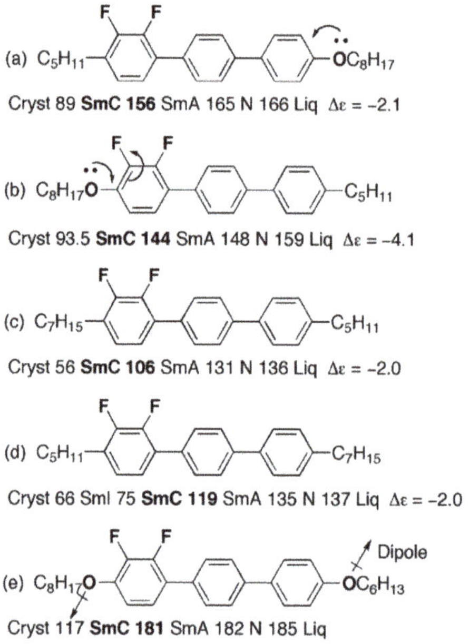

(a) C_5H_{11}–[structure]–OC_8H_{17}

Cryst 89 **SmC** 156 SmA 165 N 166 Liq $\Delta\varepsilon = -2.1$

(b) $C_8H_{17}O$–[structure]–C_5H_{11}

Cryst 93.5 **SmC** 144 SmA 148 N 159 Liq $\Delta\varepsilon = -4.1$

(c) C_7H_{15}–[structure]–C_5H_{11}

Cryst 56 **SmC** 106 SmA 131 N 136 Liq $\Delta\varepsilon = -2.0$

(d) C_5H_{11}–[structure]–C_7H_{15}

Cryst 66 SmI 75 **SmC** 119 SmA 135 N 137 Liq $\Delta\varepsilon = -2.0$

(e) $C_8H_{17}O$–[structure]–OC_6H_{13} / Dipole

Cryst 117 **SmC** 181 SmA 182 N 185 Liq

Figure 12. The effect of the incorporation and position of the polar lateral fluoro-substituents in disubstituted terphenyls on the dielectric anisotropy, birefringence and viscosity. The curved arrows show the potential movement of electrons from electron-rich areas to electron-deficient locations. In turn, this gives an image of electronic polarisation.

3.8. Gels and Polymers

Apart from the design, synthesis and applications in low-molar-mass rod-like systems, the use of coupling reactions was important in the creation of discotics, gels, oligomers, dendrimers and polymers. These topics covered various applications in areas such as adhesives, ferroelectrics, pyroelectrics, filters, coatings, alignment and high yield strength materials. The basic science and topics are wide ranging and so we cover only a few examples here.

Crossing over from low-molar-mass materials to polymers we enter the area of gels based on the research of Hickmet [80,81]. We prepared gels composed of a cross-linking monomer unit and a mixture of mesomorphic materials. In the design of the formulation, the compounds were selected to be of a similar size and chemical nature and polymerised to give a desired gel network [82–84]. Figure 13a shows a typical formulation for a gelating mixture of liquid crystals based on an achiral host mixture of difluoroterphenyls, a chiral dopant to impart ferroelectricity, and 10 wt % of a cross-linking monomer to produce a gel [85]. When subjected to photopolymerisation, a gel existed in a nematic phase above 98.5 °C; on cooling, a smectic A* phase was formed followed by a ferroelectric smectic C* phase at 83.0 °C. Of course, the gel did not crystallise at low temperatures below −20 °C before starting to show semblances of glassifying. At room temperature, the response time in the ferroelectric phase was just 40 ms in an electric field of 10 V mm^{-1}, i.e., a response time much faster than for the host/dopant mixture without polymerisation. Therefore, we move from conventional low-molar-mass materials to gelated forms that can be manipulated in many different ways.

As shown in Figure 12, the upper two materials (a) and (b), with a single alkoxy group, show the same phase sequences and similar transition temperatures, whereas material (c) with no alkoxy groups still exhibits a smectic C phase, but at a much lower temperature. This indicates that the difluoro-substituted unit is still contributing to the induction of the molecular tilt. With respect to the upper two materials, the dielectric anisotropies are different, with the lower material (b) having the higher value, which is associated with the conjugation between the oxygen and fluorine atoms, whereas this is not the case for material (a) where the polarisation through conjugation is weak. For compounds (a) and (c), the dielectric anisotropies are similar, indicating that the major contribution to the formation of the smectic C phase is through the molecular packing, but the lower transition temperatures for (c) indicate that the contribution is less than for (a), which has a larger outboard dipole associated with oxygen. Consider now the pair of compounds (c) and (d). They are analogues of one another with the two terminal chains swapped around. They have very similar transition temperatures indicating that the positions of the alkyl chains do not greatly affect mesomorphic behaviour. Compound (e), having two terminal outboard dipoles, has the classical structure for a smectic C material. It has the highest thermal stability for the smectic C phase. Overall, the least polar materials (c) and (d) have the lowest viscosity, the lowest tilt angle, whereas (e) has the highest tilt and the highest viscosity. Therefore, formulated mixtures need to have a balance of components to give optimal properties.

It is of course possible to take monomers such as those employed in making gels and to polymerise them on their own in order to form networks. In the example below, shown in Figure 13b, we use a novel prepolymeric system based on diallylamine to create a network. The mesogenic unit is again based on a difluoroterphenyl and exhibits smectic C and nematic phases in which photopolymerisation can take place. Preorganisation of the monomer species can retain to some degree its organisation in the photopolymerised nematic or smectic C phases. Such organised networks are of use in preparing polarisers, colour filters, and optical compensators. Without the use of a mesogenic unit between the diallylamine groups, materials can be created that are of use in coatings, adhesives, ionic liquids, etc. Two spin-off companies were formed by DERA to exploit various possible applications, one called NPS in 2002 and the other IPS in 2005.

3.9. Nematic Disc-like Materials

In the late 1970s, Hull became involved with the synthesis of disc-shaped molecules based on hexa-substituted benzenes and triphenylenes. This area of research became revisited at a similar time as research that was ongoing on terphenyls because there were some analogies in the synthetic methodologies. Research thus went into the synthesis of bowl-like molecules and oblong-shaped coordination complexes, for columnar ferroelectricity and nematic biaxiality, respectfully. However, by the mid-1990s, research had switched to a search for nematic discotic triphenylenes possessing negative birefringencies and potentially negative dielectric properties with the possibility of searching for biaxial nematic phases.

The first method used in 1978 to produce the 2,3,6,7,10,11-hexamethoxytriphenylene involved the oxidative trimerisation of veratrole with a third of an amount of chloranil. The process was tedious and produced low yields. A second method involved the cyclisation of veratrole in the presence of iron (III) chloride, which also gave low yields. However, synthetic refinements learned through collaborations with UEA in the consortium gave almost quantitative yields of the hexa-methoxytriphenylene. Demethylation with boron tribromide gave suitable yields of 2,3,6,7,10,11-hexahydroxytriphenylene, which could be derivatised to give discotic liquid crystals. Some of the more important materials that were prepared are shown in Figure 14. These materials are the hexa-benzoate derivatives of hexa-hydroxytriphenylene, where the benzoate esters could be designed to incorporate lateral groups, which might be polar (halogens) or apolar (aliphatic chains).

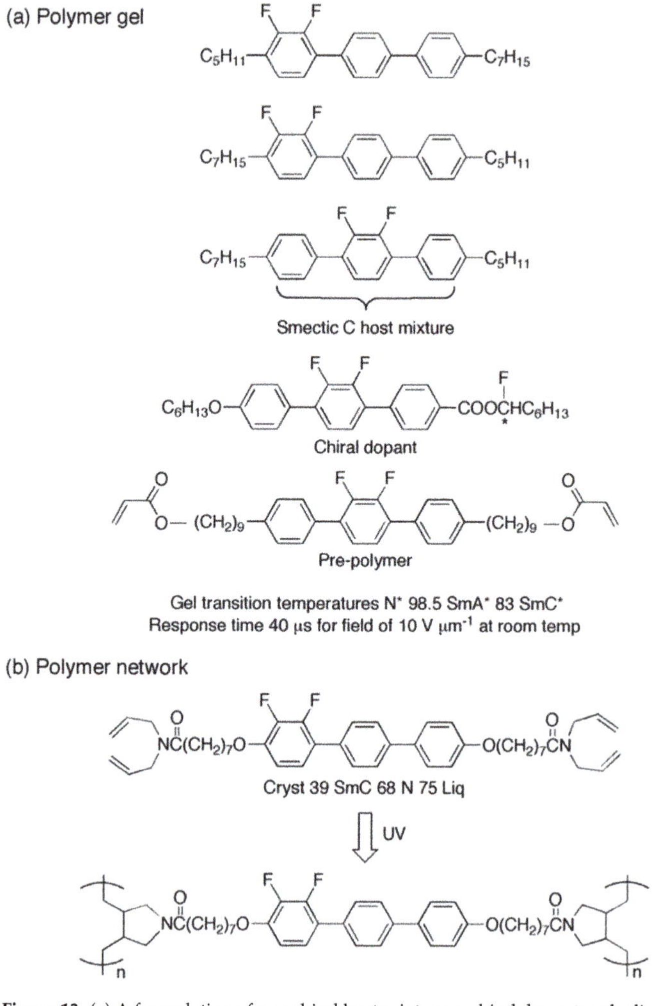

Figure 13. (a) A formulation of an achiral host mixture, a chiral dopant and a liquid crystal monomer for use in forming a liquid-crystalline gel [85] and (b) network polymeric materials formed via photopolymerisation of substituted diallylamines [86]. * indicates the location of a stereogenic centre.

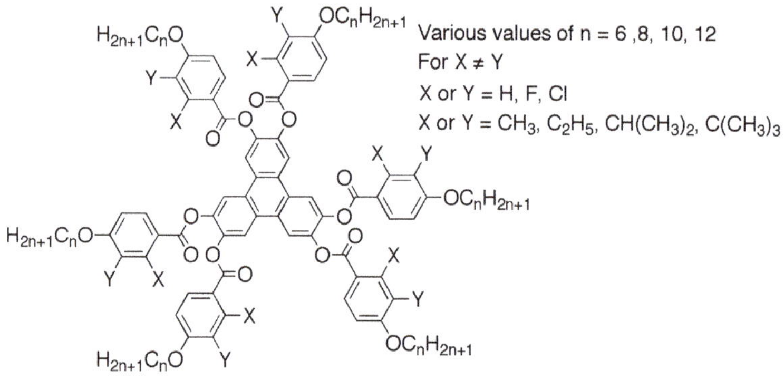

Various values of n = 6, 8, 10, 12
For X ≠ Y
X or Y = H, F, Cl
X or Y = CH_3, C_2H_5, $CH(CH_3)_2$, $C(CH_3)_3$

Figure 14. Template for targeting nematic discotic liquid crystals.

Stacking of triphenylene units together is more likely to form columns of molecules, and therefore, the incorporation of benzoates seemed a likely way to generate discotic nematic phases. To achieve this, substitution that prises apart the molecular discs was postulated, for which a wide variety of substitutions were tested, as shown in Figure 15. It was found that if the substituents in the benzoate were too large (C_2H_5, etc.) or too polar (F, etc.), nematic phases were not achieved. A methyl substituent, in the two or three positions of the outer benzoate moieties, were the only material types that would support nematic phase formation. Figure 15 shows the aromatic region of benzoate-substituted triphenylenes. The left-hand model has no substituent in the external benzoate ring, whereas the right-hand model has a large tert-butyl substituent; in this way, the separation of adjacent triphenylene rings can be illustrated. The large tert-butyl moiety sterically hinders the internal packing of the discs and twists the benzoate rings out of the plane of the triphenylene core, the benzoate rings then prising apart the molecules, thereby preventing columnar formation (and in this particular case actually preventing nematic phases forming). It appears that the methyl substituent is the right size to partially prise the discs apart allowing for the formation of the nematic phase.

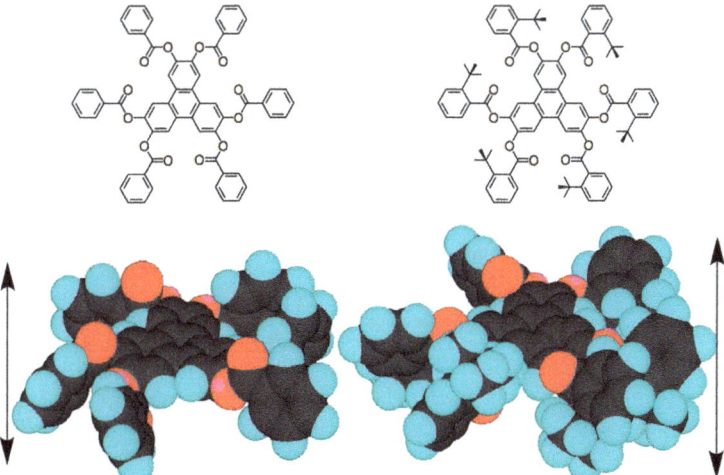

Figure 15. Steric hinderance by the benzoate esters around the central core region of triphenylene prises apart the molecules so that they look like disordered piles of pennies in the nematic discotic phase. Modelling simulations performed using ChemDraw 3D.

Table 3 shows the structures of triphenylene-2,3,6,7,10,11-hexahexayl hexakis(4-alkoxy-2 or 3-methylbenzoate(s) where the methyl group is either pointing inwards towards the triphenylene core (inner) or away from the core (outer), where the exterior alkoxy chain is also varied in length. All of the examples shown exhibit discotic nematic phases, with melting points around 100 °C and clearing points for the most part beginning around 200 °C. These temperatures are much lower than those of the nonmethyl-substituted analogues, which have greater tendencies in also forming columnar mesophases. These property–structure correlations show that lateral substitution in external benzoate rings is of practical use in the design of nematic-discotic materials for a variety of applications [87–90].

Table 3. Comparison of transition temperatures (°C) as a function of terminal alkoxy chain length and lateral substitution in the Triphenylen-2,3,6,7,10,11-yl hexa-4-alkoxybenzoates.

R'	X	Y	Cryst		DT		Drd		ND		Iso Liq
C_6H_{13}	H	H	•	186	•	193			•	274	•
C_8H_{17}	H	H	•	152			•	168	•	244	•
$C_{10}H_{21}$	H	H	•	142			•	191	•	212	•
$C_{12}H_{13}$	H	H	•	146			•	174			•
C_6H_{13}	CH_3	H	•	126					•	233	•
C_8H_{17}	CH_3	H	•	134					•	223	•
$C_{10}H_{21}$	CH_3	H	•	103					•	192	•
$C_{12}H_{13}$	CH_3	H	•	99					•	163	•
C_6H_{13}	H	CH_3	•	172					•	230	•
C_8H_{17}	H	CH_3	•	126					•	198	•
$C_{10}H_{21}$	H	CH_3	•	109					•	165	•
$C_{12}H_{13}$	H	CH_3	•	105					•	129	•

Abbreviations: Cryst—crystal, DT—tilted columnar, Drd—rectangular disordered discotic, Iso Liq—isotropic liquid.

The control over the synthetic pathways developed between Hull, UEA and Leeds Universities for such materials, and the high yields which could be achieved for their production, meant that they would be of practical use. As with the ferroelectric materials described earlier, it was possible to create polymeric networks (EU Orchis network 1989), with interest being in various areas of molecular electronics. In other areas of R&D their properties of negative birefringence of nematic discotics were of interest in optical films, in particular for applications in optical compensation films for various devices. The beautiful work at Fuji film on these films and related materials resulted in the invention and development of films for wide viewing angles in nematic displays [91].

3.10. Chiral Materials—Liquid Crystals and Dopants

Various forms of chirality permeate throughout liquid crystals from molecules [92] to mesophase structures [93]. Knowing the connectivity and relationships between the left-hand and the right-hand can be invariably important. An amusing story about this was told by George Gray: "BDH Ltd. sold commercially both Hull's right- and left-handed compounds. Customers could choose which to use. One customer decided to do better than everyone else and to use some of each additive. Of course, the two cancelled out and the effect was zero- and he complained most bitterly that our products were no good. He had to be gently educated".

Gray's story is amusing given that the first chiral cyanobiphenyl was prepared in 1973, and at that time there was no relationship between stereogenic architecture and mesophase macrostructure other than the physical properties of enantiomers would be opposite to one another. However, for mesomorphic materials, there was a need for the purposes of mixture formulation to either reduce helical pitch length or expand it. In 1976, relationships between stereochemistry, molecular structure and helical twist direction were developed for materials with single stereogenic centres, with the following relationships:

$$Rel \rightarrow Sed$$

$$Rod \rightarrow Sol$$

where R and S are the Chan, Ingold, Prelog systematic labelling systems for asymmetric centres, o and e are the parities (odd and even) for the number of atoms the centre is removed from the central rigid molecular core, and d and l (dextro and laevo) being related to the optical rotation direction for the helical structure in a chiral nematic phase. These Gray and McDonnell rules [55] were applied for formulations that did not necessarily have single enantiomers in their mixtures [94].

For smectic C ferroelectric liquid crystals, they too are dependent on the relationships between stereochemistry and broken symmetries on the macroscale [95], but in this case extra relationships are needed for the development of formulations. These include the direction of the spontaneous polarisation (Ps+ and Ps−) and the direction of the dipole at the stereogenic centre (+I, −I), thereby giving us similar rules to those of Gray and McDonnell, but with two extra terms as shown below [96].

$$(+I) \text{ Rod } (Ps-) \rightarrow (-I) \text{ Rol } (Ps+)$$

$$(+I) \text{ Sol } (Ps+) \rightarrow (-I) \text{ Sod } (Ps-)$$

$$(+I) \text{ Rel } (Ps+) \rightarrow (-I) \text{ Red } (Ps-)$$

$$(+I) \text{ Sed } (Ps-) \rightarrow (-I) \text{ Sel } (Ps+)$$

Many property–structure correlations were drawn up for ferroelectric smectic liquid crystals that related helical twist and spontaneous polarisation directions to molecular stereochemistry in an attempt to formulate mixtures where the values of the spontaneous polarisation and the helical pitch length were maximised. For devices, this optimised situation meant that helicity did not affect alignment, and a high polarisation reduced the switching voltage. A simple correlation between molecular architecture and properties is shown in Figure 16, which can be applied to some degree to nematogens and smectogens that have single stereogenic centres.

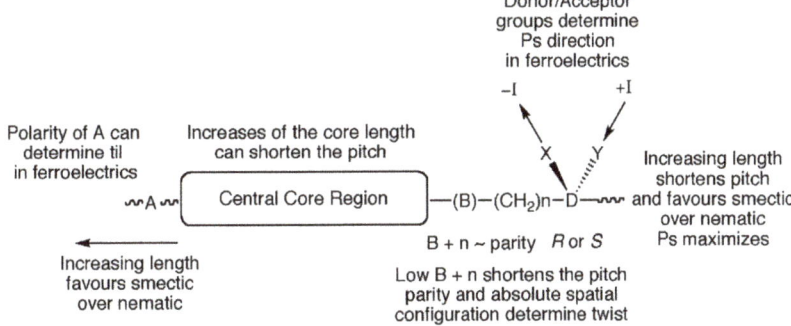

Figure 16. Property–structure correlations for chiral rod-like mesogens that exhibit either chiral nematic or ferroelectric smectic C* phases.

3.11. Serendipity—Polar Nematics

As the consortium network was nearing to its end, York (our new location) was asked to participate in two new projects; one on low birefringent ferroelectric materials that operated at ambient temperatures and another on extremely polar nematics. The first, we have somewhat discussed earlier on in this article, and as usual, to lower the birefringence required the inclusion of alicyclic ring systems, but the downside to this approach was tilted smectic phases were not usually favoured. Using an end-group design and new synthetic methodologies, this was achieved.

For the second project, achieving extremely high polarities meant using singular or multiples of strongly polar substituents. Discussions at the start of the programme focused on the use of nitro units as the polar moieties, see Figure 17. Therefore, to generate a high polarity meant having as high a proportion of nitro units as possible in the molecular structure of the target. A small molecule with a single nitro unit that mimicked cyanobiphenyls was considered, but it was already known from Gray's results, shown in Table 1, that the nitro analogues were poor nematogens. However, from Gray's 1962 textbook [42], it was also known that nitro-substituted methoxyterphenyls were mesomorphic. Furthermore, having previously synthesised re-entrant nematic materials, such as the DBnNO$_2$'s, [97] based on the research of Hardouin et al. [98–102], we additionally knew that nitro-substituted three-ring reversed and normal esters would support nematic phase formation. However, the problem with this approach was that the melting points were expected to be high and well above room temperature.

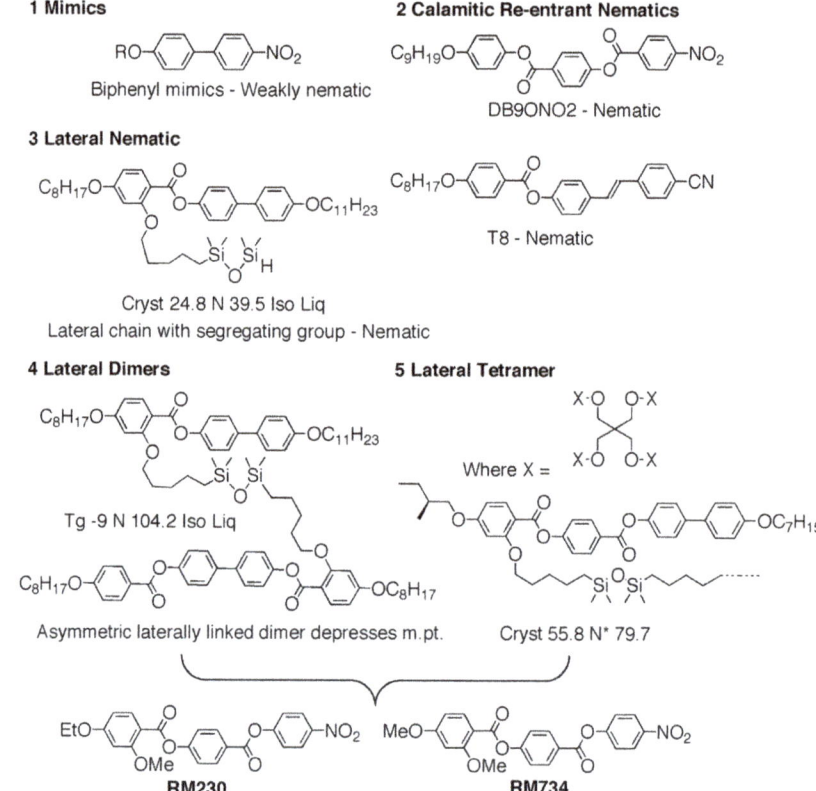

Figure 17. Outline plans for the design of highly polar nitro-, and polynitro-substituted mesogens. * indicates the location of a stereogenic centre.

To lower the melting points, we had to take a number of well-known pathways in molecular design by using dimers, trimers, tetramers, etc., [103]. Using property–structure correlations, we knew that the incorporation of lateral aliphatic chains in three- and four-ring rod-like molecules would not only depress the melting point, but would also depress smectic mesophase formation, thereby favouring nematics [104]. By joining a lateral aliphatic chain of varying lengths to a similar rod-like molecule in order to create a lateral supermolecule, we could depress crystallisation to temperatures below 0 °C. For example, by joining two laterally substituted dimers together to give a tetramer, we had also shown that we could produce materials that were nematic with useful transition temperatures via having differing arms on dimers and tetramers [105,106]. For all of the materials types, there was also the possibility of improving on the number density of nitro groups.

There was one problem with all of these concepts, and that was high viscosities and that switching would be slow, but this was not seen as a problem for the potential applications. Thus, a strategy was put in place for the start of the project (see Figure 17), but it was not long before en-route novelty appeared in the presence of an extra nematic phase for compounds RM230 and RM734.

Compounds RM230 and RM734 were materials that were made during the exploratory research and which were originally thought to exhibit nematic re-entrancy [107] after formation of a smectic C phase from a nematic phase on cooling from the liquid. It was only later that it was concluded that the materials did not have smectic C phases, and that we were on the cusp of a direct nematic-to-nematic phase transition [108,109], with the lower temperature nematic phase later classified as a new splay nematic phase [110].

In the original article concerning compound RM734 [108], the electrical field studies, using a triangular waveform and a frequency in the range of 0.1–20 Hz (ACLT property tester), gave the dielectric anisotropy in the upper temperature nematic phase to be approximately 8.5, and 6.2 in the lower phase, with Kirkwood factors of g ~0.262 and ~0.117, respectively, indicating that there was a greater antiparallel pairing in the lower temperature nematic phase. In addition, the higher temperature nematic phase also had a reasonable degree of pairing. Electrical field studies were further conducted by Clark et al. [111] and they concluded that RM734 exhibited the first ferroelectric nematic phase. In the Physics World magazine, the discovery of a ferroelectric nematic phase was listed among the "Breakthroughs of the Year" finalists for 2020, (10th Dec issue, see citation below). At a similar time to our report, Nishikawa et al. [112] also saw such behaviour in completely different materials. In these days of fast-moving discoveries, Merck reported on a material that exhibited a ferroelectric nematic phase on cooling from the liquid state and was still in that phase at room temperature [113]. Thus, we seemed to be coming around full circle from the 1980s–1990s when the work at Hull focused on ferroelectrics for microdisplays.

> *First observation of a ferroelectric nematic liquid crystal (From Physics Today)*
>
> *To Noel Clark and colleagues at the University of Colorado Boulder and the University of Utah in the US, for observing a ferroelectric nematic phase of matter in liquid crystals more than 100 years after it was predicted to exist. In this phase, all the molecules within specific patches, or domains, of the liquid crystal point in roughly the same direction—a phenomenon known as polar ordering that was first hypothesized by Peter Debye and Max Born back in the 1910s. Clark and colleagues found that when they applied a weak electric field to an organic molecule known as RM734, a striking palette of colours developed towards the edges of the cell containing the liquid crystal. In this phase, RM734 proved far more responsive to electric fields than traditional nematic liquid crystals. Although further work is required to identify materials that display the phenomenon at room temperatures, ferroelectric nematics could find applications in areas from new types of display screens to reimagined computer memory.*

As George Gray would say "For the greatest effect, an invention has to be timely—again bringing in something of the element of luck or change".

4. Conclusions

Over the time of the existence of the consortium and the networks of government facilities, industry and academic institutions, many aspects of combined and individual successes were recognized. The first recognition of the consortia members was the Queen's Award for Technological Achievement as described earlier. This was followed by the Rank Prize Award to a number of individual members. For the contributions made by chemistry research, recognition was made by the Royal Society of Chemistry to the Department of Chemistry at the University of Hull, as follows below in a part of the University press release.

Royal Society of Chemistry Landmark Award

> The University's chemistry department has received a prestigious National Historic Chemical Landmark Award for its role in the development of liquid crystals.
>
> Over the past five decades, Hull has played a crucial role in the development of liquid crystals. Initially undertaken by Professor George Gray, and continued with distinction by Professor John Goodby until recently, the work has led to major developments in liquid crystal technology. It is now being used in everything from cameras and mobile phones to the latest flat-panel computer monitors and TVs.
>
> The award was presented to the Vice Chancellor by Professor Jim Feast, President Elect of the Royal Society of Chemistry, at a special ceremony on 7 November 2005.

The highest individual award was the Kyoto Prize made in 1995 by the Inamori Foundation of Japan to Professor George Gray. Congratulatory messages were sent to the three recipients of different categories by the President of the United States, Bill Clinton, and the Prime Minister of the United Kingdom, John Major. The contents of their messages are as follows [3,48]:

Bill Clinton—President of the United States of America

> Greetings to everyone gathered for the presentation of the 1995 Kyoto Prizes. I am pleased to congratulate this year's distinguished recipients for their contributions to the betterment of humanity.
>
> This year the Inamori Foundation marks the beginning of its second decade of honoring lifetime achievements in the fields of Advanced Technology, Basic Sciences, and Creative Arts and Moral Sciences. The 1995 honorees have enriched our fundamental understanding of the universe, increased our ability to apply scientific knowledge to achieve technological progress, and advanced the conception and impact of art in our society.
>
> Dr George William Gray's seminal contributions to liquid crystal research and development have provided the basis for the liquid crystal display technology essential to virtually all contemporary computer and electronic products. Dr Chushiro Hayashi's theories on the birth and evolution of the stars and on the formation of the solar system have made him one of the giants of twentieth century astrophysics. Mr Roy Lichtenstein has formed the symbols and artifacts of contemporary society into potent artworks that redefine both the nature and purposes of art.
>
> Each of these extraordinary individuals exemplifies the deepest resources of the human spirit. For what they have given us—and continue to give—we are immensely grateful. Best wishes to all for a memorable event.

John Major—Prime Minister of the United Kingdom

> I am delighted to have this opportunity to send my warmest congratulations to the 1995 Kyoto Prize Laureates: Dr George William Gray for his contribution to research and development of liquid crystal materials; Dr Chushiro Hayashi for his contribution to the maturation of modern astrophysics; and Mr Roy Lichtenstein for his influence on contemporary fine art.
>
> I am particularly pleased and proud that a British scientist is among those honored.
>
> I warmly commend the excellent work of the Inamori Foundation to support and encourage research and for its contribution through the highly prestigious Kyoto Prizes to the recognition of outstanding achievements in Advanced Technology, Basic Sciences and Creative Arts and Moral Sciences.

In addition, to these recognitions a number of individuals received awards for their various contributions, in part or in full to the consortia. In 2005, Goodby moved the research team to the University of York, and in 2012, the remaining research contract ended. Over 40 years, Gray and Goodby with their post-doctoral researchers and doctoral research students produced in excess of 150 patents and 800 research papers on liquid crystals. On his internal move at York to become "Emeritus" Goodby commented to Cowling,

"Bernard Shaw said two things—Imagination is the beginning of creation. You imagine what you desire, you will what you imagine, and at last you create what you will, and I observed that nine out of every ten things I did were failures, so I ended up doing ten times more work".

5. Patents

All patents cited in this article are included in the list of references as they are related directly to the scientific publications reported.

Author Contributions: Both authors have made a substantial, direct and intellectual contribution to the work and approved it for publication. All authors have read and agreed to the published version of the manuscript.

Funding: This research received no external funding.

Institutional Review Board Statement: Not applicable.

Informed Consent Statement: Not applicable.

Data Availability Statement: All of the data and methodologies associated with the research described are available through the references.

Acknowledgments: We thank the contributions made over the many years that this research has been ongoing. In particular, we send our gratitude to the outstanding researchers and staff in all of the establishments, companies and universities involved. The chemistry would not have happened if it were not for them and their commitment.

Conflicts of Interest: The authors declare no conflict of interest with respect to this review.

References

1. Stonehous, J. *John Stonehouse, My Father: The True Story of the Runaway MP*; Icon Books: London, UK, 2021; 384p.
2. Hilsum, C. Flat-panel electronic displays: A triumph of physics, chemistry and engineering. *Phil. Trans. R. Soc.* **2010**, *368*, 1027–1082. [CrossRef] [PubMed]
3. Goodby, J.W.; Raynes, P. George William Gray CBE, MRIA, FRSE. *Biogr. Mems. Fell. R. Soc.* **2016**, *62*, 187–211. [CrossRef]
4. Hilsum, C. The Anatomy of a Discovery. In *Technology of Chemicals and Materials for Technology*; Howells, E.R., Ed.; Prentice Hall: Hoboken, NJ, USA, 1984; pp. 43–109.
5. George Gray Interview by Sir Harry Kroto. Available online: http://www.vega.org.uk/video/programme/25 (accessed on 9 May 2022).
6. Gray, G.W. Stable liquid crystal materials for display devices. *CVD News.* **1979**, *19*, 1–2.
7. Gray, G.W. The Clifford Paterson Lecture 1985 Liquid crystals: An arena for research and industrial collaboration among chemists, physicists and engineers. *Proc. R. Soc. Lond. A* **1985**, *402*, 1–36.
8. Gray, G.W. Reminiscences from a life with liquid crystals. *Liq. Cryst.* **1998**, *24*, 5–13. [CrossRef]
9. Goodby, J.W. The nanoscale engineering of Nematic Liquid Crystals for displays. *Liq. Cryst.* **2011**, *38*, 1363–1387. [CrossRef]
10. Barton, L.A. Dynamic scattering—A new electrooptic effect in certain classes of nematic liquid crystals. *Proc. IEEE* **1968**, *56*, 1162–1171.
11. Heilmeier, G.H.; Castellano, J.; Zanoni, L.A. Guest-host interactions in nematic liquid crystals. *Mol. Cryst. Liq. Cryst.* **1969**, *8*, 293–304. [CrossRef]
12. Heilmeier, G.H.; Zanoni, L.A.; Barton, L.A. Further studies of dynamic scattering mode in nematic liquid crystals. *Trans. IEEE* **1970**, *17*, 22–26. [CrossRef]
13. Heilmeier, G.H.; Zanoni, L.A.; Barton, L.A. Dynamic scattering in nematic liquid crystals. *Appl. Phys. Lett.* **1968**, *13*, 46–47. [CrossRef]
14. Heilmeier, G.H.; Goldmacher, J.E. Electric-field-induced cholesteric-nematic phase change in liquid crystals. *J. Chem. Phys.* **1969**, *51*, 1258–1260. [CrossRef]
15. Kelker, H.; Scheuerle, B. A liquid-crystalline (Nematic) phase with a particularly low solidification point. *Angew. Chem. Int. Ed.* **1969**, *8*, 884–885. [CrossRef]
16. Gray, G.W. A heating instrument for the accurate determination of mesomorphic and polymorphic transition temperatures. *Nature* **1953**, *172*, 1137–1140. [CrossRef]
17. Gray, G.W. *Stanton Redcroft Information Sheet*; Stanton Redcroft Ltd.: London, UK, 1972.
18. Schadt, M.; Helfrich, W. Voltage-dependent optical activity of a twisted Nematic liquid crystal. *Appl. Phys. Lett.* **1971**, *18*, 127–128. [CrossRef]
19. Goodby, J.W. Phase transitions: General and fundamental aspects. In *Handbook of Liquid Crystals*; Goodby, J.W., Collings, P.J., Kato, T., Tschierske, C., Gleeson, H.F., Raynes, P., Eds.; Wiley-VCH: Weinheim, Germany, 2014; Volume 1, pp. 59–76.

20. Helfrich, W.; Schadt, M. Lichtsteuerzelle. CH Patent 532,261, 4 December 1970.
21. Heilmeier, G.H. Electro-Optical Device. U.S. Patent 3,499,112, 3 March 1970.
22. Fergason, J.L. Display Devices Using Liquid Crystal Light Modulation. U.S. Patent 3,731,986, 22 April 1971.
23. Gray, G.W.; Chemistry Department, University of Hull, Hull, UK. Consortium Reports. Personal communication, 1972.
24. Gray, G.W.; Kelly, S.M. Bicyclo(2.2.2)octane esters exhibiting wide-range Nematic phases. *J. Chem. Soc. Chem. Commun.* **1979**, *21*, 974–975. [CrossRef]
25. Dewar, M.J.S.; Goldberg, R.S. Role of p-phenylene groups in nematic liquid crystals. *J. Am. Chem. Soc.* **1970**, *92*, 1582–1586. [CrossRef]
26. Gray, G.W.; Kelly, S.M.; McDonnell, D.G.; Mosley, A. Alkylbicyclo(2,2,2)octyl- or -polyphenyl nitriles, Their Use as Liquid Crystal Compounds and Materials and Devices Containing Them. UK Patent No 2,027,027A, 2 August 1979.
27. Gray, G.W.; Kelly, S.M. Low-melting bicyclo(2.2.2)octane esters with wide range Nematic phases. *J. Chem. Soc., Chem. Commun.* **1980**, *11*, 465–466. [CrossRef]
28. Gray, G.W.; Kelly, S.M. The synthesis of 1,4-disubstituted bicyclo(2.2.2)octanes exhibiting wide-range, enantiotropic Nematic phases. *J. Chem. Soc. Perkin II* **1981**, 26–31. [CrossRef]
29. Carr, N.; Gray, G.W.; Kelly, S.M. A comparison of the properties of some liquid crystal materials containing benzene, cyclohexane, and bicyclo(2.2.2)octane rings. *Mol. Cryst. Liq. Cryst.* **1981**, *66*, 267–282. [CrossRef]
30. Gray, G.W.; Kelly, S.M.; McDonnell, D.G.; Mosley, A. Liquid Crystal Compounds and Materials and Devices Containing Them. U.S. Patent Application No 4,261,652, 14 April 1981.
31. Gray, G.W.; Kelly, S.M.; McDonnell, D.G.; Mosley, A. Substituted Bicyclo-octane Compounds Having Liquid Crystal Properties and Materials and Devices Containing Them. UK Patent Application No 2,027,708A, 27 February 1980.
32. Deutscher, H.J.; Kuschel, F.; Schubert, H.; Demus, D. Nematic Liquid Crystal Substances for Electro Optical Assemblies—Modulation of Transmitted Light Using 4-n Alkyl Cyclohexane Carboxylic Acid-4'-Substituted Phenyl Esters. DDR Patent WP 105701, 18 June 1974.
33. Deutscher, H.-J.; Körber, M.; Schubert, H. *Advances in Liquid Crystal Research and Applications*; Bata, L., Ed.; Pergamon Press: Oxford, UK, 1980; p. 1075.
34. Deutscher, H.-J.; Laaser, B.; Dölling, W.; Schubert, H. Liquid-crystalline phenyl-trans-4-n-alkylcyclohexane carboxylates. *J. Prakt. Chem.* **1978**, *320*, 191–205. [CrossRef]
35. Osman, M.A.; Revesz, L. Mesomorphic properties of some alkyl-amino substituted phenylcyclohexanes. *Mol. Cryst. Liq. Cryst.* **1980**, *56*, 157–161. [CrossRef]
36. Eidenschink, R.; Erdman, D.; Kause, J.; Pohl, L. Substituted phenylcyclohexanes—New class of liquid-crystalline compounds. *Angew. Chem.* **1977**, *89*, 103. [CrossRef]
37. Eidenschink, R.; Kause, J.; Pohl, L. Liquid Crystalline Cyclohexane Derivatives. DE-OS Patent 2,636,684, 19 December 1978.
38. Pohl, L.; Eidenschink, R.; Kause, J.; Erdman, D. Physical-properties of nematic phenylcyclohexanes, a new class of low melting liquid-crystals with positive dielectric anisotropy. *Phys. Lett.* **1977**, *60*, 421–423. [CrossRef]
39. Eidenschink, R.; Kause, J.; Pohl, L. Liquid Crystalline Hexahydroterphenyl Derivatives. DE-OS Patent 2,701,591, 20 December 1978.
40. Eidenschink, R.; Erdman, D.; Kause, J.; Pohl, L. Substituted bicyclohexyls—New class of nematic liquid-crystals. *Angew. Chem. Int. Ed. Engl.* **1978**, *17*, 133–134. [CrossRef]
41. Eidenschink, R.; Haas, G.; Römer, M.; Scheuble, B.S. Liquid-crystalline 4-bicyclohexylcarbonitriles with extraordinary physical-properties. *Angew. Chem. Int. Ed. Engl.* **1984**, *23*, 147. [CrossRef]
42. Gray, G.W. *Molecular Structure and the Properties of Liquid Crystals*; Academic Press Incorporated: London, UK; New York, NY, USA, 1962.
43. Castellano, J.A.; Goldmacher, J.E.; Barton, L.A.; Kane, J.S. Effects of terminal group substitution on mesomorphic behaviour of some benzykideneanilines. *J. Org. Chem.* **1968**, *33*, 3501–3504. [CrossRef]
44. Boller, A.; Scherrer, H. Flussigkristalline Schiff'sche. Basen. Patent DE-OS 2,306,738, 3 October 1973.
45. Gray, G.W.; Harrison, K.J.; Nash, J.A. New family of Nematic liquid crystals for displays. *Electron Lett.* **1973**, *9*, 130–131. [CrossRef]
46. Gray, G.W.; Harrison, K.J. Substituted Biphenyl and Polyphenyl Compounds and Liquid Crystal Materials and Devices Containing Them. UK Patent 1,433,130, 30 March 1976.
47. Gray, G.W.; Harrison, K.J. Liquid Crystal Materials and Devices. U.S. Patent 3,947,375, 30 March 1976.
48. The Inamori Foundation. *Kyoto Prizes and Inamori Grants*; The Inamori Foundation: Kyoto, Japan, 1995; pp. 97–119.
49. Mandle, R.J.; Cowling, S.J.; Goodby, J.W. Isomeric trimesogens exhibiting modulated Nematic mesophases. *RSC Adv.* **2017**, *7*, 40480–40485. [CrossRef]
50. Gray, G.W.; Brynmor Jones Marson, F. Mesomorphism and chemical constitution. Part VIII. The effect of 3'-substituents on the mesomorphism of the 4'-n-alkoxydiphenyl-4-carboxylic acids and their alkyl esters. *J. Chem. Soc.* **1957**, *71*, 393–401. [CrossRef]
51. Gray, G.W.; Goodby, J.W. *Smectic Liquid Crystals—Textures and Structures*; Leonard Hill: Glasgow, Scotland; London, UK, 1984; pp. 68–81.
52. Etherington, G.; Leadbetter, A.J.; Wang, X.J.; Gray, G.W.; Tajbakhsh, A. Structure of the smectic D phase. *Liq. Cryst.* **1986**, *1*, 209–214. [CrossRef]

53. Gray, G.W.; Harrison, K.J.; Nash, J.A. Recent developments concerning biphenyl mesogens and structurally related compounds. *Pramana* **1975**, 381–396.
54. Gray, G.W.; McDonnell, D.G. Synthesis and liquid crystal properties of chiral alkyl-cyano-biphenyls (and -p-terphenyls) and of some related chiral compounds derived from biphenyl. *Mol. Cryst. Liq. Cryst.* **1976**, *37*, 189–211. [CrossRef]
55. Gray, G.W.; McDonnell, D.G. The relationship between helical twist sense. absolute configuration and molecular structure for non-sterol cholesteric liquid crystals. *Mol. Cryst. Liq. Cryst. Lett.* **1977**, *34*, 211–217. [CrossRef]
56. Gray, G.W. The liquid crystal properties of some new mesogens. *J. Phys.* **1975**, *36*, 337–347.
57. Hulme, D.S.; Raynes, E.P.; Harrison, K.J. Eutectic mixtures of nematic 4′-substituted-4-cyanobiphenyls. *J. Chem. Soc. Chem. Commun.* **1974**, 98–99. [CrossRef]
58. Demus, D. *Chemical Composition and Display Performance in Non-Emissive Electro-Optic Displays*; Kmetz, A.R., von Willisen, F.K., Eds.; Plenum Press: New York, NY, USA; London, UK, 1976; pp. 83–119.
59. Raynes, E.P. Mixed Systems, Phase Diagrams and Eutectic Mixtures. In *Handbook of Liquid Crystals*; Goodby, J.W., Collings, P.J., Kato, T., Tschierske, C., Gleeson, H.F., Raynes, P., Eds.; Wiley-VCH: Weinheim, Germany, 2014; Volume 1, pp. 351–363.
60. Khoo, I.C. Liquid Crystal Nonlinear Optics. In Proceedings of the International Conference "Quantum Optics III", Szczyrk, Poland, 3–9 September 1993.
61. Khoo, I.C.; Wu, S.-T. *Optics and Nonlinear Optics of Liquid Crystals*; Series in Nonlinear Optics Volume 1; World Scientific: Singapore, 1993.
62. Seed, A.J.; Toyne, K.J.; Hird, M.; Goodby, J.W. Synthesis and mesomorphic behaviour of high polarisability materials for non-linear optical applications. *Liq. Cryst.* **2012**, *39*, 403–414. [CrossRef]
63. Seed, A.J.; Cross, G.J.; Toyne, K.J.; Goodby, J.W. Novel, highly polarizable thiophene derivatives for use in nonlinear optical applications. *Liq. Cryst.* **2003**, *30*, 1089–1108. [CrossRef]
64. Seed, A.J.; Toyne, K.J.; Goodby, J.W. The synthesis, phase transitions and optical properties of novel tetrathiafulvalene derivatives with extremely high molecular polarizabilities. *Liq. Cryst.* **2001**, *28*, 1047–1055. [CrossRef]
65. Seed, A.J.; Toyne, K.J.; Goodby, J.W.; Hird, M. Synthesis, transition temperatures, and optical properties of various 2,6-disubstituted naphthalenes and related 1-benzothiophenes with butylsulphanyl and cyano or isothioocyanato terminal groups. *J. Mater. Chem.* **2000**, *10*, 2069–2080. [CrossRef]
66. Cross, G.J.; Seed, A.J.; Toyne, K.J.; Goodby, J.W.; Hird, M.; Artal, M.C. Synthesis, transition temperatures, and optical properties of compounds with simple phenyl units linked by double bond, triple bond, ester or propiolate linkages. *J. Mater. Chem.* **2000**, *10*, 1555–1663. [CrossRef]
67. Seed, A.J.; Toyne, K.J.; Goodby, J.W. The synthesis of the series of monofluoro-substituted 4-butylsulphanyl-4′-cyanobiphenyls and the effect of the position of fluorine within the core on the refractive indices, optical anisotropies, polarisabilities and order parameters. *J. Mater. Chem.* **1995**, *5*, 2201–2208. [CrossRef]
68. Seed, A.J.; Toyne, K.J.; Goodby, J.W. Synthesis of some 2,4- and 2,5-disubstituted thiophene systems and the effect of the pattern of substitution on the refractive indices, optical anisotropies, polarisabilities and order parameters in comparison with those of the parent biphenyl and dithienyl systems. *J. Mater. Chem.* **1995**, *5*, 653–661.
69. Hird, M.; Seed, A.J.; Toyne, K.J.; Goodby, J.W.; Gray, G.W.; McDonnell, D.G. The Synthesis, Transition Temperatures and Optical Anisotropy of Some Isothiocyanato-substituted Biphenyls. *J. Mater. Chem.* **1993**, *3*, 851–860. [CrossRef]
70. Heck, R.F.; Nolley, J.P. Palladium-catalyzed vinylic hydrogen substitution reactions with aryl, benzyl, and styryl halides. *J. Org. Chem.* **1972**, *37*, 2320–2322. [CrossRef]
71. Miyaura, N.; Yamada, K.; Suzuki, A. A new stereospecific cross-coupling by the palladium-catalyzed reaction of 1-alkenylboranes with 1-alkenyl or 1-alkynyl halides. *Tetrahedr. Lett.* **1979**, *20*, 3437–3440. [CrossRef]
72. Miyaura, N.; Suzuki, A. Stereoselective synthesis of arylated (E)-alkenes by the reaction of alk-1-enylboranes with aryl halides in the presence of palladium catalyst. *J. Chem. Soc. Chem. Commun.* **1979**, 866–867. [CrossRef]
73. Reiffenrath, V.; Krause, J.; Plach, H.J.; Weber, G. New liquid-crystalline compounds with negative dielctric anisotropy. *Liq. Cryst.* **1989**, *5*, 159–170. [CrossRef]
74. Gray, G.W.; Hird, M.; Lacey, D.; Toyne, K.J. The synthesis and transition temperatures of some 4,4″-dialkyl- and 4,4″-alkoxyalkyl-1,1′:4′,1″-terphenyls with 2,3- or 2′,3′-difluorosubstituents and of their biphenyl analogues. *J. Chem. Soc. Perkin Trans. 2* **1989**, 2041–2053. [CrossRef]
75. Schiekel, M.F.; Fahrenschon, K. Deformation of nematic liquid crystals with vertical orientation in electrical fields. *Appl. Phys. Lett.* **1971**, *19*, 391–393. [CrossRef]
76. Bremer, M.; Klasen-Memmer, M.; Pauluth, D.; Tarumi, K. Novel liquid crystal materials with negative dielectric anisotropy for TV applications. *J. Soc. Inform. Disp.* **2006**, *14*, 517–521. [CrossRef]
77. Hird, M.; Goodby, J.W.; Toyne, K.J. Nematic materials with negative dielectric anisotropy for display applications. In *Liquid Crystal Materials, Devices, and Flat Panel Displays*; Shashidhar, R., Gnade, B., Eds.; SPIE: Washington, DC, USA, 2000; Volume 3955, pp. 15–23.
78. McMillan, W.L. Simple molecular theory of the smectic C phase. *Phys. Rev. A* **1973**, *8*, 1921–1929. [CrossRef]
79. Gasowska, J.S.; Cowling, S.J.; Cockett, M.C.R.; Hird, M.; Lewis, R.A.; Raynes, E.P.; Goodby, J.W. The influence of an alkenyl terminal group on the mesomorphic behaviour and electro-optic properties of fluorinated terphenyl liquid crystals. *J. Mater. Chem.* **2010**, *20*, 299–307. [CrossRef]

80. Hikmet, R.A.M.; Boots, H.M.J.; Michielsen, M. Ferroelectric liquid crystal gels network stabilized ferroelectric display. *Liq. Cryst.* **1995**, *19*, 65–76. [CrossRef]
81. Hikmet, R.A.M.; Michielsen, M. Anisotropic network stabilized ferroelectric gels. *Adv. Mater.* **1995**, *7*, 300–304. [CrossRef]
82. Goodby, J.W.; Toyne, K.J.; Hird, M.; Styring, P.; Lewis, R.A.; Beer, A.; Dong, C.C.; Glendenning, M.E.; Jones, J.C.; Lymer, K.P.; et al. Ferroelectric liquid crystalline materials: Hosts, dopants and gels for display applications. *Mol. Cryst. Liq. Cryst.* **2000**, *346*, 169–182. [CrossRef]
83. Goodby, J.W.; Toyne, K.J.; Hird, M.; Styring, P.; Lewis, R.A.; Beer, A.; Dong, C.C.; Glendenning, M.E.; Jones, J.C.; Lymer, K.P.; et al. Chiral liquid crystals for ferroelectric, electroclinic and antiferroelectric displays and photonic devices. In *Liquid Crystal Materials, Devices and Flat Panel Displays*; Shashidhar, R., Gnade, B., Eds.; SPIE: Washington, DC, USA, 2000; Volume 3955, pp. 2–14.
84. Beer, A.; Goodby, J.W.; Kelly, S.M.; Hird, M. Liquid Crystal Materials. EU Patent EP 1,141,172, 10 October 2001.
85. Hird, M.; Goodby, J.W.; Toyne, K.J. The development of materials, mixtures and gels for ferroelectric displays. *Mol. Cryst. Liq. Cryst.* **2001**, *360*, 1–15. [CrossRef]
86. Blackwood, K.M.; Goodby, J.W.; Hall, A.W.; Milne, P.E.Y. Use of Poly(diallylamine) Polymers. EU Patent EP 1,265,942, 26 January 2005.
87. Hindmarsh, P.; Watson, M.J.; Hird, M.; Goodby, J.W. Investigation of the effect of bulky lateral substituents on the discotic mesophase behaviour of triphenylene benzoates. *J. Mater. Chem.* **1995**, *5*, 2111–2124. [CrossRef]
88. Hindmarsh, P.; Hird, M.; Styring, P.; Goodby, J.W. Lateral substitution in the peripheral moieties of triphenylen-2,3,6,7,10,11-yl hexa-4-alkoxybenzoates: Dimethyl-substituted systems. *J. Mater. Chem.* **1993**, *3*, 1117–1128. [CrossRef]
89. Beattie, D.R.; Hindmarsh, P.; Goodby, J.W.; Haslam, S.D.; Richardson, R.M. Triphenylene Hexa-n-alkylcyclohexanoates—A new series of disc-Like liquid crystals. *J. Mater. Chem.* **1992**, *2*, 1261–1266. [CrossRef]
90. Beattie, D.R.; Goodby, J.W.; Gray, G.W.; Jones, J.C.; McDonnell, D.G.; Phillips, T.J.; Hindmarsh, P.; Hird, M. Discotic Compounds for Use in Liquid Crystal Mixtures. EU Patent EP 0,702,668, 26 August 1998.
91. Bushby, R.J.; Kawata, K. Liquid crystals that affected the world: Discotic liquid crystals. *Liq. Cryst.* **2011**, *38*, 1415–1426. [CrossRef]
92. Taugerbeck, A.; Booth, C.J. Design, Synthesis of Chiral Nematic Liquid Crystals. In *Handbook of Liquid Crystals*; Goodby, J.W., Collings, P.J., Kato, T., Tschierske, C., Gleeson, H.F., Raynes, P., Eds.; Wiley-VCH: Weinheim, Germany, 2014; Volume 3, pp. 429–492.
93. Davis, E.J.; Goodby, J.W. Symmetry and Chirality in Liquid Crystals. In *Handbook of Liquid Crystals*; Goodby, J.W., Collings, P.J., Kato, T., Tschierske, C., Gleeson, H.F., Raynes, P., Eds.; Wiley-VCH: Weinheim, Germany, 2014; Volume 1, pp. 197–230.
94. Gray, G.W.; McDonnell, D.G. New low-melting cholesterogens for electro-optical displays and surface thermography. *Electron Lett.* **1975**, *11*, 556–557. [CrossRef]
95. Hird, M. Ferroelectricity in liquid crystals—Materials, properties and applications. *Liq. Cryst.* **2011**, *38*, 1467–1493. [CrossRef]
96. Goodby, J.W.; Chin, E.; Leslie, T.M.; Geary, J.M.; Patel, J.S. Helical twist sense and spontaneous polarization direction in ferroelectric smectic liquid crystals 1. *J. Am. Chem. Soc.* **1986**, *108*, 4729–4735. [CrossRef]
97. Gray, G.W.; Goodby, J.W. *Smectic Liquid Crystals—Textures and Structures*; Leonard Hill: Glasgow, Scotland; London, UK, 1984; pp. 143–149.
98. Sigaud, G.; Hardouin, F.; Achard, M.F.; Levelut, A.M. A new type of smectic A phase with long range modulation in the layers. *J. Phys.* **1981**, *42*, 107–111. [CrossRef]
99. Levelut, A.M.; Tarento, R.J.; Hardouin, F.; Achard, M.F.; Sigaud, G. Number of SmA phases. *Phys. Rev. A* **1981**, *24*, 2180–2187. [CrossRef]
100. Hardouin, F.; Levelut, A.M.; Sigaud, G. Anomalies of periodicity in some liquid crystalline cyano derivatives. *J. Phys.* **1981**, *42*, 71–77. [CrossRef]
101. Tinh, N.-H.; Hardouin, F.; Destrade, C.; Gasparoux, H. New phase transitions SmC-SmC$_2$ and SmA$_d$-SmC$_2$ in pure mesogens. *J. Physique Lett.* **1982**, *43*, 739–744. [CrossRef]
102. Hardouin, F.; Tinh, N.-H.; Achard, M.F.; Levelut, A.M. A new thermotropic smectic phase made of ribbons. *J. Physique Lett.* **1982**, *43*, 327–331. [CrossRef]
103. Saez, I.M.; Goodby, J.W. Supermolecular Liquid Crystals. In *Liquid Crystalline Functional Assemblies And Their Supramolecular Structures, Structure And Bonding*; Mingos, D.M.P., Kato, T., Eds.; Springer: Berlin/Heidelberg, Germany, 2008; Volume 128, pp. 1–62.
104. Saez, I.M.; Goodby, J.W. Supermolecular Liquid Crystals. *J. Mater. Chem.* **2005**, *15*, 26–40. [CrossRef]
105. Saez, I.M.; Goodby, J.W. Design and properties of Janus Supermolecular Liquid Crystals. *Chem. Commun.* **2003**, 1726–1727. [CrossRef]
106. Saez, I.M.; Goodby, J.W. "Janus" Supermolecular Liquid Crystals—Giant molecules with hemispherical architectures. *Chem. Eur. J.* **2003**, *9*, 4869–4877. [CrossRef] [PubMed]
107. Mandle, R.J. The Nitro Group in Liquid Crystals. Ph.D. Thesis, University of York, York, UK, 2012.
108. Mandle, R.J.; Cowling, S.J.; Goodby, J.W. A Nematic to Nematic Transformation Exhibited by a Rod-like Liquid Crystal. *Phys. Chem. Chem. Phys.* **2017**, *19*, 11429–11435. [CrossRef] [PubMed]
109. Mandle, R.J.; Cowling, S.J.; Goodby, J.W. Rational design of rod-like Liquid Crystals exhibiting two Nematic phases. *Chem. Eur. J.* **2017**, *23*, 14554–14562. [CrossRef] [PubMed]

110. Mertelj, A.; Cmok, L.; Sebastián, N.; Mandle, R.J.; Parker, R.R.; Whitwood, A.C.; Goodby, J.W.; Čopič, M. Splay Nematic Phase. *Phys. Rev. X* **2018**, *8*, 041025. [CrossRef]
111. Chen, X.; Korblova, E.; Dong, D.; Wei, X.; Shao, R.; Radzihovsky, L.; Glaser, M.A.; Maclennan, J.E.; Bedrov, D.; Walba, D.M.; et al. First-principles experimental demonstration of ferroelectricity in a thermotropic nematic liquid crystal: Polar domains and striking electro-optics. *Proc. Natl. Acad. Sci. USA* **2020**, *117*, 14021–14031. [CrossRef] [PubMed]
112. Nishikawa, H.; Shiroshita, K.; Higuchi, H.; Okumura, Y.; Haseba, Y.; Yamamoto, S.I.; Sago, K.; Kikuchi, H. A Fluid Liquid-Crystal Material with Highly Polar Order. *Adv. Mater.* **2017**, *29*, 1702354. [CrossRef] [PubMed]
113. Manabe, A.; Bremer, M.; Kraska, M. Ferroelectric nematic phase at and below room temperature. *Liq. Cryst.* **2021**, *48*, 1079–1086. [CrossRef]

Article

Anatomy of a Discovery: The Twist–Bend Nematic Phase

David Dunmur

Christ Church, University of Oxford, Oxford OX1 2JD, UK; d.dunmur@gmail.com

Abstract: New fluid states of matter, now known as liquid crystals, were discovered at the end of the 19th century and still provide strong themes in scientific research. The applications of liquid crystals continue to attract attention, and the most successful so far has been to the technology of flat panel displays; this has diversified in recent years and LCDs no longer dominate the industry. Despite this, there is plenty more to be uncovered in the science of liquid crystals, and as well as new applications, novel types of liquid crystal phases continue to be discovered. The simplest liquid crystal phase is the nematic together with its handed or chiral equivalent, named the cholesteric phase. In the latter, the aligned molecules of the nematic twist about an axis perpendicular to their alignment axis, but in the 1970s a heliconical phase with a tilt angle of less than 90° was predicted. The discovery of this phase nearly 40 years later is described in this paper. Robert Meyer proposed that coupling between a vector order parameter in a nematic and a splay or bend elastic distortion could result in spontaneously splayed or bent structures. Later, Ivan Dozov suggested that new nematic phases with splay–bend or twist–bend structures could be stabilised if the appropriate elastic constants became negative. Theoretical speculation on new nematic phases and the experimental identification of nematic–nematic phase transitions are reviewed in the paper, and the serendipitous discovery in 2010 of the nematic twist–bend phase in 1″,7″-bis(4-cyanobiphenyl-4′-yl)heptane (CB7CB) is described.

Keywords: liquid crystal; nematic; twist-bend phase

1. Introduction

Much has been written about the twist–bend nematic phase since its experimental identification was published more than 10 years ago. In a recent review [1], Rebecca Walker observed that:

"The prediction [2,3] and subsequent experimental discovery [4] of the twist–bend nematic phase, N_{TB}, is undeniably one of the most significant recent developments in the field of liquid crystals."

Like many advances in science, this particular discovery has not been without controversy [5], and there are still different views on the structure and nature of the twist–bend nematic phase. Additionally, there are differences of opinion as to who discovered what and when, and to whom any credit is due, if appropriate, for the scientific advance. Our knowledge of the natural world has accumulated through incremental steps due to the collaborative and interactive research of scientists, and this is true for the discovery of the twist–bend nematic liquid crystal phase. That is not to dismiss individual claims for the first or significant breakthrough, and not unnaturally we all want to be recognised for our ground-breaking discoveries. The case of the twist–bend nematic phase is no exception, and at one time there were three universities around the world all claiming the exclusive credit for the discovery of this new type of liquid crystal. Such stories are part of the process and deserve to be recorded, and this article reviews the development of the science behind the discovery of the twist–bend nematic phase. The author admits to an interest since he was one of those involved in the initial discovery, but it is the intention to present as fair and documented account as possible, recognising the work of many scientists who contributed to the initial discovery.

Citation: Dunmur, D. Anatomy of a Discovery: The Twist–Bend Nematic Phase. *Crystals* **2022**, *12*, 309. https://doi.org/10.3390/cryst12030309

Academic Editor: Ingo Dierking

Received: 4 February 2022
Accepted: 17 February 2022
Published: 22 February 2022

Publisher's Note: MDPI stays neutral with regard to jurisdictional claims in published maps and institutional affiliations.

Copyright: © 2022 by the author. Licensee MDPI, Basel, Switzerland. This article is an open access article distributed under the terms and conditions of the Creative Commons Attribution (CC BY) license (https://creativecommons.org/licenses/by/4.0/).

To set the discovery of the twist–bend nematic phase in context, traditionally liquid crystal phases were classified as nematic with no positional order, smectic with one degree of translational order, and columnar with two degrees of translational order. In recent times various sub-categories have been described, such as "banana" phases formed from bent-core shaped molecules, and our knowledge of the detailed structures of increasingly diverse liquid crystal phases has vastly increased. Within the established classes of liquid crystals, a variety of different structures have been identified, but for the nematic phase there have been only two types—the simple nematic formed from rod-like or disc-like molecules, and the chiral nematic or cholesteric phase formed from optically active or chiral (handed) molecules lacking a centre or plane of symmetry. There have been reports over the years of other nematic phases, of which the biaxial nematic phase with two optic axes but no translational order is one example. The optical differences between the nematic and chiral nematic phases are dramatically clear using a polarising microscope, and indeed it was the chiral nematic or cholesteric phase of cholesteryl benzoate that was the first liquid crystal to be identified [6]. The difference between a nematic phase and its chiral equivalent is that in the former, the orientation of the optic axis, or director, is distributed randomly through a non-aligned sample, whereas in a chiral nematic a helix forms, such that the director is at right angles to the helix axis. This rather remarkable structural difference does not seem to have much effect on the physical properties of the chiral and achiral nematic phases, but it does have a big effect on the optical properties.

The close packing of chiral or handed molecules inevitably produces a twist between adjacent molecules, and so the formation of a twisted structure for a chiral nematic phase is expected. Twisted structures are also found in some smectic phases formed from chiral molecules, such as twist–grain boundary phases and chiral tilted smectic C and related phases, and for such layered phases the director is not usually at 90° to the helix axis or layer normal but tilted at some smaller angle. Introducing another symmetry axis such as a helix raises the possibility of a biaxial phase. Such a biaxial nematic phase was proposed in 1970 by Freiser [7], and there has been a considerable research effort to identify such a phase. For a review, see [8]. A biaxial nematic phase should be distinguishable through various physical properties, and potentially has a number of applications.

For nematic phases there is no layer normal, but in chiral nematics there is a helix axis of molecular twist, to which the director axis is perpendicular. Macroscopically, the phase remains uniaxial, but locally, within a single period of the helix twist, the structure is biaxial. The essence of this local biaxiality is restricted rotation about the long axes of the molecules, which drives the formation of a macroscopic twist perpendicular to these long axes, but there is no reason to exclude a chiral structure in which the nematic director is tilted to the helix axis at less than 90°. The formation of twisted nematic phases seems to be dependent on the molecules being chiral, but the induction of twist through a local molecular tilt is an unanticipated development in the structure of fluid phases. The introduction of molecular tilt to a nematic phase is another way of lowering the symmetry and generating new and perhaps interesting properties. There are no clues as to how to engineer a molecule to tilt with respect to an axis of twist. Furthermore, the properties of such a structure are hard to fathom. It is perhaps for this reason that, in contrast to the biaxial nematic, there was no concerted research effort to find a tilted nematic phase, and its appearance had to await its serendipitous discovery. This is the story we wish to relate here.

2. Structures of Liquid Crystal Phases

A traditional approach to the understanding of the microscopic structure of liquid crystal phases is from the perspective of molecular shape. Thus, rod-like molecules result in nematic and smectic phases, while disc-like molecules can form nematic and columnar phases. Other molecular shapes can introduce new features to the molecular organisation, and in the recent past attention has focussed on bent-core molecules [9]. The latter have not disappointed, and a host of so-called "banana" phases B1 to B8 have been discovered with a variety of phase structures and properties. A banana shape representing a bent-core molecule is of lower symmetry than rods and discs, and it can have an associated dipole, electric and/or steric, perpendicular to its major axis, which can influence the molecular assembly in condensed phases. The interactions driving molecular organisation in liquid crystals are a combination of repulsive (shape) forces and attractive forces of dispersion, electrostatic (polar) and hydrogen-bonding, and all of these are linked through the molecular structure, which may itself have intrinsic flexibility.

Molecules do not have to be polar, i.e., possess a dipole moment, to form a liquid crystal phase, though many are. The polar group or groups in a molecule have a direct effect on the dielectric properties, but in structurally organised phases such as liquid crystals, polar interactions can influence the local structure [10]. Intermolecular interactions may cause the dipole moments on adjacent molecules to favour a parallel or anti-parallel arrangement, and this can be dependent on the overall molecular shape. Thus, rod-like molecules with longitudinal dipoles favour anti-parallel correlations, while those with transverse dipoles have a tendency for a net parallel alignment of molecular dipoles [11]. The reverse arrangements apply to disc-like molecules, and for molecular shapes such as bent-core or flexible species, the situation becomes more complicated. These effects are often magnified in smectic phases where there can be monolayer or bilayer modulation of the molecular polarisation, and for some symmetries a macroscopic ferroelectric polarisation can develop.

It was Pierre Gilles de Gennes (Nobel Prize for Physics, 1993) who provided the basis for our physical understanding of liquid crystals through his publication in 1974 of "The Physics of Liquid Crystals" [12], in which the existence of different types of liquid crystal phases is discussed. As far as we know, there is only one type of gas, unless one counts an ionised gas, a plasma, as a different phase. Similarly, there is only one liquid phase, although many different liquids are immiscible, but their fluid properties are similar. Different crystalline structures are distinguishable by topological and optical symmetries, and for some materials there can be phase transitions between them. For liquid crystals, distinguishable types were identified in the earliest research, but until recently only two species of nematic phase have been experimentally acknowledged: the nematic phase and the chiral nematic phase. This article focuses on the theoretical background and experimental evidence for an additional nematic phase, which has been identified as a twist–bend nematic.

De Gennes, in the first edition of his book [12], discusses the possibility of another nematic phase with a biaxial symmetry, and also (p. 244) a transition to a conical phase from a cholesteric phase induced by a magnetic field. Such a transformation could be possible if the bend elastic constant is anomalously low, and this prediction of a conical phase in which there is a component of the director parallel to the helix axis is attributed to R. B. Meyer [13], although it is noted that such a phase had not been observed. The idea of a heliconical nematic phase is developed in a subsequent publication by Meyer [2], except that the driving force for the formation of a conical phase is identified as spontaneous polarisation. A perpendicular component of the polarisation couples with the bend distortion of the director to give a nematic state with non-zero bend. Meyer suggested that non-chiral bent molecules might be able to form a state of finite torsion and bend, and he identified this as a helical twist–bend nematic phase. A similar coupling between the parallel component of the polarisation with a splay distortion could result in a non-uniform splay–bend structure. Spontaneous splay or bend distortions can only exist in phases with vector order and the existence of polarisation can in principle stabilise splay–bend or twist–bend phases, as shown by Meyer.

A number of examples of nematic–nematic transitions have been experimentally identified. Such transitions were observed in liquid crystals formed from poly-ethers [14,15], but the structure of the nematic phases, labelled n_2, involved in the phase change were not fully characterised. The transition was tentatively attributed to a competition between the rigid and flexible parts of the polymer, which could result in two uniaxial nematic phases of different order parameters.

So-called dimeric liquid–crystal compounds consist of two rigid rod-like molecular fragments connected by a linking group and have been the subject of many investigations. The linking group may have a fixed molecular geometry, which gives rigid bent-core mesogens (molecules that form mesophases) or may be a flexible unit such as an alkyl chain, which allows the liquid crystal-like fragments to adopt a number of orientations. A study of dimeric phenyl alkoxybenzoates linked by flexible diiminoalkylene spacers [16] revealed a nematic–nematic phase transition in a single compound, which the authors attributed to changes in local structure resulting in the N_X phase, a precursor to forming a B6 phase. Further studies of structurally related rigid symmetric dimers of phenyl alkylbenzoates linked through an oxadiazole unit [17] suggested the existence of an unknown X phase. It was reported that these investigations using calorimetry, X-ray scattering, and optical microscopy provided evidence for the formation of molecular clusters, and their segregation into domains of opposite chiral handedness. This remarkable feature of symmetry-breaking was observed in an achiral fluid with chiral separation, such that domains of opposite chirality formed. Clark and others [18] had already shown spontaneous separation of equal and opposite chiral domains in an achiral bent core smectic C phase, but not in a nematic phase. Similarly, Lagerwall [19] proposed that achiral molecules could organise in domains of left- and right-handed helices in a smectic C phase to give a twist–bend structure. Spontaneous symmetry breaking has also been seen in a planar-aligned nematic device containing a bent-core liquid crystal [20].

The speculation of Meyer concerning other nematic phases has been added to by other predictions from theory. For nematic liquid crystals composed of bent-core molecules, Lubensky and Radzihovsky [21] predict a number of lower symmetry nematic and chiral nematic phases, but do not mention a heliconical nematic. Similarly, Mettout [22], using group theoretical arguments, proposes a variety of novel nematic phases of molecules of differing symmetries. Using Landau theory, a twisted conical phase following the melting of a hexagonal columnar phase of long polymer molecules was proposed by Kamien [23]. If the condition is satisfied that the twist elastic constant (k_2) is greater than the bend elastic constant (k_3) then the conical phase may be stabilised, but this experimental condition is rarely satisfied.

An asymmetric distribution of electric charge in a particle is conveniently quantified in terms of an electric dipole. Asymmetry in particle or molecular shape is sometimes described as a steric dipole, which represents a skewed structure having an imbalanced distribution of mass along a particular direction. This steric dipole is a vector, but there is no generally accepted definition of such a quantity. In 2001, Dozov [3] proposed that nematic phases of finite torsion and bend, or splay and bend, similar to those suggested by Meyer, could arise for achiral banana-shaped molecules if the bend elastic constant became negative. For these shapes, the steric dipole (d) can be represented by the radius of curvature (r) of the bent shape, $d = (\ell/r^2)r$ where ℓ is the length of the banana-shaped molecule. The steric dipole d can couple with the bend distortion of the director, giving rise to nematic states of permanent twist–bend or splay–bend, and for a negative bend elastic constant, these can become the stable ground states. The pathological behaviour of the bend elasticity constant for banana molecules approaching zero, or going negative, causes spontaneous symmetry-breaking and the "splay–bend" and "twist–bend" phases become stable. In achiral systems, the twist–bend is two-fold degenerate, and domains of left- and right-handed twist are expected to develop, but Dozov does not explore the details of the domain structure or the defects which might stabilise it.

Phase transitions are macroscopic phenomena, and so attract macroscopic interpretation rather than a molecular model. Phase properties are conveniently categorised by symmetry, of which the simplest, $O(3)$, is an isotropic fluid such as a gas or liquid. When two phases have the same symmetry, the transition between them must be first order with a non-zero entropy of transition. Liquid crystals are characterised by orientational phase transitions associated with rotational symmetry breaking. In the absence of external influences such as electric, magnetic, mechanical, or surface forces, spontaneous symmetry breaking derives from molecular interactions resulting in structures with various degrees of orientational order. Changes of temperature and/or pressure may cause phase transitions between these phases, which are usually second order or weakly first order. One transition that is strictly symmetry forbidden is from a non-chiral state to a handed or chiral state, but it does happen with certain liquid crystals in which domain formation of opposite chirality preserves the overall achiral symmetry of the phase. Spontaneous deracemization in achiral smectic phases of liquid crystals has been observed, with domains of opposite chirality appearing as stripes tilting in different directions, and this has been compared [24] with the spontaneous deracemization of tartaric acid salts observed by Pasteur.

Fluid phases, and indeed mobile "crystal" phases, have structures which are subject to thermal fluctuations. Such changes in the local structure can be detected through a variety of physical properties, most especially through light scattering, and the opaqueness of liquid crystals is a consequence of these fluctuations. A convenient representation of the fluctuation modes is as "normal" modes, such that the normal coordinates represent squared terms in a free-energy expression. Under appropriate conditions, the equipartition of energy principle can be applied to determine the contributions of different normal modes to the free energy. In uniaxial nematics, the normal modes for director fluctuations are "splay–bend" and "twist–bend", while in chiral nematics they are "planar helical" and "heliconical modes". The contributions of the chiral fluctuation modes to optical properties have been evaluated theoretically [25,26]. Contributions of fluctuations to the free energy of phases will influence the stability of the phases and the nature of transitions between phases. For example, unfrozen fluctuations can prevent the formation of more asymmetric ordered states [22], and symmetry breaking second order phase transitions may become weak first order transitions [27].

For some researchers, the next best thing to a confirmatory experiment is a computer simulation. This is as true in liquid crystals as in other areas of condensed matter physics, and computer simulation studies [28] have been a valuable technique in understanding the structures and properties of different mesophases. The essence of computer simulations is a mathematical algorithm, which may be adaptable, that carries all the physics of local interactions responsible for the structures and properties of the phases of interest. Such studies can confirm experimental observations or theoretical predictions, but they cannot probe outside the limitations of the mathematical model that controls the simulation. Memmer [29] was the first to publish a simulated structure for the nematic twist–bend phase, using a model of two connected Gay–Berne particles with an included angle of 140° to represent a banana-shaped mesogen. There were no centres of charge, and so no dipole–dipole interactions, but Memmer notes that there is a steric dipole defined by the connected G–B particles. The computer simulation of twisted structures is influenced by the periodic boundary conditions and sample size, and Memmer states that the system studied generated a right-handed helix. For achiral particles, a left-handed structure should appear with equal probability, but could be separated by a large enthalpy barrier in the simulations.

3. The Discovery of the Twist–Bend Nematic Phase

The extensive introduction above records what, in a patent application, would be described as prior art up to around 2010. A number of publications had appeared that reported a new nematic phase, though structural studies and characterisations had not been sufficient for proper identification. In this section we will review claims of possible new nematic phases, and the serendipitous discovery of the twisted nematic phase at the end of the first decade of the 21st century.

Liquid crystal phases formed from bent-core or banana-shaped molecules have augmented the range of identified phase types, and many of these have layered structures and are related to the traditional smectic phases. In 2003, the famous liquid crystal group from Halle identified [30] a new nematic phase formed from asymmetric rigid bent-core mesogens, which they labelled as the N_X phase. This was thought to be a nematic columnar phase formed from bundles of bent-core molecules arranged in a nematic-like order. The N_X

phase exhibited a fan-like optical texture, and under some conditions displayed domains of opposite handedness, but no suggestion of a twisted structure was advanced. New nematic phases were reported in a number of other bent-core mesogenic compounds [31,32], and these were explained in terms of clusters or cybotactic groups forming with nematic order, often as a precursor to a smectic or columnar phase.

The existence of liquid crystalline order in biological systems has long been recognised, though the phase structures formed are much less studied than for low molecular weight materials. A suspension of helical flagella isolated from *Salmonella typhimurium* is reported [33] as forming a chiral conical nematic phase, though the twist is intrinsic to the flagella rather than formed through molecular interactions. The helices of the flagellae intermingle, and their polydispersity prevents the formation of layered structures. It is proposed that the director follows the helical structure set by the constituent molecules. Preparations of the flagella suspensions when viewed under a polarising microscope show a striped texture of alternating birefringence, and the authors concluded that their conical phase was similar to that proposed earlier by Meyer [2].

Dimeric liquid crystals are a class of compounds in which two mesogenic units are linked either through a rigid connecting unit or through a flexible alkyl or alkoxy chain of methylene units. For flexibly connected dimers there has been considerable focus on the effect of the parity of the alkyl chain, odd or even, on the liquid crystalline phase behaviour of the compounds [34]. In an extensive study of α, ω diiminoalkylene-linked alkoxyphenyl benzoates, Šepelj et al. [16] found that one compound exhibited a new monotropic nematic phase which was labelled N_X. The characteristic feature identified for this phase was an optical texture having spiral domains with alternating handedness, the origin of which could not be explained.

To establish a new liquid crystal phase requires the characterisation of its symmetry and its optical properties together with a model for the molecular arrangement in the ordered fluid, usually based on X-ray scattering. Studies of other physical properties can give an indication of the internal structure of a fluid, but invariably need some theoretical model to interpret the measurements. A review of the relationship between the dielectric properties of liquid crystals and the shapes of the constituent mesogens [35] revealed that the permittivity components of the dimeric mesogen 1″,7″-bis(4-cyanobiphenyl-4′-yl)heptane (CB7CB), the structure of which is represented in the figure below, showed a discontinuity at a temperature of 12 °C below the nematic to isotropic transition: this was interpreted as indicative of a phase transition to a new type of nematic phase.

NC—⌬—⌬—$(CH_2)_7$—⌬—⌬—CN

Chemical structure of CB7CB

The compound CB7CB was first synthesised in the Southampton liquid crystal group [36], who made a preliminary examination of the phase properties, noting a nematic phase and a lower temperature phase tentatively identified as a smectic C phase. Measurements of the dielectric relaxation in flexible dimeric liquid crystals [37] revealed unusual behaviour and were explained [38] in terms of a model that took account of different conformational states of the flexible dimeric molecules. It was found that lowering the temperature of the nematic phase of these dimers caused substantial changes to their average molecular shape, as represented by contributions from different conformations of the linking flexible alkyl chain. The observed transition to a lower temperature nematic phase could be attributed to shape changes of the flexible dimeric molecules. Further studies of the proposed new nematic phase of CB7CB were carried out, and a key observation [39] was made that the deuterium NMR quadrupole splitting measured for deuterated CB7CB d_4 bifurcated at the transition from the high temperature nematic phase to the lower temperature nematic phase. This result is indicative of a symmetry-breaking transition and the formation of equal domains of left- and right-handed chiral molecules. The liquid crystal dimer investigated was on average achiral, but particular conformers stabilised by the flexible alkyl chain could be chiral with equal concentrations of left- and right-handed species. The origin of the symmetry-breaking was not apparent and clearly needed further investigation.

Every second year the liquid crystal community gathers for its International Conference, and in 2010 the 23rd such meeting was held in Kraków, Poland. There were about a dozen talks and posters on possible new nematic phases, including four posters on nematic–nematic phase transitions in flexible liquid crystal dimers. The author of this review contributed a talk entitled "A liquid crystal dimer with a bent nematic phase", which reported on a collaborative project involving 13 researchers working in six different institutions around Europe. The talk presented results on the identification and characterisation of a new nematic phase in CB7CB labelled as a twist–bend nematic phase. After the talk there were a few questions including one from J. K. Vij of Trinity College Dublin, Eire, concerning details of one of the optical textures presented. Vij had contributed a poster to the conference on a related compound CB11CB, which exhibited an additional phase at a lower temperature than the conventional nematic phase. The work by Vij was subsequently published later in 2010 [40], while the presentation on the twist–bend phase of CB7CB appeared in 2011 [4].

These publications gave slightly conflicting views of the structure of the new nematic phase observed in homologues of the α,ω-bis[(4-cyanobiphenyl)-4′]alkanes. The paper from Vij and collaborators from the Universities of Dublin and Hull confirmed the nematic nature of the phase in CB11CB using X-ray scattering and concluded from a Landau de Gennes calculation that the observed periodic deformation in thin films was a result of at least one of the elastic constants for splay or twist becoming negative. Their investigations failed to detect any evidence of symmetry breaking due to chirality. On the other hand, the paper [4] on CB7CB presented as a talk at the Kraków conference gave evidence from ^{2}H NMR that, for particular methyl protons in the alkyl chain, there were two non-equivalent sites in the oriented mesogen, consistent with the presence of chiral symmetry-breaking giving left and right enantiomers. Dielectric measurements indicated that at the transition from the high-temperature nematic to the unidentified lower temperature phase a macroscopic tilt developed with respect to the rubbing direction, and furthermore a calculation of the bend elastic constant of CB7CB predicted that it could be negative. The conclusion of the paper was that the low temperature nematic phase shared many of the characteristics of the twist–bend nematic phase proposed by Dozov [3] for materials of negative bend elastic constant. However, there were still unresolved questions concerning the identification of the twist–bend nematic phase. Although chiral symmetry breaking in CB7CB had been demonstrated as a possibility, further optical confirmation was lacking, and if the phase was heliconical, then it should be possible to determine the pitch of the helix and the tilt angle. Tilt had been observed at the transition from the nematic to the twist–bend nematic phase, but no estimate of its magnitude was given.

These and other outstanding questions concerning the new nematic phase were resolved to some extent in subsequent papers by a number of authors from different institutions. In the decade 2010 to 2020 there was an explosion of more than a thousand publications concerning many aspects of the twist–bend nematic phase. This is illustrated by the graph of numbers of papers containing "twist–bend nematic" in their titles. Because of the vagaries of titles, the total number of papers on the topic is much greater than the numbers given in graph. A few papers became identified as "highly-cited", and one [4] was selected by the editors of the American Physical Society Physical Review E as the milestone paper for 2011.

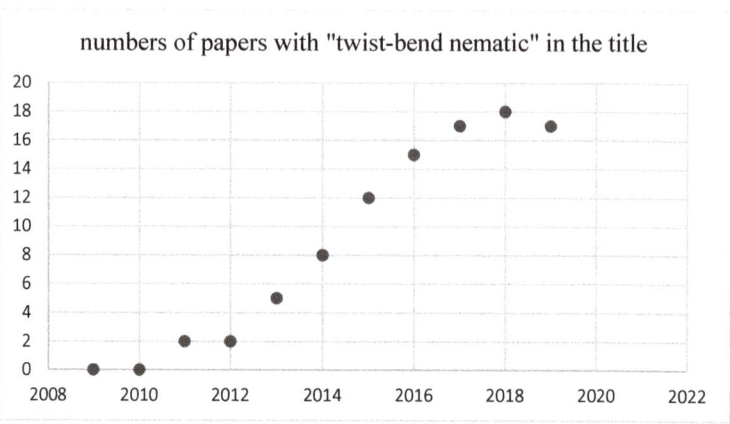

4. The Structure of the Twist–Bend Nematic Phase

The explosion of interest in the twist–bend nematic phase that followed from 2011 included key papers providing additional evidence for the proposed structure of the heliconical phase. In particular, a paper [41] from a large group at the University of Colorado Boulder presented freeze-fracture measurements on CB7CB, showing a periodic structure consistent with a helix having a pitch of 8.3 nm with a director cone angle of 25°. Freeze-fracture was also among the techniques used by a combined group of researchers from the Liquid Crystal Institute at Kent State University and the universities of Dublin, Aberdeen and Hull, who reported further results [42] on the twist–bend phases of CB7CB and a related mixture confirming the heliconical structure of the phase. Other examples of the twist–bend nematic phase also were reported [43] from studies of related homologues of CB7CB.

The absence of X-ray diffraction peaks corresponding to the helix is due to its glide symmetry, which means that there is no electron density modulation associated with the heliconical arrangement of molecules. However, diffraction can be observed from the periodicity of the helix for suitable samples using the method of resonant X-ray scattering which utilises polarised radiation. A group from the University of Colorado Boulder used the Advanced Light Source at the Lawrence Berkeley National Laboratory to confirm the helical structure of the twist–bend phase of CB7CB [44], and further demonstrated that the pitch increases rapidly as the transition to the normal nematic phase is approached from lower temperatures. A more detailed resonant X-ray scattering study of the twist–bend nematic phase of a mixture, the component molecules of which were structurally similar to CB7CB, gave measurements of the helical pitch [45], and demonstrated its increase with increasing temperature. Further analysis of the scattering pattern revealed that the flexible molecules modified their structure in the nematic twist–bend environment to adapt to the helical arrangement in the phase.

The chirality of the twist–bend phase is a consequence of its heliconical structure, but in an overall achiral system, symmetry-breaking must result in the formation of equal contributions from left- and right-handed helices. Optical textures of the twist–bend nematic phase provide some confirmation of domain formation of opposite handed structures, and in the characteristic striped textures, originally labelled as "rope-like domains", the birefringence of adjacent domains alternates, with the maximum axis of birefringence deviating alternately to the left and right by about 10°. Although consistent with optically chiral domains, these observations are not definitive. An indirect indicator of chirality in a sample can be revealed by NMR spectroscopy, since in a chiral phase of flexible molecules, the equivalence between certain nuclear sites can be removed by symmetry-breaking and detected through an appropriate NMR diagnostic. This was done for the material CB7CB, and the results demonstrated [46] that the twist–bend nematic phase of CB7CB was chiral, with inequivalent C−H bonds in the alkyl chain related by left-right symmetry and contributing equal intensities to the NMR signal. Optical chirality in homeotropically aligned thin films of the twist–bend phase nematic phase of CB7CB and other homologues has been confirmed using circular dichroism [47].

Much of the investigative research on the structure of the twist–bend mesophase has been conducted on the material CB7CB, which was the first low molecular weight compound to be identified as exhibiting a twist–bend nematic phase [4]. However, it will be recalled that the first report of a nematic–nematic transition was described [14,15] in a series of main-chain liquid crystal copolyethers with semi-flexible biphenylethane groups linked by odd-numbered oligomethylene spacers. The structure of the liquid crystal phases of these materials has been reinvestigated [48] using techniques of grazing incidence X-ray diffraction and polarised infra-red spectroscopy. This work has unequivocally shown that the phases originally identified as low temperature nematic phases are in fact twist–bend nematic phases, and so were probably the first materials to be revealed exhibiting such a phase.

5. Features of the Twist–Bend Nematic Phase

This commentary on the discovery of the twist–bend nematic phase focuses only on a class of materials that exhibit the characteristic features identified in the low temperature phase of CB7CB. Thus, other low-temperature nematic phases that may or may not exist in other bent-core or banana-shaped molecules are not considered, neither are they excluded as possible new nematic phases. The main identifying feature of the twist–bend phase is a weak first order phase transition on lowering the temperature to a phase showing no X-ray scattering characteristic of long-range positional order. The higher temperature phase may be a normal nematic phase or sometimes an isotropic phase [49], and the low temperature phase should have features showing the tilt or twist of the director.

The compounds that have been identified in this article as having twist–bend nematic phases are dimers, oligomers, or polymers, with mesogenic units linked by flexible alkyl chains. The flexibility of the constituent molecules may be a stabilising aspect of the twist–bend phase which enables the molecular structure to adapt to the local environment, self-selecting conformations that result in director tilt or a preferred handed-twist (chirality). The importance of the molecular structure adapting to the twist–bend phase is signalled by the dielectric measurements on the material CB7CB, which has two dipolar groups linked by a methylene chain. In the isotropic phase, the mean permittivity (~10) is what would be expected of a fluid composed of monomer units of a short alkyl chain cyanobiphenyl, that is the mesogenic end groups are responding dielectrically as though they are not connected. On cooling into the normal nematic phase, over a small temperature interval there is a rise in the parallel permittivity and a fall in the perpendicular permittivity, which would be expected for a monomeric dipolar mesogen. Then, as the nematic order develops, there is a dramatic change in behaviour with both components and the average permittivity falling as the temperature decreases and the nematic order increases. This indicates that the flexible link is favouring an extended conformation of the connected mesogenic groups, and the average dipole moment of the mesogenic dimer falls. It continues to reduce on lowering the temperature until the transition to the twist–bend nematic is reached, when there is a sharp increase in the rate of decline of the average dipole moment. A few degrees into the twist–bend nematic, the dielectric anisotropy becomes zero. A careful X-ray study of a model dimer [45] has provided evidence that molecular conformations do adjust to the helical structure, such that the gain in translational entropy along the axis exceeds any enthalpy penalty from a change in the conformational distribution.

A key feature of the twist–bend nematic phase is chirality. Direct evidence of twist may be manifest by intrinsic optical chirality, but other techniques can observe evidence of tilt or twist. The optical textures or patterns shown by thin films of liquid crystal viewed under a polarising microscope have become valuable diagnostics for many phase types. Some phases show characteristic textures, and for the twist–bend nematic phase, stripped textures having an apparent alternating twist, rope-like domains, have been frequently identified, although other textures can also be seen for different sample preparations. In achiral systems, the formation of the twist–bend phase results in equal left-handed and right-handed domains, but the matter of chiral separation has yet to be fully explored.

A model for a twist–bend nematic phase is easily imagined, and the driving force for its manifestation has been suggested as due to unusual elastic properties and/or ferroelectric organisation of constituent molecules. Understanding the formation of a twist–bend phase can only be obtained through careful measurement and interpretation of its physical properties. Since the materials to be studied are anisotropic fluids, their symmetry will determine the nature of the properties to be measured, and there is usually a requirement for samples to be macroscopically aligned. Being liquid crystals, elasticity and viscosity are key properties, but the complexity of the structure and symmetry of N_{TB} phase make definitive measurements and interpretation difficult.

The symmetry of a helix can be defined in various ways, but for a sample of achiral twist–bend nematic it is expected that both left- and right-handed helical structures will be present with defects between the different chiral domains. To define an elasticity or viscoelastic tensor for such a system is clearly difficult, and there is the added practical difficulty of ensuring a controlled alignment to provide measurement of appropriate components of the tensor properties. For these reasons, measurements of the physical properties of compounds forming twist–bend nematic phases have mostly been made in the nematic phases of the materials. There is some evidence that macroscopic alignment in the high-temperature nematic phase is preserved in the twist–bend phase, so it is possible to obtain information on the likely structure of the material.

The significance of the bend elastic constant in determining the formation of a twist bend nematic phase or heliconical phase really emerged from the work of Meyer [13,50,51]. He found that if a magnetic field is applied perpendicular to the helix axis of a chiral nematic phase, then the pitch increased. De Gennes expands on this [52] to suggest that for a small bend elastic constant, the orientation of the director could make an angle of less than 90° to the magnetic field, and so an induced heliconical phase could be formed. This was the basis of Meyer's original conjecture about the formation of a twist–bend phase. Dozov extended this idea to show that a negative bend elastic constant could, not surprisingly, result in a permanently bent structure, as represented by the twist–bend phase. Further theoretical work [53,54] established that a negative bend elastic constant (k_3) was not mandatory, only that k_3 was small and less than the twist elastic constant. As explained above, measurements of elastic properties in the twist–bend phase present difficulties, but it has been shown [55] that in CB7CB the bend elastic constant as measured in the nematic phase is smaller than both the splay and twist constants and decreases with temperature through the nematic phase as the transition to the twist–bend phase is approached. An alternative or additional driving force proposed for the formation of a twist–bend nematic phase was the local organisation of transverse molecular dipoles of bent-core molecules resulting in a macroscopic polarisation, and there have been a number of theoretical and experimental studies of related phenomena such as flexoelectricity. An electro-clinic effect has been observed [56] in the twist–bend phase of CB7CB, which proves the chirality of the phase, and analysis of the results provided an estimate of the short pitch of the helix and a value for the heliconical tilt angle.

Over the past 50 years, nematic liquid crystals have been transformational in the display industry, so it is natural to look for new applications of the twist–bend nematic phase in the area of opto-electronics. It is possible to electrically switch the optical appearance of thin films of N_{TB}, and preliminary measurements [57] indicate that Fréedericksz transitions occur, but the structures of the switched states need further study. It has been suggested [58] that the electric field-induced distortions of the twist–bend nematic phase are similar to those observed in chiral smectic A phases and include a sub-microsecond flexoelectric-induced rotation of the optic axis. There are many aspects of the structure and properties of the twist–bend nematic phase that remain to be understood, and as has often been the case, a determining factor in the research will be the availability of suitable materials exhibiting the phase at convenient temperatures. However, this constraint on experiments does not inhibit theoretical studies. For example, a model for bend distortions in the twist–bend phase based on the Frank free energy has been developed [59], which might add to our understanding of complex structures in thin films.

6. Conclusions

A liquid crystal phase with a heliconical structure is a concept that follows from the models that we have for the chiral nematic phase. Robert Meyer recognized this around 1969 when investigating [50,51] the effect of a magnetic field on the structure of a chiral nematic phase formed from a mixture of p-azoxyanisole and cholesteryl acetate, as related in the book by de Gennes [52]. Such a field-induced heliconical phase occurring in a chiral material would require an anomalously small bend elastic constant. Meyer later proposed [2] that a helical state of finite torsion and bend is possible in a nematic if there is a non-zero polarisation perpendicular to the local director but noted that no helical structure had ever been reported in a non-chiral nematic. Thus, the factors identified for the stability of a heliconical nematic phase were a small bend elastic constant and/or a local perpendicular polarisation, yet the appearance of a chiral structure in an achiral material seemed unlikely.

In the decades following Meyer's observations, there was no research effort to find a heliconical nematic phase, although there were occasional reports of possibly novel but unidentified nematic phases [15–17]. In 2001, Dozov explored the consequences of a negative bend elastic constant in the context of Landau theory and predicted [3] that a state of continuous conical twist–bend could be stabilized with a two-fold degenerate twist left and right.

Like so many liquid crystal discoveries, the experimental identification of a twist–bend nematic phase was dependent on serendipitous studies of a particular material. A class of liquid crystals known as dimeric mesogens includes the compound CB7CB, which had presented some unexplained properties [37,38]. A careful study over a few years by a number of collaborating research groups through Europe established that this liquid crystal did indeed exhibit the features of a twist–bend nematic phase proposed by the theoretical work of Meyer and Dozov. The publication [4] of these results stimulated a flurry of research activity around the globe, and confirmation of the twist–bend structure was rapidly established by a number of research groups [41,42] in a variety of compounds. There followed a great increase in the investigation of novel nematic phases, a number of which were revealed to have characteristics of the twist–bend phase. Science does not provide a template for nature, and the structures of liquid crystal phase variants are determined by particular intermolecular interactions in different materials.

The emergence of the twist–bend nematic phase has been a consequence of the research of many scientists. Collaboration between groups has aided the evaluation of the structure of the new phase and should enhance the investigation of new properties with possible applications. Liquid crystals are examples of soft matter, and the variety of structures observed, including the twist–bend nematic phase, might be important in understanding the structure and growth of natural materials.

Funding: This research received no external funding.

Acknowledgments: The author wishes to thank all those involved in the discovery of the twist–bend nematic phase. Particular mention is made of those from collaborating laboratories in Europe who contributed to the characterization and identification of the twist–bend nematic phase in 1″,7″-bis(4-cyanobiphenyl-4′-yl)heptane CB7CB [4]. The liquid crystal group from the Polytechnic University of Barcelona carried out calorimetric measurements to establish the nature of the phase transition from nematic to twist–bend nematic, and showed that the proximity to a tricritical point suggested an additional degree of molecular ordering in the low-temperature phase. Dielectric measurements carried out in the Applied Physics Department of the University of the Basque Country, Bilbao, gave further information on the macroscopic structure of the phases and strongly indicated that a tilt of the director away from the rubbing direction occurred at the phase transition to a twist–bend nematic. Confirmation of the nematic nature of the twist–bend phase was provided by the liquid crystal group of the HH Wills Physics Laboratory at the University of Bristol, which determined an intercalated structure with a weak tendency to layer formation i.e., not a smectic. The theory group from the Department of Chemical Science, University of Padua, used a surface interaction (SI) model to take account of molecular flexibility in the analysis of dielectric measurements, which showed the inadequacy of the rotational isomeric state model to describe the properties of flexible dimeric mesogens. This group also carried out calculations using the SI model to predict that the bend elastic constant of CB7CB could be negative. Evidence of chiral symmetry breaking came from magnetic resonance measurements for which selectively deuterated samples of CB7CB were prepared by the group from the Department of Biophysics, Max-Planck Institute for Medical Research, Heidelberg. The compound CB7CB was originally prepared in the liquid crystal group of the School of Chemistry, University of Southampton. Magnetic resonance measurements, ESR and NMR, established that chiral symmetry breaking occurred at the phase transition from nematic to twist–bend nematic. This was dramatically illustrated by a bifurcation in the quadrupolar splitting for the 1″ and 7″ deuterons in CB7CB. Optical textures characteristic of the twist–bend nematic phase were identified in the Southampton group, which also coordinated the research from other contributing laboratories. Finally, the author would like to acknowledge the role of Professor Geoffrey Luckhurst as leader of the Southampton Liquid Crystal Group, the research of which resulted in the discovery of the twist–bend nematic phase.

Conflicts of Interest: The author declares no competing interest.

References

1. Walker, R. The twist-bend phases: Structure–property relationships, chirality and hydrogen-bonding. *Liq. Cryst. Today* **2020**, *29*, 2–14. [CrossRef]
2. Meyer, R.B. *Structural Problems in Liquid Crystal Physics, in Molecular Fluids*; Balian, R., Weill, G., Eds.; Vol. XXV-1973 of Les Houches Summer School in Theoretical Physics; Gordon and Breach: New York, NY, USA, 1976; pp. 273–373.
3. Dozov, I. On the spontaneous symmetry breaking in the mesophases of achiral banana-shaped molecules. *Eur. Lett.* **2001**, *56*, 247–253. [CrossRef]
4. Cestari, M.; Diez-Berart, S.; Dunmur, D.A.; Ferrarini, A.; de la Fuente, M.R.; Jackson, D.J.B.; Lopez, D.O.; Luckhurst, G.R.; Perez-Jubindo, M.A.; Richardson, R.M.; et al. Phase behaviour and properties of the liquid crystal dimer 1″,7″-bis(4-cyanobiphenyl-4′-yl)heptane: A twist-bend nematic liquid crystal. In Proceedings of the 23rd International Liquid Crystal Conference, Kraków, Poland, 11–16 July 2010; Volume 84, p. 031704. [CrossRef]
5. Dozov, I.; Luckhurst, G.R. Setting things straight in "The twist-bend nematic: A case of mistaken identity. *Liq. Cryst.* **2020**, *47*, 2098–2115. [CrossRef]
6. Reinitzer, F. Contributions to the understanding of cholesteryls. *Monat. f. Chem.* **1888**, *9*, 421–441. [CrossRef]
7. Freiser, M.J. Ordered States of a Nematic Liquid. *Phys. Rev. Lett.* **1970**, *24*, 1041–1043. [CrossRef]
8. Luckhurst, G.R.; Sluckin, T.J. (Eds.) *Biaxial Nematic Liquid Crystals*; Wiley: Chichester, UK, 2015.
9. Jákli, A.; Lavrentovich, O.D.; Selinger, J.V. Physics of liquid crystals of bent-shaped molecules. *Rev. Mod. Phys.* **2018**, *90*, 045004. [CrossRef]
10. Dunmur, D.A. Dielectric studies of intermolecular and intramolecular interactions in nematic liquid crystals. *Proc. SPIE* **2002**, *4759*, 196–208.
11. Dunmur, D.A.; Palffy-Muhoray, P. A mean field theory of dipole-dipole correlation in nematic liquid crystals. *Mol. Phys.* **1992**, *76*, 1015–1024. [CrossRef]
12. De Gennes, P.G. *The Physics of Liquid Crystals*; Oxford University Press: Oxford, UK, 1974.
13. Meyer, R.B.; University of Harvard, Cambridge, MA, USA. Personal communication, 1969.
14. Ungar, G.; Percec, V.; Zuber, M. Liquid crystalline polyethers based on conformational isomerism. 20. Nematic-nematic transition in polyethers and copolyethers based on 1-(4-hydroxyphenyl)2-(2-R-4-hydroxyphenyl)ethane with R = fluoro, chloro and methyl and flexible spacers containing an odd number of methylene units. *Macromolecules* **1992**, *25*, 75–80. [CrossRef]
15. Ungar, G.; Percec, V.; Zuber, M. Influence of molecular structure on the nematic-nematic transition in polyethers based on 1-(4-hydroxy-phenyl)-2-(2-R-4-hydroxyphenyl ethane where R = CH3 and Cl, and flexible spacers with an odd number of methylene units. *Polym. Bull.* **1994**, *32*, 325–330. [CrossRef]
16. Šepelj, M.; Lesac, A.; Baumeister, U.; Diele, S.; Nguyen, H.L.; Bruce, D.W. Intercalated liquid-crystalline phases formed by symmetric dimers with an α,ω-diiminoalkylene spacer. *J. Mater. Chem.* **2007**, *17*, 1154–1165. [CrossRef]
17. Görtz, V.; Southern, C.; Roberts, N.W.; Gleeson, H.F.; Goodby, J.W. Unusual properties of a bent-core liquid-crystalline fluid. *Soft Matter* **2009**, *5*, 463–471. [CrossRef]
18. Link, D.R.; Natale, G.; Shao, R.; Maclennan, J.E.; Clark, N.A.; Körblova, E.; Walba, D.M. Spontaneous Formation of Macroscopic Chiral Domains in a Fluid Smectic Phase of Achiral Molecules. *Science* **1997**, *278*, 1924–1927. [CrossRef]

19. Lagerwall, S.T. Ferroelectric Liquid Crystals. In *Handbook of Liquid Crystals*, 1st ed.; Demus, D., Goodby, J.W., Gray, G.W., Spiess, H.-W., Vill, V., Eds.; Wiley-VCH: New York, NY, USA, 1999; Volume 2B, p. 515.
20. Salter, P.S.; Benzie Reddy, R.A.; Tschierske, C.; Elston, S.J.; Raynes, E.P. Spontaneously chiral domains of an achiral bent-core nematic liquid crystal in a planar aligned device. *Phys. Rev. E* **2009**, *80*, 031701. [CrossRef] [PubMed]
21. Lubensky, T.C.; Radzihovsky, L. Theory of bent-core liquid-crystal phases and phase transitions. *Phys. Rev. E* **2002**, *66*, 031701. [CrossRef]
22. Mettout, B. Macroscopic and molecular symmetries of unconventional nematic phases. *Phys. Rev. E* **2006**, *74*, 041701. [CrossRef]
23. Kamien, R.D. Liquids with Chiral Bond Order. *J. Phys. II* **1996**, *6*, 461–475. [CrossRef]
24. Takezoe, H. Spontaneous Achiral Symmetry Breaking in Liquid Crystalline Phases. *Top. Curr. Chem.* **2011**, *318*, 303–330. [CrossRef]
25. Filev, V.M. Temperature-induced inversion of rotation of the polarization-plane of light in liquid crystals. *JETP Lett.* **1983**, *37*, 703–706.
26. Dolganov, V.K.; Demikhov, E.I. Pretransitional effects near blue phases of a cholesteric liquid-crystal. *JETP Lett.* **1983**, *38*, 445–447.
27. Kats, E.I.; Abalyan, T.V. Ordering and phase transitions in liquid crystals. *Phase Transit.* **1991**, *91*, 237–268. [CrossRef]
28. Pasini, P.; Zannoni, C. (Eds.) *Advances in the Computer Simulations of Liquid Crystals*; Kluwer: Dordrecht, The Netherlands, 1998.
29. Memmer, R. Liquid crystal phases of achiral banana-shaped molecules: A computer simulation study. *Liq. Cryst.* **2002**, *29*, 483–496. [CrossRef]
30. Schröder, M.W.; Diele, S.; Pelzl, G.; Dunemann, U.; Kresse, H.; Weissflog, W. Different nematic phases and a switchable SmCP phase formed by homologues of a new class of asymmetric bent-core mesogens. *J. Mater. Chem.* **2003**, *13*, 1877–1882. [CrossRef]
31. Tamba, M.G.; Baumeister, U.; Pelzl, G.; Weissflog, W. Banana-calamitic dimers: Unexpected mesophase behaviour by variation of the direction of ester linking groups in the bent-core unit. *Liq. Cryst.* **2010**, *37*, 853–874. [CrossRef]
32. Keith, C.; Lehmann, A.; Baumeister, U.; Prehm, M.; Tschierske, C. Nematic phases of bent-core mesogens. *Soft Matter* **2010**, *6*, 1704–1721. [CrossRef]
33. Barry, E.; Hensel, Z.; Dogic, Z.; Shribak, M.; Oldenbourg, R. Entropy-Driven Formation of a Chiral Liquid-Crystalline Phase of Helical Filaments. *Phys. Rev. Lett.* **2006**, *96*, 018305. [CrossRef]
34. Imrie, C.T.; Luckhurst, G.R. Liquid Crystal Dimers and Oligomers. In *Handbook of Liquid Crystals*, 1st ed.; Demus, D., Goodby, J.W., Gray, G.W., Spiess, H.-W., Vill, V., Eds.; Wiley-VCH: New York, NY, USA, 1999; Volume 2B, Chapter X; p. 801.
35. Dunmur, D.A.; de la Fuente, M.R.; Perez-Jubindo, M.A.; Diez, S. Dielectric studies of liquid crystals: The influence of molecular shape. *Liq. Cryst.* **2010**, *37*, 723–736. [CrossRef]
36. Barnes, P.J.; Douglass, A.; Heeks, S.K.; Luckhurst, G.R. An enhanced odd-even effect of liquid crystal dimers orientational order in the α,ω-bis(4′-cyanobiphenyl-4yl) alkanes. *Liq. Cryst.* **1993**, *13*, 603. [CrossRef]
37. Dunmur, D.A.; Luckhurst, G.R.; De La Fuente, M.R.; Diez, S.; Jubindo, M.A.P. Dielectric relaxation in liquid crystalline dimers. *J. Chem. Phys.* **2001**, *115*, 8681–8691. [CrossRef]
38. Stocchero, M.; Ferrarini, A.; Moro, G.J.; Dunmur, D.A.; Luckhurst, G.R. Molecular theory of dielectric relaxation in nematic dimers. *J. Chem. Phys.* **2004**, *121*, 8079. [CrossRef] [PubMed]
39. Luckhurst, G.R.; Tropea, Italy. Personal communication, 2009.
40. Panov, V.P.; Nagaraj, M.; Vij, J.K.; Panarin, Y.P.; Kohlmeier, A.; Tamba, M.G.; Lewis, R.A.; Mehl, G. Spontaneous Periodic Deformations in Nonchiral Planar-Aligned Bimesogens with a Nematic-Nematic Transition and a Negative Elastic Constant. In Proceedings of the 23rd International Liquid Crystal Conference, Kraków, Poland, 11–16 July 2010; Volume 105, p. 167801. [CrossRef] [PubMed]
41. Chen, D.; Porada, J.H.; Hooper, J.B.; Klittnick, A.; Shen, Y.; Tuchband, M.R.; Korblova, E.; Bedrov, D.; Walba, D.M.; Glaser, M.A.; et al. Chrial heliconical ground state of nanoscale pitch in a nematic liquid crystal of achiral molecular dimers. *Proc. Natl. Acad. Sci. USA* **2013**, *110*, 15931–15936. [CrossRef] [PubMed]
42. Borshch, V.; Kim, Y.-K.; Xiang, J.; Gao, M.; Jakli, A.; Panov, V.P.; Vij, J.K.; Imrie, C.T.; Tamba, M.G.; Mehl, G.H.; et al. Nematic twist-bend phase with nanoscale modulation of molecular orientation. *Nat. Commun.* **2013**, *4*, 2635. [CrossRef] [PubMed]
43. Henderson, P.A.; Imrie, C.T. Methylene-linked liquid crystal dimers and the twist-bend nematic phase. *Liq. Cryst.* **2011**, *38*, 1407–1414. [CrossRef]
44. Zhu, C.; Tuchband, M.R.; Young, A.; Shuai, M.; Scarbrough, A.; Walba, D.M.; Maclennan, J.E.; Wang, C.; Hexemer, A.; Clark, N.A. Resonant CarbonK-Edge Soft X-Ray Scattering from Lattice-Free Heliconical Molecular Ordering: Soft Dilative Elasticity of the Twist-Bend Liquid Crystal Phase. *Phys. Rev. Lett.* **2016**, *116*, 147803. [CrossRef] [PubMed]
45. Stevenson, W.D.; Ahmed, Z.; Zeng, X.B.; Welch, C.; Ungar, G.; Mehl, G.H. Molecular organisation in the twist-bend nematic phase by resonant x-ray scattering at the Se K-edge and by SAXS, WAXS and GIXRD. *Phys. Chem. Chem. Phys.* **2017**, *19*, 13449. [CrossRef]
46. Beguin, L.; Emsley, J.W.; Lelli, M.; Lesage, A.; Luckhurst, G.R.; Timimi, B.A.; Zimmermann, H. The Chirality of a Twist–Bend Nematic Phase Identified by NMR Spectroscopy. *J. Phys. Chem. B* **2012**, *116*, 7940–7951. [CrossRef] [PubMed]
47. Stevenson, W.D.; Zeng, X.; Welch, C.; Thakur, A.K.; Ungar, G.; Mehl, G.H. Macroscopic chirality of twist-bend nematic phase in bent dimers confirmed by circular dichroism. *J. Mater. Chem. C* **2020**, *8*, 1041–1047. [CrossRef]
48. Stevenson, W.D.; An, J.; Zeng, X.B.; Xue, M.; Zou, H.X.; Liu, Y.S.; Ungar, G. Twist-bend phase in biphenylethane-based copolymers. *Soft Matter* **2018**, *14*, 3003. [CrossRef]

49. Dawood, A.A.; Grossel, M.C.; Luckhurst, G.R.; Richardson, R.M.; Timimi, B.A.; Wells, N.J.; Yousif, Y.Z. On the twist-bend nematic phase formed directly from the isotropic phase. *Liq. Cryst.* **2016**, *43*, 2–12. [CrossRef]
50. Meyer, R.B. Effects of electric and magnetic fields on the structure of cholesteric liquid crystals. *Appl. Phys. Lett.* **1968**, *12*, 281–282. [CrossRef]
51. Meyer, R.B. Distortion of a cholesteric structure by a magnetic field. *Appl. Phys. Lett.* **1969**, *14*, 208–209. [CrossRef]
52. De Gennes, P.G.; Prost, J. *The Physics of Liquid Crystals*, 2nd ed.; Oxford University Press: Oxford, UK, 1993; p. 292.
53. Virga, E.G. Elasticity of Twist-Bend Nematic Phases. In *Differential Geometry and Continuum Mechanics*; Chen, G.Q., Grinfeld, M., Knops, R., Eds.; Springer Proceedings in Mathematics & Statistics; Springer: Cham, Switzerland, 2015; Volume 137, pp. 363–380.
54. Barbero, G.; Evangelista, L.R.; Rosseto, M.P.; Zola, R.S.; Lelidis, I. The Physics of Liquid Crystals. *Phys. Rev. E* **2015**, *92*, 030501. [CrossRef] [PubMed]
55. Babakhanova, G.; Parsouzi, Z.; Paladugu, S.; Wang, H.; Nastishin, Y.A.; Shiyanovskii, S.V.; Sprunt, S.; Lavrentovich, O.D. Elastic and viscous properties of the nematic dimer CB7CB. *Phys. Rev. E* **2017**, *96*, 062704. [CrossRef] [PubMed]
56. Meyer, C.; Luckhurst, G.R.; Dozov, I. Flexoelectrically Driven Electroclinic Effect in the Twist-Bend Nematic Phase of Achiral Molecules with Bent Shapes. *Phys. Rev. Lett.* **2013**, *111*, 067801. [CrossRef]
57. Panov, V.; Song, J.-K.; Mehl, G.; Vij, J. The Beauty of Twist-Bend Nematic Phase: Fast Switching Domains, First Order Fréedericksz Transition and a Hierarchy of Structures. *Crystals* **2021**, *11*, 621. [CrossRef]
58. Meyer, C.; Dozov, I.; Davidson, P.; Luckhurst, G.R.; Dokli, I.; Knezevic, A.; Lesac, A. Electric-field effects in the twist-bend nematic phase. *Proc. SPIE Emerg. Liq. Cryst. Technol.* **2018**, *10555*, 105550Z. [CrossRef]
59. Binysh, J.; Pollard, J.; Alexander, G.P. Geometry of Bend: Singular Lines and Defects in Twist-Bend Nematics. *Phys. Rev. Lett.* **2020**, *125*, 047801. [CrossRef]

Article

Reappraisal of The Optical Textures of Columnar Phases in Terms of Developable Domain Structures with Relaxed Constraints and a Rationale for The Striated Texture

John E. Lydon

Faculty of Biological Sciences, The University of Leeds, Leeds LS2 9JT, UK; j.e.lydon@leeds.ac.uk

Abstract: Optical textures pictured in the seminal 1974 textbook, *The Microscopy of Liquid Crystals*, by Norman Hartshorne, have been reappraised. Some of these, which were described by Hartshorne (and many others) as *confused focal conics*, were of chromonic and discotic phases, which had not been identified at that time—and would now be recognized as developable domain structures of columnar phases. It is suggested that the rigorous constraint of isometry in these is relaxed in regions of the director field under high stress. A rationale for the characteristic striated appearance of columnar textures is proposed, in which the molecular columns are bundled together, forming twisted ropes within the domains. It is also suggested that the regular alternation of opposing domains in M ribbons minimizes the slippage of columns required as the mesophase develops, and an explanation of the characteristic multi-pole appearance of the brushes in the optical textures of columnar structures is proposed.

Keywords: Hartshorne; Bouligand; liquid crystals; optical textures; developable domains; columnar hexagonal phases

Citation: Lydon, J.E. Reappraisal of The Optical Textures of Columnar Phases in Terms of Developable Domain Structures with Relaxed Constraints and a Rationale for The Striated Texture. *Crystals* **2022**, *12*, 1180. https://doi.org/10.3390/cryst12081180

Academic Editors: Ingo Dierking and Charles Rosenblatt

Received: 16 May 2022
Accepted: 16 August 2022
Published: 22 August 2022

Publisher's Note: MDPI stays neutral with regard to jurisdictional claims in published maps and institutional affiliations.

Copyright: © 2022 by the author. Licensee MDPI, Basel, Switzerland. This article is an open access article distributed under the terms and conditions of the Creative Commons Attribution (CC BY) license (https://creativecommons.org/licenses/by/4.0/).

1. Preface

This article is divided into three numbered parts, Section 1, which includes the Introduction and background, Section 2, which describes the reappraisal of Norman Hartshorne's inferences drawn from his observed optical textures in light of subsequent published work, and Section 3, which contains original material: new tentative hypotheses for the striated texture and the multi-nucleated appearance of discontinuities in the optical textures of columnar structures.

In the years between 1930 and his death in 1982, Norman Hartshorne was among a small handful of internationally recognized experts on optical microscopy in general and of liquid crystals in particular. In 1974, he wrote the first book dealing specifically with the optics of mesophases, *The Microscopy of Liquid Crystals*, Volume 48 in *The Microscope Series* [1]. This comprehensive small volume covered the optical microscopy of the then-known world of liquid crystals, lyotropic and thermotropic. It included descriptions of mesophases which would now be recognized as developable domain structures of columnar phases [2,3]. Previously, his textbook, *Crystals and the polarizing microscope*, co-authored by the geologist, Allan Stuart, [4] was regarded as the definitive volume on the subject (and ran to seven editions). In his later life, after retiring as Reader in Physical Chemistry at the University of Leeds, he ran courses in optical microscopy under the banners of both the Royal Microscopical Society and the London branch of the McCrone Research Institute.

The came at a time of rapid expansion of the subject [5]. Prior to this, things had been relatively simple and neatly compartmentalized into lyotropic mesogens, which were largely the possession of the soap and detergent industry, and the thermotropics, which were synthesized by organic chemists and studied by a few physicists. Their only property, which was seen as being exploitable at that time, was the thermochromic nature, of cholesteric phases, for temperature mapping, in medicine (for detecting cancerous growths) and for military night vision devices [6]. There were occasional far-sighted papers such as Bernal and Fankuchen's article on the mesophase formed by the tobacco mosaic virus in 1941 [7], but these were isolated and were in no way mainstream liquid crystal literature works. Reviews of liquid crystals rarely mentioned carbonaceous phases or biological materials other than lipids. The aggregation of dye molecules in solution was not mentioned in the context of liquid crystalline phases. Each area had its own terminology, and the papers dealing with them were usually published in different journals. The first specialist journal dealing with liquid crystals, *Molecular Crystals* (which was to evolve into Molecular Crystals and Liquid Crystals), appeared in 1966. Its rival journal, *Liquid Crystals*, started in 1986, and *Liquid Crystals Today* in 1990.

In the following decades, the situation became rapidly more complex. Chromonic phases were identified [8–10], and discotic thermotropic phases appeared [11–13], in spite of Vorländer apparently "proving" that they were impossible [14]. Liquid crystal displays had started to cover the world. There was increased funding for liquid crystal research, and the rate of publication of articles about mesophases began to increase exponentially. The fascinating complexities of blue phases, banana-shaped mesogens, twist grain boundaries, and liquid crystalline elastomers were all well below the horizon, waiting to be discovered.

Over the years since the 1970s, there has been a sequence of invaluable textbooks on the optical textures of liquid crystals, including those listed in [15–17], but Hartshorne's book is of particular historical importance because of the work it stimulated. It was written when discotic and chromonic phases had been produced and studied, but not yet characterized and recognized as radically new families of liquid crystals. He had accurately described their optical textures and optical properties (i.e., their optical textures, birefringence, and refractive indices) in detail, but he was not able to deduce their structures. Like everyone else at that time, he was attempting to define them in terms of the known mesophase structures, i.e., the nematic and smectic thermotropic states and the lyotropic neat and middle phases.

The frontispiece of Hartshorne's book (the only colour illustration in the volume) is shown in Figure 1a. His caption reads: "These batonnets and spherulites are characteristic of the lyotropic mesophase, M, of disodium chromoglycate (INTAL), Fisons new anti-asthma drug. They are formed on cooling an aqueous solution saturated at about 65–70° to about 50°".

The features identified here as "spherulites" would now be recognized as developable domain structures [2,3] formed by M phase columns curving into circular arcs. I suggest that the richly decorated straight-sided polygon (resembling a miniature item of art deco jewellery), described as a "batonnet", is actually a single crystal of Intal, with added surface detail formed from epitaxially positioned strands and circles of the M phase. Note that the larger circular decorations tend to occur above the edges and corners of the crystal, whereas the linear and rectangular features lie in lines on the horizontal surface of the crystal, usually along prominent crystallographic axes. I interpret the complete anulus, shown on the lower right of the photograph, as a ring of chromonic M phase, formed around an air bubble.

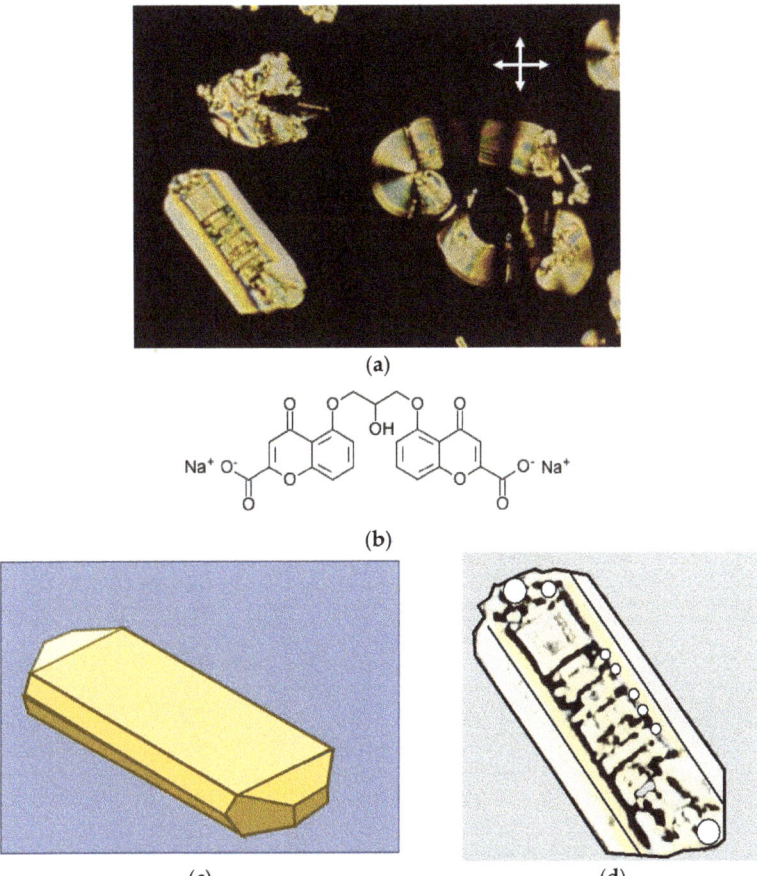

Figure 1. (a) A reproduction, with the permission of Microscope Publications, of the frontispiece in Hartshorne's 1974 textbook, *The Microscopy of Liquid Crystals*, [1] showing regions of the mesophase of Intal forming in the isotropic liquid [1]. (b) The molecular structure of the Intal. (c) The habit of a single crystal of Intal. (d) Enhanced image of the decorated "batonnet" in (a), with some circular features outlined.

2. Introduction

2.1. Diisobutylsilane Diol and Intal

I described previously how I came to be given an early copy of Hartshorne's book in 1974, at one of his courses in London (where, to my considerable embarrassment, I had to give the introductory lecture when he became ill) [8].

The early chapters of this book contained the standard groundwork for that time: basic microscopy, hot stages, birefringence, lyotropics and thermotropics, optical textures—schlieren patterns of nematics and focal conics—director fields of smectics, cholesteric phases, etc., with familiar drawings, such as those of Dupin cyclides and the polygonal texture of smectic phases. However, the final chapter was surprising and certainly not what you would expect to find in a textbook intended for beginners in the field. It was entitled simply *Some unusual mesophase systems*. Norman Hartshorne was highlighting material for further investigation. Its fifteen pages covered two topics. The first was the optical textures of the mesophases of the small thermotropic mesogen, di-isobutyl silane diol, shown in Figure 2 [12]. Its compact molecules did not resemble those of any known mesogen and did not look in any way suitable for mesophase formation. It occurred to me that the mesogenic units of the discotic phase might actually be the hydrogen-bonded dimers shown in Figure 2b.

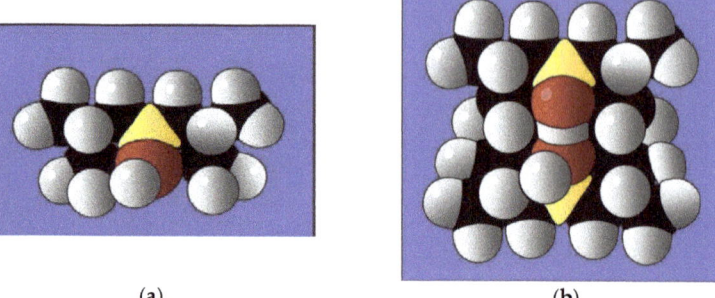

(a)   (b)

Figure 2. Molecular models of the monomer of diisobutyl silane diol (**a**) and the dimer (**b**) that was proposed as the structural unit of the discotic phase [12].

The second half of the chapter described the lyotropic phases formed by aqueous solutions of the then new anti-asthmatic drug, produced by Fisons. This was disodium chromoglycate, commercially known in the U.K. as Intal (from **I**nterference with **Al**lergy) and in the USA as Chromolyn. The first compound was arguably the first discotic mesogen to be characterised [12], and the second became the archetypal chromonic liquid crystalline mesogen [11,12].

In 1974, reading and re-reading Hartshorne's book, it dawned on me that although his observations would almost certainly be faultless, his interpretations of them might not necessarily be correct and that he might not have taken into consideration the possibility that he was dealing with radically new types of the mesophase that did not conform to any of the then-known structures. Perhaps the familiar terms such as *spherulites, batonnets*, and *focal conics*, which he had used, were not appropriate. Puzzlingly, in his comments about the two mesophases formed by Intal, he described the textures of the more dilute N phase as resembling those formed by small-molecule **thermotropic** nematic phases and the textures of the M phase as resembling those of **lyotropic** middle phases, with hexagonal arrays of cylindrical micelles, as shown in Figure 3. In this, he was perfectly correct, of course, but he did not comment on the apparently paramorphotic relationship between their two textures nor suggest that both mesophases were built out of the same stacks of molecules, nor appreciate that there was something unusually precise about the curvature of the director field within the domains in the optical texture. In addition, he specifically commented on the unusual fine detail of striations that develop with time, but did not suggest any explanation.

Bearing in mind the similarity of the optical texture of the M phase of Intal and the middle phase of conventional lyotropic systems, I wondered if they were both composed of cylindrical columns, as sketched in Figure 3.

It seemed almost heresy to question the work of such a monumental figure, and it was not until I had talked the matter over with George Gray at Hull that I decided to risk proposing new models for both systems [8,11] and eventually coined the term "chromonic".

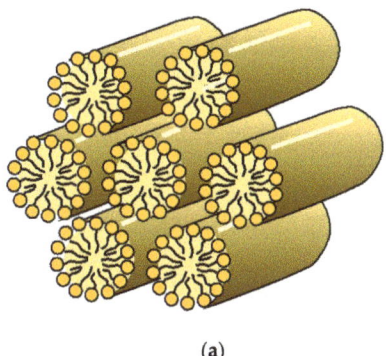 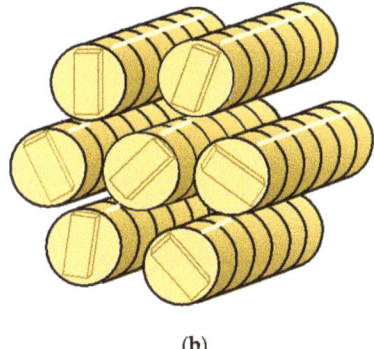

(a) (b)

Figure 3. Stylized representations of hexagonal columnar phases. (**a**) The M1 middle phase of an amphiphile/water system, showing a hexagonal array of cylindrical micelles in an aqueous continuum. The wavy radial lines represent more-or-less fluid aliphatic chains, and the small peripheral circles around each cylinder represent the hydrophilic head groups, with ionic or hydrogen bonding properties. (**b**) The hexagonal columnar structure of the chromonic M phase, showing the stacked aromatic molecules held face-to-face by π–π interactions. The rectangles drawn on the face of the discs indicate the random orientation of the stacks of aromatic groups in the columns. The peripheral hydrophilic groups attached to the aromatic cores make the columns soluble in the water continuum.

2.2. Developable Domains

The crucial factor determining the structures of developable domains is that columnar structures *splay* requires relatively high energy and the easiest distortion is *bend*. This leads to textures composed of units termed *developable domains* (discussed below), with isometric packing, of a hexagonal array, where the closest separation of the columns is constant throughout the structure.

A developable surface is a smooth surface that can be flattened onto a plane in some way by folding, bending, rolling, cutting, or gluing, without distortion (i.e., it can be bent without stretching or compression). Cylinders and cones have developable surfaces. All developable surfaces are ruled surfaces, but the converse is not necessarily true. The sphere is not a developable surface, and cartography is forced to be a two-stage process. First, part of the Earth's surface is projected onto a developable surface (a cylinder in the case of the Mercator projection) and, hence, onto a plane.

The cylinder-to-plane transformation is sketched in Figure 4a; industrial examples occur in large-scale engineering, such as shipbuilding, where plates (curved in one dimension) for the hulls of vessels are formed from flat sheets of metal in a rolling mill. An ancient example of the use of developable domains is the cylindrical seal used for authenticating official documents, shown in Figure 4b. This was rolled across wet clay, sealing a document, to identify the contents.

Isometry: Structures are said to be isometric if the distances of the closest approach between all neighbouring columns are constant and equal. The sketches shown Figures 5 and 6 illustrate the distinction between isometric and equidistant arrangements.

(a) (b)

Figure 4. (**a**) An example of a developable surface, where a cylinder is unrolled into a plane. This surface is also ruled, and the pattern of parallel straight lines drawn on the cylinder surface is preserved during the transformation. Figure (**b**) shows an ancient Mesopotamian cylindrical seal (dating to about 5000 BC) and its imprint. [18].

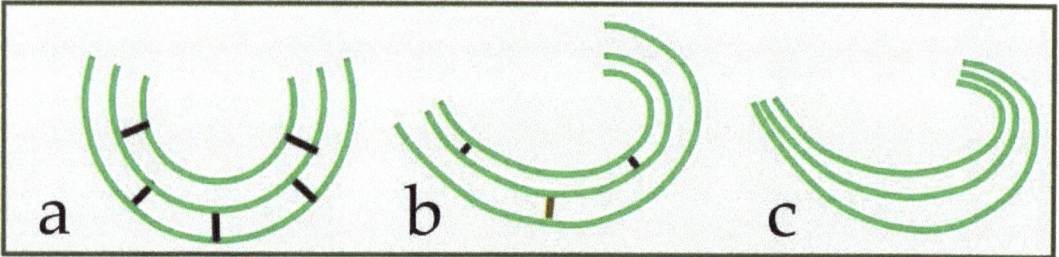

Figure 5. Isometric and equidistant structures in two dimensions. (**a**) shows an isometric arrangement, where the distances of the closest approach between all neighbouring columns are constant and equal. Note that the columns lie in a family of concentric circular arcs. (**b**) shows an equidistant array, where the distances of the closest approach are constant along the length of a pair of neighbouring fibrils, but are not equal throughout a domain. The pattern shown in (**c**) is neither isometric nor equidistant. Redrawn from Atkinson et al. [19].

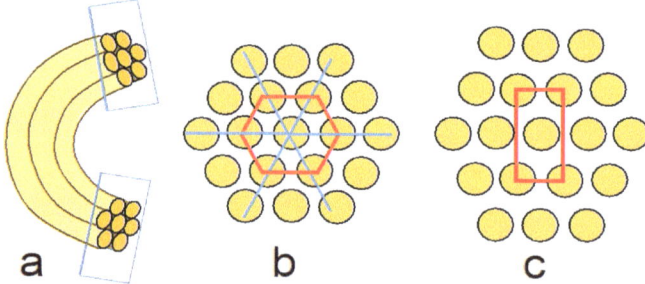

Figure 6. Isometric and equidistant structures in three dimensions. (**a**) A fragment of a developable domain structure, where all columns are curved in arcs of circles, but not splayed or twisted, and where the distribution of columns in any perpendicular plane shows the same hexagonal lattice of column axes. If the hexagonal distribution shown in (**b**) is distorted, for example into a centred rectangular lattice, as shown in (**c**), the structure becomes equidistant, but is no longer isometric. Redrawn from Atkinson et al. [19].

2.3. Compression Structures

The middle phase (M) of lyotropic systems was so-called because it occurs in the concentration range between that of the neat phase and of the isotropic liquid. It shows a range of paramorphotic optical textures, depending on how the mesophase had been created. In particular, the herringbone texture, shown in Figure 7a, appears when the concentrated isotropic solution is cooled. Until about 1980, it was widely assumed that every optical texture not immediately recognizable as a schlieren texture of a nematic phase was a focal conic image from a smectic phase. The explanation given for textures of this kind, was that the solute molecules are at their lowest energy state when they are packed in layers. This view was abandoned when it became accepted that the middle phases were columnar. It was then argued that it was energetically favourable for an isolated molecule to become intercalated within an existing column. When elongating columns become crowded in a confined space, they are forced to meander and, ultimately, to buckle sharply into zig-zag layers. As sketched in Figure 8, they are pictured as compressed springs, with the energy gain from the molecular intercalation being sufficient to cope with the energy cost of any concomitant long-range distortion of the director field that this may cause.

Herringbone textures, produced by internally generated compressive forces or by externally imposed stress, occur over a wide range of dimensions from μm in liquid crystalline materials to meters in geological strata, which have been subjected to massive tectoidal forces. Figure 7a shows the herringbone texture of the M phase of Intal, formed by cooling a concentrated solution of the isotropic liquid. Figure 7b shows a geological formation with strata distorted into chevron bands by external compression [20].

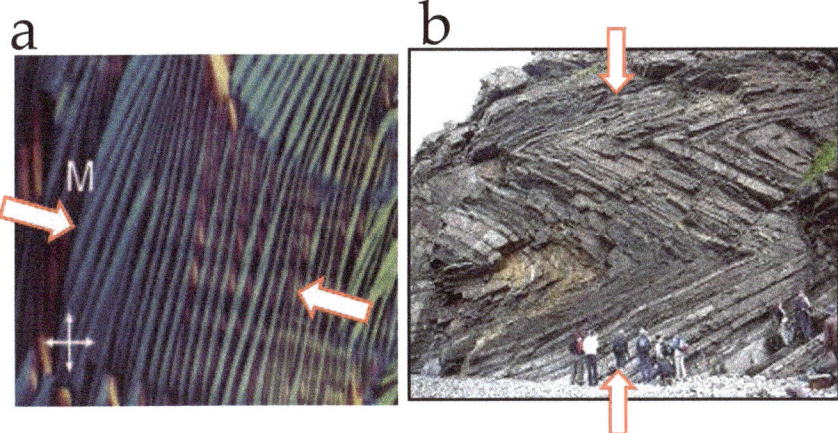

Figure 7. Herringbone textures. (**a**) shows the optical texture of the M phase of Intal. (**b**) shows the chevron distortion of strata in the cliff at Millhook Haven in North Cornwall Cornwall [20]. The arrows indicate the direction of the compressive stress.

Sketches like that in Figure 8a give the impression that the meandering pattern is a continuous sine wave, distinct from the zig-zag herringbone pattern, which is clearly divided into domains. However, this is not the case for narrow stripes of the mesophase, where the meandering wave is not sinusoidal and is composed of alternating domains, as will be described below.

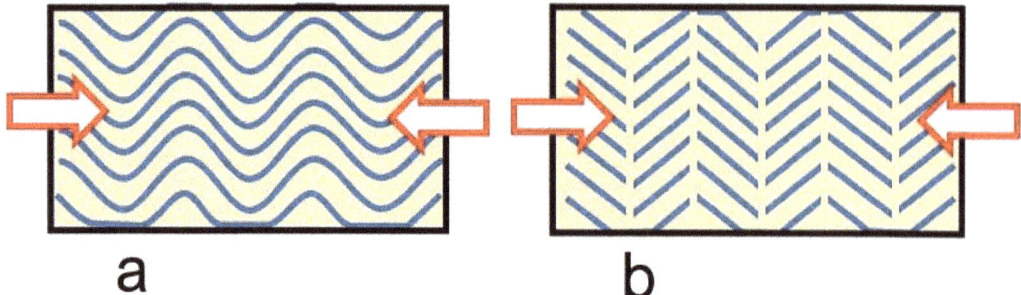

Figure 8. The development of the herringbone texture from the smoothly curved meandering pattern (**a**) to the broken zig-zag drawn in (**b**).

2.4. Optical Textures of Columnar Mesophase Stripes

For columnar phases confined between a slide and cover slip, the only plane in which the director field can meander in response to the compressive forces is that perpendicular to the viewing direction. In a recent publication by Bramble et al. [21], a technique for producing straight narrow stripes of the mesophase is described. This uses a prepared substrate, with alternating surface bands of a wetting and de-wetting nature, and when the mesophase is spread over the surface, it separates into distinct bands. In this case, the mesophase is constrained in both width and thickness, as sketched in Figure 9. The variety of director field patterns spontaneously created is shown in Figure 10. Although the stripes are not all identical in appearance (presumably because of minor irregularities in the dimensions of the prepared surface and the thickness of the mesogen stripe), all the textures appear to consist of developable domains. Note the interesting "egg and dart" texture of stripes **e** and **f** and the overspill region of **g**.

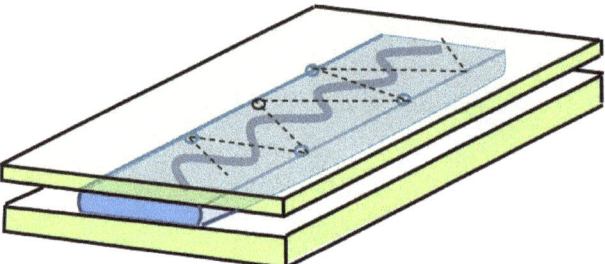

Figure 9. The developable domain structure of a thin band of a columnar mesophase, held between a slide and cover slip. Note that the only plane in which the director field can meander in response to the compressive forces is perpendicular to the viewing direction, causing all the domains to appear as nested arcs of concentric circles. The broken lines indicate domain boundaries, and the small circles show the positions of the poles of the concentric director fields. Redrawn from [21].

Reference [21] also contains a description of a different preparation, which has a preferred normal alignment of the columns at the mesophase/air boundary, as shown in Figure 11a. This shows a region of a stripe containing a single disclination viewed between crossed polars, with a red 1λ plate inserted. Figure 11b shows the inferred director field. Note the range of the yellow and blue areas, which indicates that there are regions of gentle splay extending on both sides of the singularity, over a range of about 200 µm, which gradually bring the director field back to a normal alignment at the edge of the mesophase stripe. Clearly, the constraint against splay, which gives rise to developable domains, can be overridden when the opportunity arises.

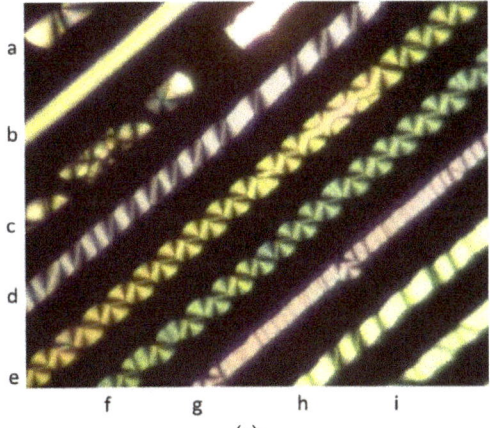
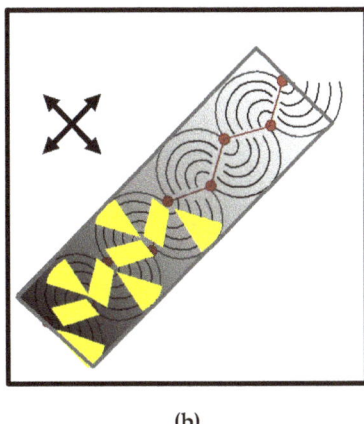

Figure 10. Taken from Bramble et al. [21], photographed at 150° in reflection mode, between crossed polars. (**a**) Stripes of a columnar Col$_h$ mesophase spontaneously aligned on a patterned silicon substrate with alternate 25 µm-wide bands of a wetting and non-wetting surface. (**b**) An analysis of the director field of the egg and dart patterns in stripes e and f, in terms of the developable domain. The red lines joining them indicate the domain boundaries.

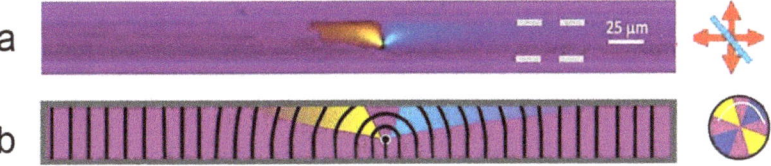

Figure 11. (**a**) shows the appearance of a 25 µm-wide stripe of a columnar mesophase around a single disclination (extending upwards from the centre line of the stripe), when viewed between crossed polars with a red 1λ plate inserted. In Figure (**b**), the black lines have been added to indicate the inferred director field. Note the distribution of the yellow and blue areas, indicating the extent of the regions of splay extending from the singularity. The orientation diagram shown on the right shows the alignments corresponding to the blue, magenta, and yellow regions of the image. The white arc indicates the range of orientations within the director field of the sample. Taken from Bramble et al. [21].

3. Discussion

3.1. Slippage

The rationale for the developable domain concept, proposed by Yves Bouligand [2], assumes that the molecular columns are effectively smooth and featureless and that they are easily bent and can slide past each other when a domain is growing. This picture must be reasonably valid; otherwise, the structure would not be able to assemble with columns bending with such a wide range of radii of curvature within each domain. However, it is perhaps also reasonable to assume that there is at least some small periodic lateral interaction between the columns, which causes them to prefer to assemble in register and to stop them building up a cumulative offset strain along the stripe. This could explain why developable domains are usually paired, as indicated in Figure 12. In these two highly stylized sketches, the dashed lines indicate columns of molecules in a hexagonal array, showing how sequences of alternating domains could build up within a parallel region of the director field, without requiring any large-scale slippage between the columns.

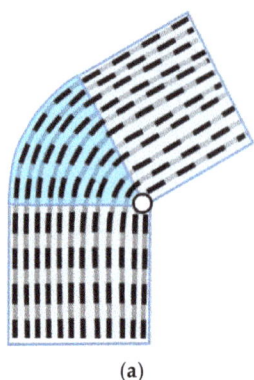 

(a)   (b)

Figure 12. In these highly stylized sketches, the broken lines indicate columns of molecules in an M phase structure. The lines have been drawn in this fashion to make the lateral positioning clear. Note the way in which a pairing of domains maintains the long-range register between the columns. In Figure (**a**), the columns are drawn in register at the bottom of the figure and slide out of phase as they curve through the wedge of the developable domain. In Figure (**b**), the addition of an apposed domain cancels the effect of the first and brings the columns back into register. Note that the long stripes of the developable domain structure shown in Figure 10 contain a balanced pairing of segments.

3.2. Striations

Lyotropic middle (hexagonal) phases, chromonic M ribbons, and the triangulated plates and ribbons of columnar discotics have all been found decorated with patterns of striations, which follow the course of the director field, as shown in Figure 13. No other optical texture is categorized by features of this kind. Although the striations appear to be at more-or-less equal spacing, the blocks of colour in the optical textures often show abrupt and apparently random changes, as shown in Figure 1b. (They are very different from the regular banding patterns associated with the helicoidal layered structures of cholesteric, N* phases, where there is a smooth sinusoidal variation of intensity.) The presence of these striations has been commented on repeatedly, but, as far as I am aware, no detailed explanation has been offered for their presence.

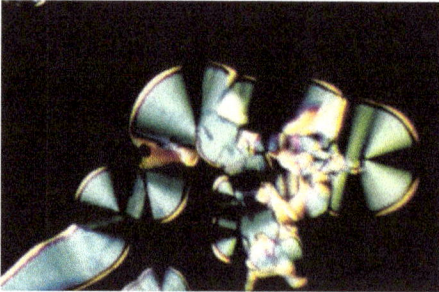

Figure 13. A ribbon of developable domains of the columnar phase of Intal, formed when the isotropic solution is cooled. Note the pattern of striations following the alignment of the director field.

Hartshorne comments that:

"The well-developed striated textures shown by M phases are frequently displayed by M phases between crossed polars, particularly when they have formed slowly or have aged somewhat". [1] page 114

3.3. Helical Bundles

In their beautifully illustrated geometrical analysis of assemblies of one-dimensional filaments, Atkinson et al. comment:

> "The relatively restrictive geometry of equidistant fields raises interesting questions about the relationship between the problem of packing finite versus infinite equidistant curves . . . . The structure of finite equidistant bundles may be much less constrained than equidistant fields. Discrete equidistant bundles of this sort have ready applications to physical systems, from collagen triple helices and other dense packed biological systems". [19]

They discussed the geometry of "almost equidistant" arrays, by relaxing the strict isometric constraints of Bouligand's developable domain structure [2,3]. The examples they discussed were for toroids, which are of relevance to the structures of "spherical" viruses, where the closed loop of nucleic acid is constrained within an icosahedral capsid. Here, space is very constricted, and there is no room for the nucleic acids to adopt an unstrained developable domain structure; furthermore, there is considerable distortion in addition to bending. I suggest that in the region near the poles in director fields such as those pictured in Figure 10, pressure from the outer regions of the director field causes the inner regions near the dislocations, where the director field is highly curved, to become highly distorted.

Bearing this in mind, I suggest that the relatively slow reorganization process, which produces striations, involves the consolidation of the structure into twisted bundles of columns—perhaps with equidistant rather than isometric structures—as sketched in Figure 14.

Each individual strand within a twisted multistrand helix is a coiled coil structure, sometimes called a *gyre*. The formations of helices and gyres are both ways of relieving twist strain—as is apparent when an elastic band is twisted under tension and then, relaxed. If the molecular units are non-chiral, there may be an alternation, or random mixture of clockwise and anti-clockwise helices, as Bernal and Fankuchen found in the mesophase formed by concentrated dispersions of the tobacco mosaic virus [7].

As mentioned by Atkinson et al. [19], multiple helices are common in biological polymer systems. They listed the double chains of DNA, actin fibrils, which spontaneously wrap around each other to form a double helix, and collagen fibres, which are three-stranded helices. To this list, one could add xylan (a linear polysaccharide in the cell walls of green algae), which is also a three-stranded helix [22].

Figure 14. A sketch of a helical bundle of columns, which, it is proposed, slowly forms within columnar developable domain structures, giving rise to the striated appearance. Redrawn from Atkinson et al. [19].

3.4. Multi-Pole Nucleation

There is another unexplained characteristic feature of developable domain optical textures. When columnar phases are formed from the isotropic liquid, they often appear to grow from a small nucleus and expand outwards. However, as shown in Figure 15, at higher magnification, the nucleus does not appear to be a single sharp point, as seen in the optical textures of small-molecule nematic schlieren patterns. It is revealed as two pairs of points about 1 µm apart joined by a short length of a straight line. This is evident in the optical textures of the columnar phases of nucleic acids, shown in Figure 15 [23].

The postulated way in which the multiple nuclei are formed is shown in Figure 15c. In this, the director field is a consequence of the same mechanical properties of a mesophase as those that give rise to developable domains. To create a radiating pattern from a single point would involve an energetically prohibitive level of splay, and multi-pole nucleation provides a way of avoiding this.

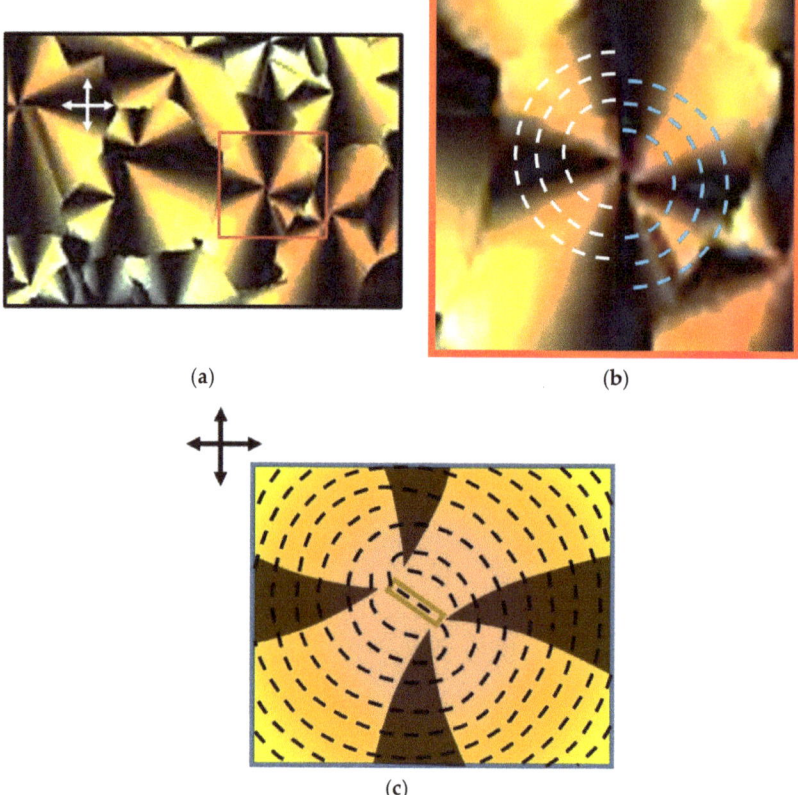

Figure 15. (**a**). The optical texture of the chromonic liquid crystalline phase formed by a concentrated aqueous solution of short-stranded DNA. Taken from [23]. Note the faint, straight line in the middle of the outlined square, which appears to join the two pairs of poles centring the pattern of striations. Furthermore, note the way in which the four dark brushes curve as they approach the poles (a pattern that is repeated in other parts of the micrograph). (**b**) An enlarged view of the square outlined in (**a**), with the pattern of striations superimposed. (**c**) A stylized sketch of the optical texture in the region of the multipole nucleus, stressing the curvature of the dark extinguished areas of the optical texture and showing the proposed pattern of the director field around the complex nucleus.

The helical bundle and the developable domain structure are the only two patterns of columns that are strictly equidistant [19], and the sketch in Figure 16 shows the postulated director field between the helical germ and the outer region (which might be first established as a developable domain and, then, revert to a striated array of bundles of helical columns).

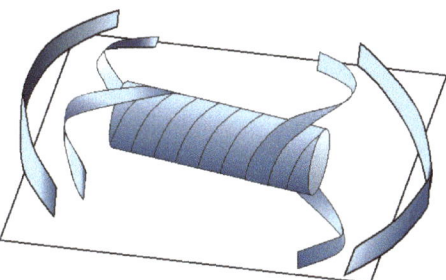

Figure 16. Taken from [21]. A stylized sketch of the proposed director field pattern when the mesophase region is extending outwards from a helically wound core. Note that the ribbons drawn in this figure are intended to give a three-dimensional impression of the alignment within the director field. They are not intended to imply that the striations are caused by ribbon-like fibrils.

3.5. Helical Fibrils

Some additional support for the hypothesis of helical bundles comes from the study by Khan et al. of the triphenylene mesogen 2,3,6,7,10,11-hexakishexyloxytriphenylene (HAT6) [24]. This compound is a bona fide columnar mesogen, exhibiting a nematic and a hexagonal columnar mesophase with the familiar developable domain optical textures. However, when it is mixed with an acetonitrile-based solvent, it forms a gel state in a thermally reversible "physical" process. (In this context, the word "physical" means without any covalent bonding being involved, and the forces involved in the cross-linking are considered to be hydrogen bonding or π–π interactions.) The gel is composed of an open network of strands of columnar material, as shown in Figure 17. Note the banded appearance of the strands viewed between crossed polars, indicating helical twisting.

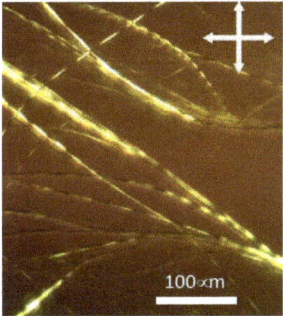

Figure 17. The open mesh of the gel phase of HAT 6 fibres dispersed in an acetonitrile-based solvent, viewed between crossed polars. Note the appearance of a helical gyre structure in the fibres. From Khan et al. [24]. I am indebted to Professor Richard Bushby for drawing my attention to this work.

4. Retrospect

In 1974, based on its optical texture as described by Norman Hartshorne [1], page 127, I concluded that the mesophase of diisibutyl silane diol might be composed of an array of columns of dimeric units. I observed that this phase appeared to have the mechanical properties of a sheet of paper, which can be curved in one direction, but which cannot then be curved in a different direction without being torn [12]. This picture falls short of the concept of developable domains, but points in that direction. I did not possess the geometrical insight of Yves Bouligand, and his publication did not to appear until six years later [2,3]. Furthermore, I had not considered the significance of the striations decorating the domains in the optical textures of columnar mesophases nor noticed that, in the M ribbons of Intal, the director tends to follow either straight lines or is curved into apparently precise arcs of circles.

The great fictional detective created by Arthur Conan Doyle solved cases that baffled everyone else, by his powers of close observation of apparently inconsequential detail:

"It has long been an axiom of mine that the little things are infinitely the most important" Sherlock Holmes, to Dr Watson in *A case of identity* [25].

Then, he followed a process of inductive reasoning, wherever it led (ignoring any preconceptions). In retrospect, I see that I had clearly learned little from my extensive reading of the Sherlock Holmes stories in my childhood.

Funding: This research received no external funding.

Acknowledgments: I am extremely grateful to Lavrentovich for pointing out a significant number of errors in the first draft of this publication. Amongst other things, these concerned the geometrical features of focal conic structures of smectic phases (which I had included with the intention of drawing parallels with adaptable domain structures). These errors arose because I was not aware of the distinction between hyperboloids and catenates, and I thought that a smectic layer within a focal conic domain was a minimal, doubly ruled structure. (I gather that this misconception is widespread.) The relevant section was removed from the text. I also thank Richard Bushby and Chris Hammond, of the University of Leeds, for their advice and encouragement, and Andrew Lydon for his assistance in preparing this article. I am indebted to the Faculty of Biological Sciences of The University of Leeds for the fellowship, which has enabled me to continue working on liquid crystals.

Conflicts of Interest: The author declares no conflict of interest.

Dedication: to Norman Hartshorne with thanks for his patient and quietly inspiring Teaching, which gave me the same love of optical microscopy that had dominated his scientific life.

References

1. Hartshorne, N.H. (Ed.) *The Microscopy of Liquid Crystals*; Volume 48 of the Microscope Series; Microscope Publications the University of California Publications: Chicago, IL, USA, 2009; ISBN1 0904962032; ISBN2 9780904962031.
2. Bouligand, Y. Defects and textures of hexagonal discotics. *J. Phys.* **1980**, *41*, 1307–1315. [CrossRef]
3. Kleman, M. Developable domains in hexagonal liquid crystals. *J. Phys.* **1980**, *41*, 737–745. [CrossRef]
4. Hartshorne, N.H.; Stuart, A. (Eds.) *Crystals and the Polarising Microscope*, 4th ed.; First published 1934; Hodder & Stoughton Educational: London, UK, 1970; ISBN-10: 0713122560; ISBN-13: 978-713122565.
5. Dunmur, D.; Slukin, T. (Eds.) *Soap, Science, and Flat-Screen TVs: A History of Liquid Crystals*; OUP Oxford: Oxford, UK, 2014; ISBN 13-0198700838.
6. Barrall, E.M., II; Porter, R.S.; Johnson, J.F. Temperatures of Liquid Crystal Transitions in Cholesteryl Esters by Differential Thermal Analysis. *J. Phys. Chem.* **1966**, *70*, 385–390. [CrossRef] [PubMed]
7. Bernal, J.D.; Fankuchen, I. Crystallographic studies of Plant Virus Preparations. *J. Gen. Physiol.* **1941**, *25*, 111–146. [CrossRef] [PubMed]
8. Lydon, J.E. A personal history of the early days of chromonics. *Liq. Cryst. Today* **2007**, *16*, 13–27. [CrossRef]
9. Lydon, J.E. New models for the mesophases of disodium cromoglycate (INTAL). *Mol. Cryst. Liq. Cryst.* **1980**, *64*, 19–24. [CrossRef]
10. Lydon, J.E. Chromonic Liquid Crystals. *Curr. Opin. Colloid Interface Sci.* **1998**, *3*, 458–466. [CrossRef]
11. Bunning, J.D.; Lydon, J.E.; Eaborn, C.M.; Jackson, P.; Goodby, J.W.; Gray, G.W. Classification of the mesophase of diisobutylsilanediol. *J Chem. Soc. Faraday Trans.* **1982**, *1*, 713–724. [CrossRef]
12. Lydon, J.E. The Pre-history of discotic mesophases—A personal account of the study of the mesophase of diisobutylsilane diol. *Liquid Cryst.* **2015**, *425*, 666–677. [CrossRef]
13. Chandrasekhar, S.; Sadashiva, B.K.; Suresh, K.A. Liquid crystals of disc-like molecules. *Pramana* **1977**, *9*, 471–480. [CrossRef]
14. Bushby, R.J. The Prehistory of discotic liquid crystal. *Liq. Cryst. Today* **2014**, *42*, 14–17. [CrossRef]
15. Demus, D.; Richter, L. (Eds.) *Textures of Liquid Crystals*; VEB Deutscher Verlag für Grundstoffindustrie: Leipzig, Germany, 1978; ISBN 10: 3527257969. ISBN 13: 9783527257966.
16. Gray, G.W.; Goodby, J.W.; Hill, L. (Eds.) *Smectic Liquid Crystals: Textures and Structures*; The University of California: Downtown Oakland, CA, USA, 1984; ISBN 0863440258. ISBN 9780863440250.
17. Dierking, I. (Ed.) *Textures of Liquid Crystals*; Wiley-VCH Verlag GmbH & Co. KGaA: Weinheim, Germany, 2003; ISBN 9783527307258. ISBN 9783527602056. [CrossRef]
18. Collon, D. *First Impressions: Cylinder Seals in the Ancient Near East, Their History and Significance*; British Museum Publication: London, UK, 1987.
19. Atkinson, D.W.; Santangelo, C.D.; Grayson, G.M. Constant spacing of filament bundles. *New J. Phys.* **2019**, *21*, 062001. [CrossRef]

20. Pinterest. Available online: https://www.geologypage.com/2018/12/millook-haven-beach-england.html (accessed on 5 May 2022).
21. Bramble, J.P.; Tate, D.J.; Evans, S.D.; Lydon, J.E.; Bushby, R.J. Alternating defects and egg and dart textures in de-wetted stripes of discotic liquid crystal. *Liq. Cryst.* **2021**, *49*, 543–558. [CrossRef]
22. Nakata, M.; Zanchetta, G.; Chapman, B.D.; Jones, C.D.; Cross, J.O.; Pindak, R.; Bellini, T.; Clarke, N.A. DNA duplexes end-to-end stacking and liquid crystal condensation of 6– to 20–Base Pairs. *Science* **2007**, *318*, 1276–1279. [CrossRef] [PubMed]
23. Atkins, E.D.T.; Parker, K.D. The helical structure of a β-D-1,3-xylan. *J. Polym. Sci. C Polym. Symp.* **2007**, *28*, 69–81. [CrossRef]
24. Khan, A.A.; Kamarudin, M.A.; Qasim, M.M.; Wilkinson, T.D. Formation of physical-gel redox electrolytes through self-assembly of discotic liquid crystals: Applications in dye sensitized solar cell. *Electrochim. Acta* **2017**, *244*, 162–171. [CrossRef]
25. Doyle, A.C. A case of identity. In *The Adventures of Sherlock Holmes, Adventure III Sherlock Holmes*; The Strand Magazine; The Strand: London, UK, 1891.

Review

Liquid Crystal Dimers and Smectic Phases from the Intercalated to the Twist-Bend

Corrie T. Imrie [1,*], Rebecca Walker [1], John M. D. Storey [1], Ewa Gorecka [2] and Damian Pociecha [2]

1. Department of Chemistry, School of Natural and Computing Sciences, University of Aberdeen, Scotland AB24 3UE, UK
2. Faculty of Chemistry, University of Warsaw, ul. Zwirki i Wigury 101, 02-089 Warsaw, Poland
* Correspondence: c.t.imrie@abdn.ac.uk

Abstract: In this review we consider the relationships between molecular structure and the tendency of liquid crystal dimers to exhibit smectic phases, and show how our application of these led to the recent discovery of the twist-bend, heliconical smectic phases. Liquid crystal dimers consist of molecules containing two mesogenic groups linked through a flexible spacer, and even- and odd-membered dimers differ in terms of their average molecular shapes. The former tend to be linear whereas the latter are bent, and this difference in shape drives very different smectic behaviour. For symmetric dimers, in which the two mesogenic groups are identical, smectic phase formation may be understood in terms of a microphase separation into distinct sublayers consisting of terminal chains, mesogenic units and spacers, and monolayer smectic phases are observed. By contrast, intercalated smectic phases were discovered for nonsymmetric dimers in which the two mesogenic units differ. In these phases, the ratio of the layer spacing to the molecular length is typically around 0.5 indicating that unlike segments of the molecules overlap. The formation of intercalated phases is driven by a favourable interaction between the different liquid crystal groups. If an odd-membered dimer possesses sufficient molecular curvature, then the twist-bend nematic phase may be seen in which spontaneous chirality is observed for a system consisting of achiral molecules. Combining the empirical relationships developed for smectogenic dimers, and more recently for twist-bend nematogenic dimers, we show how dimers were designed to show the new twist-bend, heliconical smectic phases. These have been designated SmC_{TB} phases in which the director is tilted with respect to the layer plane, and the tilt direction describes a helix on passing between layers. We describe three variants of the SmC_{TB} phase, and in each the origin of the symmetry breaking is attributed to the anomalously low-bend elastic constant arising from the bent molecular structures.

Keywords: liquid crystal dimers; intercalated; interdigitated; twist-bend nematic; twist-bend smectic; chirality; resonant soft X-ray scattering

1. Overview

Over the last decade arguably the hottest topic in liquid crystals science has been the twist-bend nematic, N_{TB}, phase following its discovery in 2011 [1], some ten years after its prediction by Dozov [2]. We will return to the N_{TB} phase later, but widely overlooked in Dozov's seminal work was the prediction of twist-bend smectic phases, and in this review, we trace the discovery of these phases for liquid crystal dimers [3]. Although the aim of this Special Issue is to provide an overview of the state-of-the-art of current UK liquid crystals research, the work we describe would not have been possible without a close collaboration between the Universities of Aberdeen and Warsaw. Indeed, the very essence of liquid crystals research is the need for a multidisciplinary approach, and science should know no borders. In keeping with this Special Issue's aim, however, we have attempted to focus primarily on the contribution to the overall work made in Aberdeen, but note that this story is fundamentally one of collaboration.

At the root of this work are liquid crystal dimers, and we begin in Section 2 with a very brief description of the characteristic behaviour of this fascinating class of low molar mass liquid crystals. In Section 3, we consider the smectic behaviour of symmetric dimers, and follow this in Section 4 by describing the discovery of the intercalated smectic phases for nonsymmetric dimers. We then return to describe the N_{TB} phase in Section 5 and briefly consider the types of liquid crystal dimer known to exhibit the phase. Pulling together the themes developed in Sections 3–5, we describe the discovery of the newest class of smectic phases in liquid crystal dimers, the twist-bend smectic phases in Section 6. We finish this review with an outlook of what may be achieved in this area in the future.

2. Liquid Crystal Dimers

Liquid crystal dimers consist of molecules containing two semi-rigid mesogenic moieties linked though a flexible spacer normally, but not always, an alkyl chain, and can be divided broadly into two classes. In symmetric dimers the two mesogenic units are the same, whereas in nonsymmetric dimers they differ. Detailed reviews of structure-property relationships in liquid crystal dimers can be found elsewhere [4–6], whereas for the purpose of this overview we need to consider only what may be referred to as their archetypal behaviour. Figure 1a shows how the melting points and nematic-isotropic transition temperatures, T_{NI}, vary as the length of the spacer is increased for a series of symmetric liquid crystal dimers, the α,ω-bis(4-nitroazobenzene-4′-oxy)alkanes, [7]

and these are referred to using the acronym BNABO*n* in which *n* refers to the number of methylene units in the flexible spacer. It can be seen in Figure 1a that T_{NI} initially exhibits a strong alternation as the parity of the spacer is varied, but that this attenuates on increasing spacer length. In this alternation it is the even members of the series that show the higher values of T_{NI}. It is interesting to note that the melting points for this particular series also alternate on increasing the spacer length, and again the even members show the higher values, but this is somewhat less regular behaviour than that seen for T_{NI}, and is not observed for all dimer series. The dependence of T_{NI} on spacer length and parity in dimers is most often attributed to molecular shape when considering the spacer in its all-*trans* conformation. For an even-membered spacer, the two mesogenic units are more or less parallel, whereas for an odd-membered spacer they are inclined to each other and the molecule is bent (Figure 1b). The linear shape of an even-membered dimer is more compatible with the molecular organisation found in the nematic phase than the bent-shape of an odd-membered dimer, and this accounts for the higher values of T_{NI} seen for the former. Although intuitively pleasing, this interpretation does not account for the pronounced alternation also observed in the nematic-isotropic entropy change on increasing the spacer length, and to do so the inherent flexibility of the spacer must be accounted for by considering a wide range of conformations and not solely the all-*trans* form [5]. For our purposes, however, it is sufficient to remember that, on average, an even-membered dimer is essentially linear whereas an odd-membered dimer is bent, and that this difference in average shape decreases as the spacer length increases given the increasing number of conformations available to the spacer.

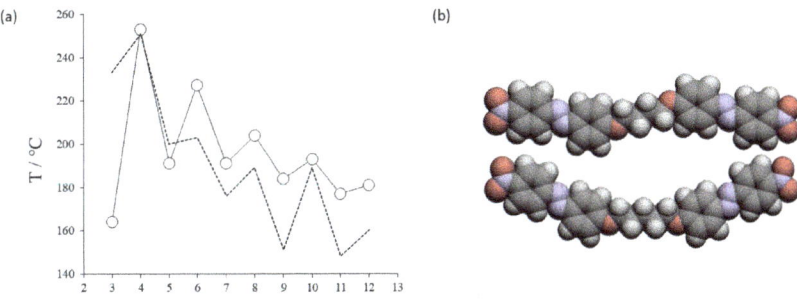

Figure 1. (a) The dependence of the nematic-isotropic transition temperatures, T_{NI}, on the number of methylene groups, n, in the flexible spacer for the BNABOn series [7]. The circles indicate T_{NI}, and the broken line joins the melting points. (b) Molecular shapes of BNABO4 (upper) and BNABO5 (lower).

3. Symmetric Liquid Crystal Dimers and Smectic Phases

The interest in liquid crystal dimers can be traced back to the early 1980s and the suggestion by Griffin and Britt that they may be used as model compounds for the technologically important semi-flexible main chain liquid crystal polymers [8] whereas their discovery, many decades earlier, by Vörlander had been largely overlooked [9]. The dimeric molecular architecture represented a marked deviation in the design of a low mass liquid crystal. The overwhelming majority of low molar mass liquid crystals up to that point consisted of molecules containing a single semi-rigid core attached to which were one or two flexible alkyl chains. In essence, the interactions between the cores accounted for the liquid crystal behaviour and the terminal chains used to reduce the melting point and drive smectic phases. In a dimer, this structure is inverted and now the core of the molecule is flexible, and we have seen already the importance of this in controlling their shape and hence, transitional behaviour.

The majority of dimers reported in the 1980s showed only nematic behaviour and Griffin and Britt attributed this to their inherent molecular flexibility suppressing smectic phase formation [9]. With hindsight, the absence of smectic behaviour was surprising. It was well-known that molecular inhomogeneity, such as the chemically distinct regions found in a dimer, drives the formation of smectic phases, and there appeared to be no fundamental reason for dimers not to exhibit smectic behaviour. It was almost a decade later, however, before the first family of liquid crystal dimers were reported that showed rich smectic polymorphism, the α,ω-bis(4-n-alkylanilinebenzylidene-4'-oxy)alkanes [10]:

$$H_{2m+1}C_m-\text{[ring]}-N=\text{[ring]}-O(CH_2)_nO-\text{[ring]}=N-\text{[ring]}-C_mH_{2m+1}$$

and these are referred to using the acronym $m.OnO.m$ in which n and m refer to the number of carbon atoms in the spacer and terminal chains, respectively. The strategy underpinning the design of this family of dimers was straightforward and centred upon the need to be able to readily vary the lengths of both the spacer and the terminal alkyl chains, a condition met by the $m.OnO.m$ molecular architecture. In addition, they may be considered to be the dimeric analogues of the N-(4-n-alkyloxybenzylidene)-4'-n-alkylanilines known to be a rich source of smectic phases [11].

The versatility in the synthetic approach used to prepare the $m.OnO.m$ series allowed for the transitional behaviour of 132 members of the family to be reported [10]. These included varying the spacer length, n = 1–12, and the terminal alkyl chain lengths, m = 0–10. These dimers did indeed exhibit rich smectic polymorphism including smectic A and C phases, hexatic smectic B and F phases, and soft crystal behaviour. Two novel modulated hexatic phases were discovered for the odd-membered dimers having the longest spacers (n = 9, 11) and terminal chains (m = 10), and in a later study, involving even longer terminal alkyl chains (m = 12, 14), further examples were found [12]. In the higher temperature phase, the smectic layers have a periodic modulation analogous to that found in the Sm C ~ ribbon phase. This modulated hexatic phase is only observed for dimers with an odd parity spacer, suggesting that the average bent molecular shape drives its formation. This is a theme that runs through this overview.

The study of the $m.OnO.m$ series revealed that increasing the spacer length for a given terminal chain length promotes nematic behaviour and this is shown for the 5.OnO.5 series in Figure 2. This was surprising and contravened the very general observations at the time that increasing the length of an alkyl chain promoted smectic behaviour in low molar mass mesogens, and that increasing the length of the spacer in a semi-flexible main chain polymer also promoted smectic behaviour. It is also interesting to note that the SmA-I transition temperatures also exhibited a pronounced odd-even effect as the length and parity of the spacer was varied (Figure 2) as seen for T_{NI} (Figure 1a). This indicates that the bent odd-membered dimers experience greater difficulty in packing into smectic structures than their linear, even-membered counterparts. Increasing the terminal chain length for a given spacer promotes smectic behaviour as seen, for example, for the $m.O5O.m$ series in Figure 3, and as would be expected.

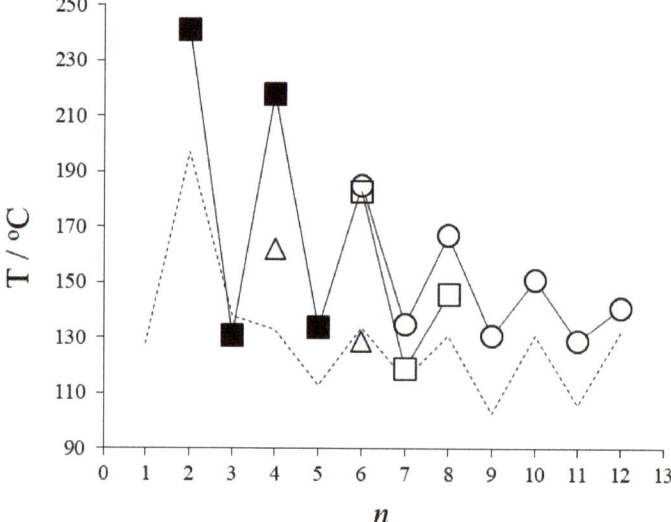

Figure 2. The dependence of the transition temperatures on the number of methylene groups, n, in the flexible spacer for the 5.OnO.5 series [10]. The broken line joins the melting points. Filled squares denote T_{SmAI}; unfilled squares T_{SmAN}; circles T_{NI}; triangles T_{SmASmB}.

A simple empirical relationship emerged from the study of the $m.OnO.m$ family of compounds relating the observation of smectic behaviour to the relative lengths of the terminal chains, m, to that of the spacers, n [10]. Thus, for smectic behaviour to be observed $m > 0.5\ n$. In addition, all the smectic phases observed for these dimers possessed a monolayer structure, i.e., the layer spacing, d, corresponded to the full molecular length, l. These observations are now established as being rather general for symmetric dimers with only a small number of exceptions known [5], and imply that in the smectic layer the mesogenic units, spacers and terminal chains may be considered to microphase separate into distinct sublayers as sketched in Figure 4a. An alternative packing arrangement of the dimers in which the terminal chains and spacers are randomly mixed to give an intercalated structure (Figure 4b), although entropically favoured, was not observed, and this was attributed to an unfavourable interaction between the terminal chains and spacers offsetting the favourable entropic term [13].

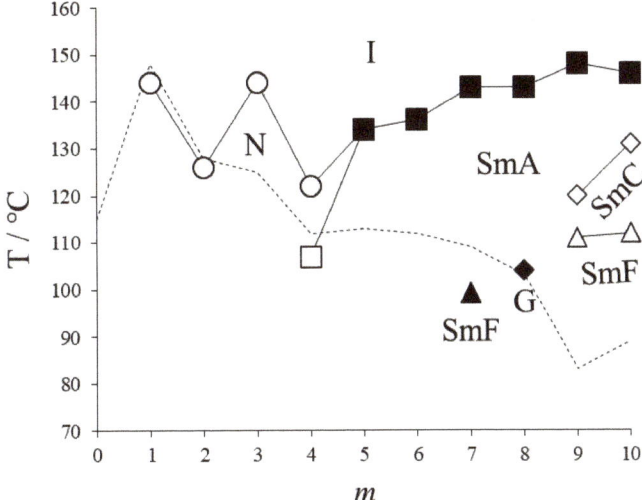

Figure 3. The dependence of the transition temperatures on the number of carbon atoms, m, in the terminal chains for the $m.\mathrm{O5O}.m$ series [10]. The broken line joins the melting points. Filled squares denote T_{SmAI}; unfilled squares T_{SmAN}; circles T_{NI}; unfilled diamonds T_{SmASmC}; filled diamonds T_{SmAG}; filled triangles T_{SmASmF}; unfilled triangles T_{SmCSmF}.

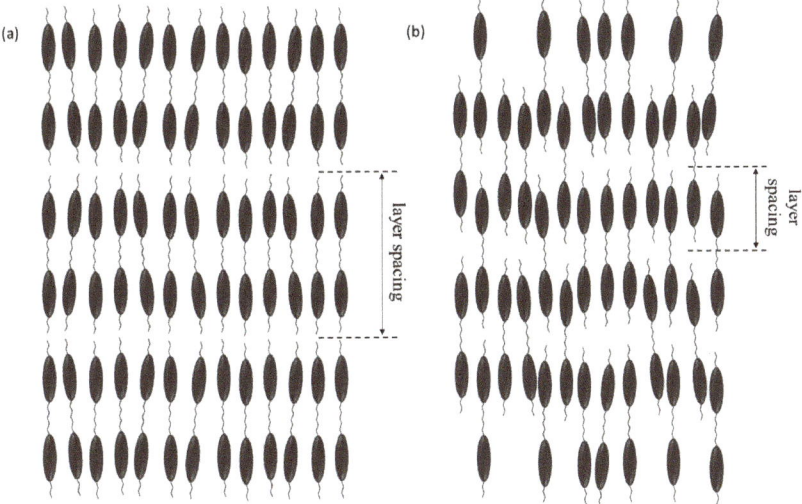

Figure 4. A sketch of (**a**) the monolayer smectic A phase ($d \sim l$) exhibited by symmetric liquid crystal dimers, and (**b**) the intercalated smectic A phase ($d \sim l/2$) observed only rarely for symmetric liquid crystal dimers.

It is noteworthy that replacing the ether links by methylene links between the spacer and mesogenic units in a dimer has a significant effect on its transitional properties. This is highlighted by a comparison of the behaviour of the α,ω-bis(4-n-alkyloxyanilinebenzylidene-4′-yl)alkanes [14,15],

$H_{2m+1}C_mO$—⟨⟩—N=⟨⟩—$(CH_2)_n$—⟨⟩=N—⟨⟩—OC_mH_{2m+1}

referred by the acronym mO-n-Om in which the hyphen is used to reflect the reversal of the Schiff's base link compared to the m.OnO.m series, with that of the corresponding ether-linked materials, the mO-OnO-Om series. Surprisingly, for short chain lengths ($m = 1, 2$), the mO-5-Om series exhibited a nematic phase and at lower temperatures, a second mesophase that exhibited a fan-like optical texture in coexistence with regions of schlieren texture and this was assigned as an anticlinic, intercalated smectic C phase although its monotropic nature precluded an unambiguous identification using X-ray diffraction (XRD). This behaviour was noted to be in contrast to that seen for the corresponding mO-OnO-Om series for which smectic behaviour was observed only if the length of the terminal chains was greater than half that of the spacer, as described earlier for the m.OnO.m series. By comparison, the even-membered mO-n-Om ($n = 4, 6$) series with short terminal chains showed solely nematic behaviour as expected. For long terminal chain lengths ($m = 9, 10$), the mO-5-Om series exhibited a G/J soft crystal phase with a modulated layer structure, whereas the corresponding members of the mO-6-Om series a G/J soft crystal phase was observed but with a simple layer structure. This difference in behaviour is similar to that seen for the corresponding members of the mO-OnO-Om series. It is important to note that the switch from an ether- to a methylene-linked spacer accentuates the bent shape of an odd-membered dimer as we will see later. The behaviour of the mO-OnO-Om series reinforced the view that the difference in shape between odd and even-membered dimers accounts, at least in part, for the differing smectic behaviour observed, with the bent odd-members having a stronger tendency to pack into tilted, alternating lamellar phases. The particularly surprising behaviour seen for mO-5-Om with $m = 1$ and 2, was later shown in fact to be a twist-bend nematic phase and we return to this in Section 5 [16].

4. Nonsymmetric Dimers and Intercalated Smectic Phases

We have seen that symmetric dimers have a strong tendency to form smectic phases having monolayer structures, and that this may reflect an unfavourable interaction between the spacers and terminal chains inhibiting intercalated arrangements. The question now arose: what if we overcame this unfavourable interaction by designing nonsymmetric dimers that exhibited a favourable specific interaction between the unlike mesogenic units, and could this drive the formation of intercalated smectic phases? To investigate this intriguing possibility, the α-(4-cyanobiphenyl-4′-yloxy)-ω-(4-n-alkylanilinebenzylidene-4′-oxy)alkanes were studied [17,18]:

NC—⟨⟩—⟨⟩—$O(CH_2)_nO$—⟨⟩—N=⟨⟩—C_mH_{2m+1}

The acronym used to describe these dimers is CBOnO.m, in which n and m refer to the number of carbon atoms in the spacer and terminal chains, respectively. This general structure was chosen because it was known that mixtures of nematogenic 4-n-alkoxycyanobiphenyls and N-(4-n-alkyloxybenzylidene)-4′-n-alkylanilines showed induced smectic behaviour implying a specific favourable interaction between the unlike mesogenic units [19,20]. The CBOnO.m family of dimers showed new patterns of liquid crystal behaviour: for example, Figure 5 shows the dependence of the transition temperatures on the length of the terminal chain for the CBO4O.m series and it is striking that smectic behaviour is observed for short and long terminal chains, but for intermediate chain lengths only a nematic phase is seen. This behaviour contravened a very general observation for conventional low molar mass mesogens that, for a given series, the smectic A-nematic transition temperature simply increases on increasing the length of a terminal chain and at some point, nematic behaviour is extinguished. To understand the behaviour seen in Figure 5, we must consider how the structure of the smectic phase changes as we increase m. For short chain lengths, the ratio of the smectic layer spacing to molecular length, d/l is about 0.5 indicating an intercalated arrangement of the dimers in which differing fragments of the molecules overlap (Figure 6a), the SmA$_c$ phase. The driving force for this arrangement was attributed to a specific interaction between the unlike mesogenic units suggested to be an electrostatic quadrupolar interaction between groups having quadrupole moments of opposite signs [21]. We should note that as sketched in Figure 6a the intercalated smectic A phase would appear to be polar, but we assumed that such molecular groupings would be randomly arranged at the macroscopic level such that ferroelectric properties would not be exhibited. This view was later supported by the failure to detect macroscopic polarisation in these phases. Returning to Figure 5, d/l for the smectic phase shown by the dimers with long terminal chains is around 1.8, implying an interdigitated arrangement of the dimers (Figure 6b), the SmA$_d$ phase, and such an arrangement is stabilised by the electrostatic interaction between the polar and polarizable cyanobiphenyl groups while the smectic phase results from the molecular inhomogeneity arising from the long terminal chains. The cross-over from an intercalated to an interdigitated structure arises from insufficient space between the layers of mesogenic units, governed by the length of the spacer, to accommodate the terminal chains and so the layers are pushed apart reducing the specific interaction between the unlike groups. The disappearance of smectic behaviour for intermediate chain lengths reflects a competition between these two incompatible arrangements, neither of which wins and hence, only a nematic phase is observed.

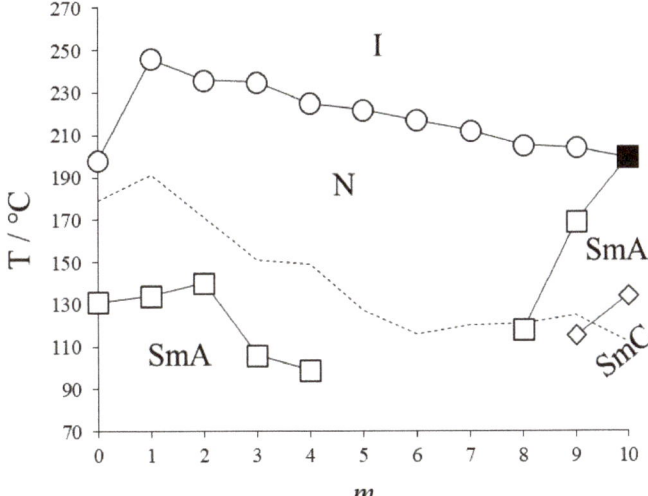

Figure 5. The dependence of the transition temperatures on the number of carbon atoms, m, in the terminal chain for the CBO4O.m series [18]. The broken line joins the melting points. Filled squares denote T$_{SmAI}$; unfilled squares T$_{SmAN}$; circles T$_{NI}$; unfilled diamonds T$_{SmCSmA}$.

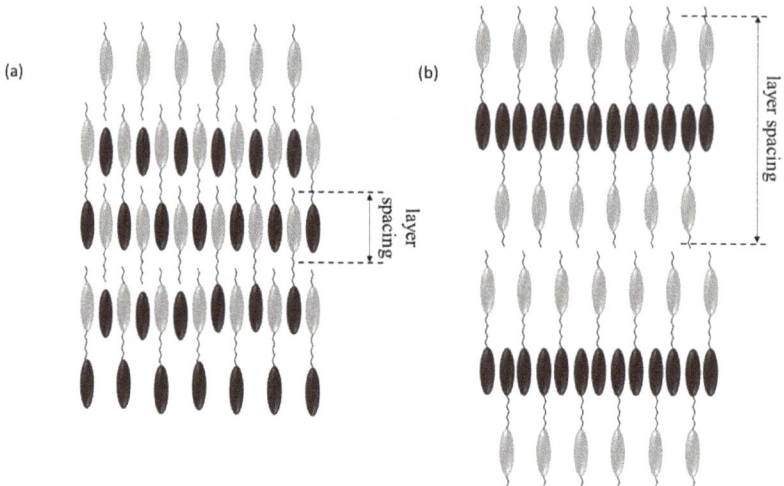

Figure 6. Sketches of (**a**) the intercalated and (**b**) the interdigitated smectic A phases shown by nonsymmetric liquid crystal dimers.

If we now consider how the phase behaviour changes on varying the length of the spacer while holding the terminal chain constant, Figure 7 shows the behaviour of the CBOnO.10 series [18]. Qualitatively similar behaviour to the CBO4O.m series is seen such that smectic phases are observed for short and long spacers, and exclusively nematic behaviour for intermediate spacer lengths. In contrast to the CBO4O.m series, however, intercalated smectic phases are observed for long spacer lengths and interdigitated phases for short spacers. This is wholly consistent with the view that the terminal chain must be accommodated within the space between the layers that is governed by the length of the spacer.

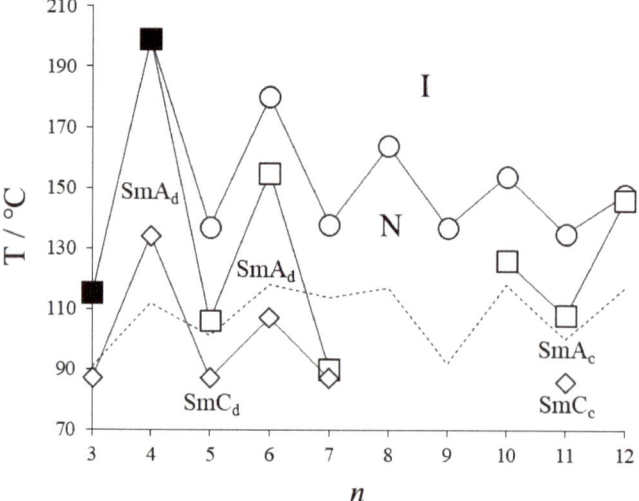

Figure 7. The dependence of the transition temperatures on the number of methylene groups, n, in the flexible spacer for the CBOnO.10 series [18]. The broken line joins the melting points. Filled squares denote T_{SmAI}; unfilled squares T_{SmAN}; circles T_{NI}; unfilled diamonds T_{SmCSmA}.

Not only was the intercalated smectic A phase shown by the CBO*n*O.*m* series but other intercalated variants including the smectic C (SmC$_c$) and I phases, as well as intercalated soft crystal phases were observed. It is important to note that tilted intercalated smectic phases were observed only for dimers containing long odd-membered spacers, and it was suggested that this reflects the difficulty that these bent dimers experience packing into an intercalated arrangement. Of particular interest was the intercalated smectic C phase shown, for example, by CBO9O.6. The anticlinic structure of the phase was identified on the basis of the observation of a schlieren optical texture containing both $s = \pm 1/2$ and $s = \pm 1$ defects [22], and the value of d/l was established using X-ray diffraction to be about 0.5 indicating an intercalated arrangement of the molecules. A sketch of the molecular arrangement proposed for the intercalated smectic C phase is shown in Figure 8 in which the tilt direction alternates between the layers such that the global tilt angle is zero but locally, within a layer, is non-zero. This structure was later confirmed using electron spin resonance spectroscopy that revealed within the intercalated smectic C phase two distinct directors with azimuthal tilt directions differing by 180° [23]. Within a layer the tilt angle was estimated to be about 18°, fully consistent with the bent geometry of these odd-membered dimers in which the mesogenic units make an angle of around 15° with the spacer. The intercalation of the dimers has been widely accepted to account for the measured values of d/l of around 0.5 although a number of issues remain to be fully resolved as recently discussed elsewhere [24].

Figure 8. Sketch of the molecular arrangement in the intercalated smectic C phase. The arrow indicates the local tilt direction.

In Figure 5, we saw how the transition temperatures varied on increasing the terminal chain length for the CBO4O.*m* series and this surprising pattern of behaviour was understood in terms of the formation of intercalated smectic phases for short terminal lengths and interdigitated smectic phases for long terminal chains. The behaviour of the CBO5O.*m* series was very different; see Figure 9. On cooling the nematic phase shown by the members with *m* = 1–9, what appeared to be a smectic phase formed, and the values of the transition temperature between this phase and the nematic phase decreased essentially linearly on increasing *m*. The structure of the lower temperature phase could not be established using X-ray diffraction due to its monotropic nature. The behaviour seen in Figure 9 was noted as being unusual and it was also noted that the CBO3O.*m* series showed broadly similar behaviour. These observations were thought to reflect the bent shape of these dimers, but its physical significance was not apparent. As we will see in the next section, however, the behaviour may now be accounted for in terms of the twist-bend nematic phase. Smectic behaviour was observed for long homologues in both these odd-membered series and we will return to this later.

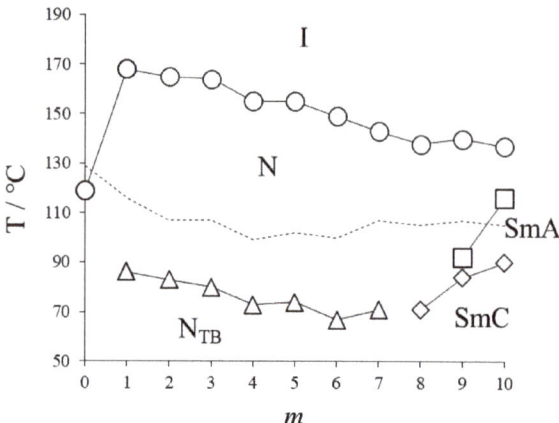

Figure 9. The dependence of the transition temperatures on the number of carbon atoms, m, in the terminal chain for the CBO5O.m series [18]. The broken line joins the melting points. Circles denote T_{NI}; triangles T_{NTBN}; squares T_{SmAN}; diamonds T_{SmCSmA}.

The intercalated smectic phases were first observed for nonsymmetric dimers and as noted earlier, the driving force for their formation was attributed to a specific favourable interaction between the unlike mesogenic units. It was known, however, that a small number of symmetric liquid crystal dimers also showed intercalated smectic phases (see, for example, [25]) although always for long odd-membered spacers. This supported the suggestion that molecular shape was an important factor in smectic phase formation. This tendency for bent odd-membered dimers to pack into tilted, intercalated smectic phases to alleviate the difficulties in packing into orthogonal arrangements has similarities to the behaviour of bent-core mesogens [26]. This view was further reinforced by the behaviour of a series of 4-decyloxy-4'-hydroxybiphenyl esters of α,ω-alkanedicarboxylic acids reported by Białecka-Florjańczyk et al. [27]:

The even members of this series showed a tilted soft crystal phase, possibly a G/J phase, whereas the odd members exhibited the B4 phase previously only observed for bent-core mesogenic materials. The B4 phase is thought to consist of bundles of twisted, rope-like smectic ribbons of finite thickness which pack to give macroscopic left- and right-handed domains [28]. The formation of helical nanofilaments in these structures is thought to be driven by the instability of the flat layers arising from the mismatch in the projections of the two crystal lattices associated with each arm of the bent core [29]. The elastic strain required to connect the two lattices may be relieved, at least in part, by bending the layers with saddle-splay curvatures. A similar model is clearly applicable to bent odd-membered dimers although would necessarily have to also include the inherent flexibility associated with the spacer. The observation of such behaviour for this particular series may be attributed to the strong lateral interactions known to exist between biphenyl fragments, a view supported by the observation of the soft crystal phase for the even members. Again, we see that molecular shape is a key factor in the pronounced differences in behaviour between odd and even-membered dimers, and the similarities between bent odd-membered dimers and the semi-rigid bent-core mesogens will be further discussed later.

5. The Twist Bend Nematic Phase

We now return to the prediction of the twist-bend nematic, N_{TB}, phase by Dozov [2] and its subsequent discovery by Cestari et al. [1]. At the root of Dozov's prediction was the assertion that bent molecules have a natural disposition to pack into bent structures, but that pure bend cannot fill space and so is forbidden. Instead, bend must be accompanied by other deformations of the director; either twist or splay. In the case of twist, this gives rise to the twist-bend nematic phase in which the director adopts a heliconical structure in which it is tilted with respect to the helical axis (Figure 10). Perhaps the most intriguing aspect of Dozov's work was the prediction of spontaneous chirality in a fluid system composed of achiral molecules having no positional order. The helices formed may be either left- or right-handed and equal amounts of both are expected.

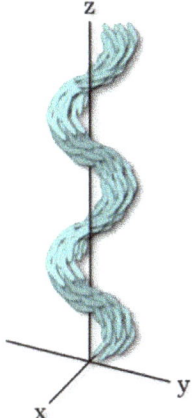

Figure 10. A schematic representation of the twist-bend nematic, N_{TB}, phase.

Odd-membered liquid crystal dimers with their bent molecular structures (Figure 1b) provided the ideal testbed for Dozov's prediction, and some ten years later, Cestari et al. reported the discovery of the N_{TB} phase in 1,7-bis(4-cyanobiphenyl-4'-yl)heptane (CB7CB) [1]:

$$NC---(CH_2)_7---CN$$

The assignment of the N_{TB} phase was confirmed using techniques including freeze fracture transmission electron microscopy [30] and resonant soft X-ray scattering (RSoXS) [31]. The events surrounding this discovery are reviewed by Dunmur in another contribution to this collection of papers [32]. A striking feature of the N_{TB} phase is that the helical pitch is very short, typically around 10 nm, corresponding to around 3–4 molecular lengths. Nematic-nematic transitions had been reported previously for other odd-membered dimers that, with the benefit of hindsight, are examples of N_{TB}-N transitions [33,34]. A common feature of these dimers is the use of methylene links to connect the spacer to the semi-rigid mesogenic units. By contrast, the majority of dimers reported prior to the discovery of the N_{TB} phase contained ether-linked spacers, i.e., $O(CH_2)_nO$, although interest in methylene-linked dimers had been triggered earlier by a prediction made by Ferrarini et al. [35,36] that systems containing a high concentration of bent conformers in the isotropic phase should exhibit a nematic-nematic transition. Again, short odd-membered methylene-linked dimers provided a testbed for these predictions. It was found that switching $O(CH_2)_nO$ for $(CH_2)_{n+2}$ saw the values of T_{NI} fall for both odd- and even-membered spacers but the reduction was greater for odd-membered spacers. This resulted in a more pronounced alternation in T_{NI} for the methylene-linked dimers on varying the spacer length and parity compared to that seen for their ether-linked counterparts [14,15]. By contrast, the entropy change associated with the nematic-isotropic transition increased for even-membered spacers but decreased for odd-membered spacers on replacing $O(CH_2)_nO$ with $(CH_2)_{n+2}$. These experimental observations were in complete agreement with the predictions of a model described by Luckhurst and co-workers in which the only difference between the dimers is the bond angle between the *para*-axis of the mesogenic unit and the first bond in the spacer [37,38]. For an ether-linked dimer this angle is about 126.4° whereas for a methylene-linked dimer it is around 113.5°. The change in this angle means that an odd-membered methylene-linked dimer is more bent than the corresponding ether-linked dimer (Figure 11). The discovery of the N_{TB} phase reignited interest in the methylene-linked dimers and previously reported dimers were shown to exhibit the new phase [16]. In addition, odd-membered ether-linked dimers were also shown to exhibit the N_{TB} phase [39] and led to the reassignment of smectic phases as N_{TB} phases as shown, for example, in Figure 9 for the CBO5O.*m* series [18,40].

Figure 11. A comparison of the shapes of odd-membered methylene- and ether-linked dimers: (a) CB7CB and (b) CBO5OCB.

There is now a large collection of odd-membered dimers known to exhibit the N_{TB} phase, and structure-property studies have focussed on the nature of the link between the spacer and mesogenic units [41–49], the length and parity of the spacer [39,50–53], the structure of the mesogenic units [54–64], and the chemical nature of the terminal groups [65–68]. Although the majority of twist-bend nematogens are odd-membered dimers, other examples include trimers and higher oligomers [69–74], hydrogen-bonded supramolecular systems [75–79], rigid-bent core systems [80,81] and polymers [82]. A recent overview of structure-property relationships in twist-bend nematogens may be found elsewhere [83].

The overarching structural requirement for the observation of the twist-bend nematic phase is molecular curvature. This is in complete agreement not only with Dozov's model [2], but also with Maier-Saupe theory for V-shaped molecules that predicts the N_{TB} phase will be formed by just a narrow range of molecular curvatures with the N-N_{TB} transition temperature being particularly sensitive to the molecular bend angle [84]. The uniformity of molecular curvature is also critical in driving the formation of the N_{TB} phase as revealed by the behaviour of the azobenzene-based dimer, CB6OABOBu [85]:

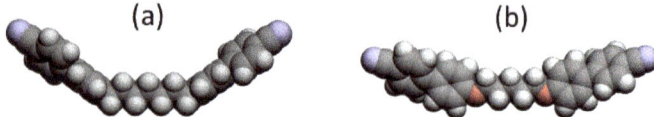

This dimer shows an isothermal N_{TB}-N transition when illuminated using UV-light and this may be attributed to the *trans-cis* photoisomerization of the azo-linkage. This transition is reversible on removing the light source, driven by the thermal *cis—trans* relaxation. The *cis* isomer is more strongly bent than the *trans* isomer (Figure 12) and a priori it may have been expected that increasing the concentration of the more bent isomer would enhance the stability of the N_{TB} phase. It is clear from this and other studies [86,87] that quite the opposite is true. This counter-intuitive observation was accounted for in terms of the shapes of the two isomers. The molecular curvature of the *trans* isomer is governed by the geometry of the spacer and hence, is spatially uniform. In contrast, the *cis* isomer contains two centres of bend, namely the spacer and the azobenzene fragment, which results in a change in bend polarity along the molecule. This spatially varying bend is not compatible with the local structure of the N_{TB} phase.

Figure 12. The shapes of the (**a**) *trans* and (**b**) *cis* isomers of CB6OABOBu.

6. The Twist-Bend Smectic Phases

Dozov, in his seminal work, not only predicted the existence of the N_{TB} phase for bent mesogens, but also noted that the same arguments could *give rise to a similar symmetry breaking even in apolar banana smectics* [2]. This aspect of the study was largely overlooked, and the search for these twist-bend smectic phases became a focus of our work. At the outset of our studies, the great majority of the N_{TB} phases reported either vitrified or crystallised on cooling, and only rarely had smectic-N_{TB} or B-N_{TB} transitions been reported [58,88–90]. In designing materials to potentially exhibit twist-bend smectic phases we identified three key structural criteria: (i) a uniform molecular curvature compatible with the N_{TB} phase, and, based on our previous studies of dimeric smectogens, (ii) a specific favourable interaction between the mesogenic units to promote smectic behaviour, and (iii) a terminal chain that could be readily varied in length to control the packing of the molecules and to promote molecular inhomogeneity.

The first question to address is how does the more pronounced molecular curvature, required to promote the formation of twist-bend phases, affect the tendency of dimers to exhibit smectic phases? Figure 13 shows the dependence of the transition temperatures on the length of the terminal chain for the CB6O.m series [40,91]:

$$\text{NC}-\bigcirc-\bigcirc-(CH_2)_6O-\bigcirc-\overset{\diagdown}{N}-\bigcirc-C_mH_{2m+1}$$

in which the hexyloxy spacer ensured the necessary molecular curvature for the N_{TB} phase to be observed [42]. Our initial study of this series included varying the terminal chain length, m, from $m = 1–10$, [40] and in a subsequent study we reported the transitional behaviour of the longer members with $m = 11–18$ [91]. The phase behaviour of the CB6O.m series with $m = 1–10$ [40] may be compared to that of the CBO5O.m series shown in Figure 9 [18]. The bent nature of both series reduces their smectic tendencies compared to, for example, the linear CBO4O.m series shown in Figure 5 [18]. For the CBO5O.m series, N_{TB} phases are observed for $m = 1–7$. N_{TB} behaviour is extinguished at $m = 8$ and smectic-nematic transitions are observed for $m = 8–10$. For the more bent CB6O.m series, N_{TB} phases are observed for $m = 1–10$, and smectic behaviour emerges at $m = 10$ which exhibits a Sm-N_{TB} transition to be discussed later [40]. This reduction in the smectic tendencies on increasing molecular bend may be quantified by comparing the scaled transition temperature, T_{SmN}/T_{NI}, for CBO5O.10 of 0.924, with T_{SmNTB}/T_{NI}, for CB6O.10 of 0.909. It is interesting to note that the value of T_{NTBN}/T_{NI} for CB6O.10 is 0.925, i.e., essentially the same as T_{SmN}/T_{NI}, for CBO5O.10. It appears, therefore, that increasing molecular curvature increases the tendency to exhibit the N_{TB} phase at the expense of smectic behaviour. Ironically, Dozov highlighted the challenge in obtaining the N_{TB} phase would be to supress the formation of smectic phases in bent-core systems [2] in which symmetry breaking had been attributed to specific polar interactions [92]. For odd-membered dimers the origin of the symmetry breaking is quite different and may be attributed to anomalously low values of the bend elastic constant arising from the bent molecular geometry. Thus, the inherent flexibility and bent shape of these odd-membered dimers suppresses smectic behaviour, and the challenge was to design odd-membered dimers in which the tendency to form smectic phases, rather than the N_{TB} phase, was enhanced.

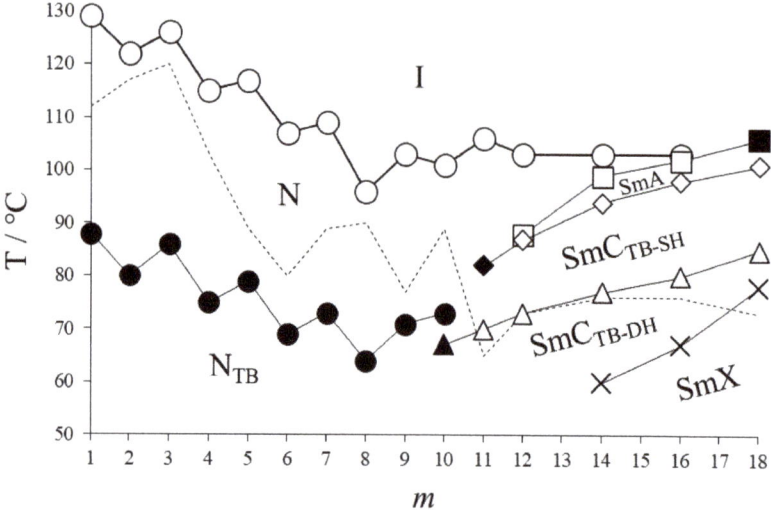

Figure 13. The dependence of the transition temperatures on the number of carbon atoms, m, in the terminal chain for the CB6O.m series [40,91]. The melting points are connected by the broken lines. Unfilled circles denote T_{NI}; filled circles T_{NTBN}; filled squares T_{SmAI}; unfilled squares T_{SmAN}; unfilled diamonds $T_{SmCTB\text{-}SHSmA}$; filled diamonds $T_{SmCTB\text{-}SHN}$; unfilled triangles $T_{SmCTB\text{-}DHSmCTB\text{-}SH}$; filled triangles $T_{SmCTB\text{-}DHNTB}$; crosses $T_{SmXSmCTB\text{-}DH}$.

In order to achieve this goal, we increased the interactions between the mesogenic moieties by increasing the structural anisotropy of the benzylideneaniline moiety in the CB6O.m series to obtain the CB6OIBeOm series [3]:

Again, we chose the hexyloxy spacer to impart the required molecular curvature and varied the terminal chain length, $m = 1\text{--}10$, to increase molecular inhomogeneity and further promote smectic behaviour. For short terminal chain lengths, $m = 1\text{--}6$, a transition between the conventional and twist-bend nematic phases was observed, whereas for $m \geq 7$, the N_{TB} phase was extinguished and up to four lamellar phases were found below the N phase (Figure 14). The layer spacings in all the smectic phases corresponded to approximately the molecular length. The lowest temperature lamellar phase, seen for most of the homologues, $m = 2, 4\text{--}10$, is a hexatic-type smectic phase, designated HexI, in which the molecules tilt towards the apex of the local in-plane hexagons. The three higher temperature smectic phases showed liquid-like ordering of the molecules within the layers, and the changes in optical texture seen on cooling from the nematic phase are shown in Figure 15a. The highest temperature smectic phase was optically uniaxial and assigned as a SmA phase. On cooling, a schlieren texture developed, and at the transition there was no change in layer spacing, only the layer thermal expansion coefficient differed between the phases, both negative in sign, suggesting both phases to be orthogonal. The observed schlieren texture suggested that this phase was the biaxial smectic A, SmA$_b$, phase, in which molecular rotation around the long axis is, to some degree, frozen and this assignment is consistent with the increase in layer spacing on cooling. The absence of a polar response under an electric field indicated that in the SmA$_b$ phase the dipole moments of the molecules must be locally compensated. On cooling the SmA$_b$ phase into the lowest temperature liquid-like smectic phase, the layer spacing decreased continuously suggesting a transition to a tilted phase, but surprisingly a homeotropic optical texture was restored (Figure 15a) ruling out the possibility of a simple all-in-one plane synclinic or anticlinic SmC phase. The observed optical uniaxiality of this tilted phase strongly suggested an averaging of the molecular orientations arising from the formation of a helical structure. This view was supported using soft X-ray resonant scattering which revealed that, at the transition from the SmA phase, a resonant Bragg peak developed corresponding to a pitch length of around 150 Å decreasing to around 117 Å at the transition to the HexI phase (Figure 15b). The pitch length was incommensurate with the layer thickness, corresponding to around 3–4 smectic layer distances. This heliconical smectic C phase was designated as the SmC$_{TB}$ phase and the molecular arrangement within the phase is shown schematically in Figure 16. The SmC$_{TB}$ phase does not appear optically active suggesting that essentially free molecular rotation occurs around the long molecular axis such that the biaxiality is too weak to give rise to detectable layer chirality, and it is known that optical activity is negligible in chiral smectic phases in which the pitch length is much shorter than the wavelength of light as is the case here. A resonant peak was also detected in the HexI phase corresponding to a pitch length of around 100 Å, around 2.2 smectic layer spacings, and this was essentially temperature independent (Figure 15b). The HexI phase showed a weakly birefringent texture (Figure 15a) with optically active domains and gave a circular dichroism signal around the absorption band of the material. This optical activity was attributed to the strongly inhomogeneous electron distribution across the layer and the tilting of the biaxial molecules, i.e., layer chirality (Figure 17); this originates given that the bent molecules are tilted with respect to the layer normal and changing the direction of the bend vector with respect to the tilt plane results in a non-superimposable mirror image. The morphology of the HexI phase was studied using atomic force microscopy revealing uniformly oriented entangled filaments with an average diameter of around 50 nm. These filaments had a uniform twist sense over areas of micron dimensions. The HexI phase, therefore, exhibits structural chirality at different length-scales, namely layer chirality, nanoscale helices and mesoscopic helical filaments, and strongly resembles the B$_4$ phase but this has an internal crystalline structure [28,93] whereas the HexI phase does not. It should be noted that although all the smectic phases are single layer type with d~l, in the HexI phase, an additional, very weak diffraction signal corresponding to $2l$ periodicity was detected [94], evidencing a tendency towards bilayer packing driven by a small imbalance in the density of the cyano groups at consecutive layer interfaces. Based on the RSoXS pattern it was deduced that the structure of HexI phase is anticlinic with a longer helix superimposed on that of the bilayer unit. A comparison of the calculated and experimental RSoXS patterns suggests that the layer chirality is coupled to the handedness of the longer superimposed helix.

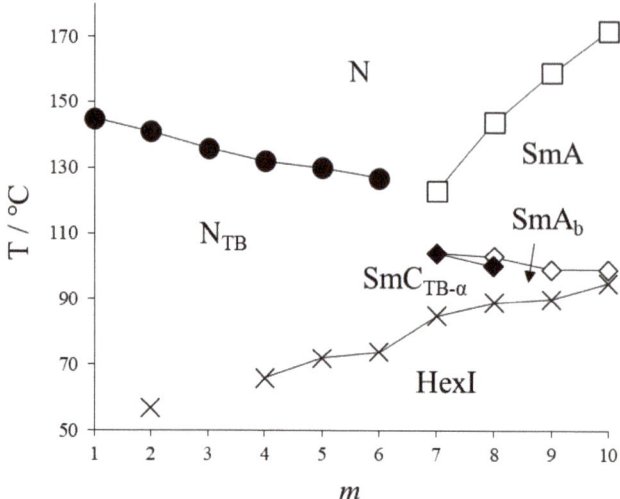

Figure 14. The dependence of the transition temperatures on the length of the terminal chain, m, for the CB6OIBeOm series [3]. The N-I transition temperatures have not been shown in order to enlarge the other phase regions. The melting points have been omitted for the sake of clarity. Filled circles denote T_{NTBN}; unfilled squares T_{SmAN}; unfilled diamonds $T_{SmAbSmA}$; filled diamonds $T_{SmCTB\text{-}\alpha SmAb}$; crosses $T_{HexI\text{-}}$.

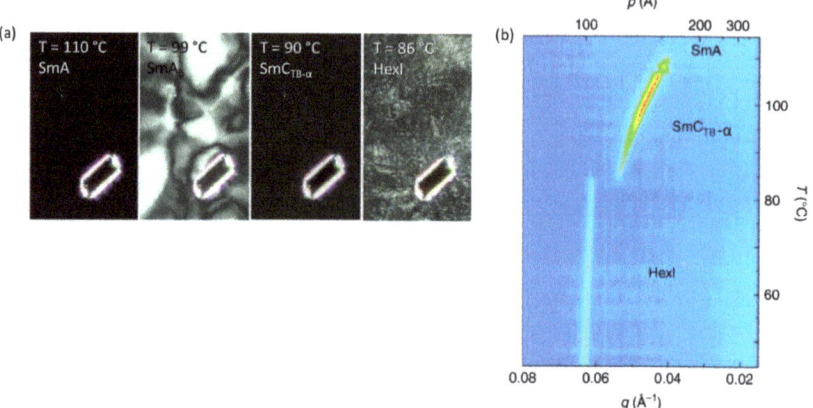

Figure 15. (a) The optical textures observed between crossed polarisers for CB6OIBeO8 on cooling in an homeotropic cell. A glass bead has been used to indicate that the textures represent the same area of the slide; (b) temperature dependence of the resonant soft X-ray diffraction signal for CB6OIBeO7 recorded on cooling (taken from [3]).

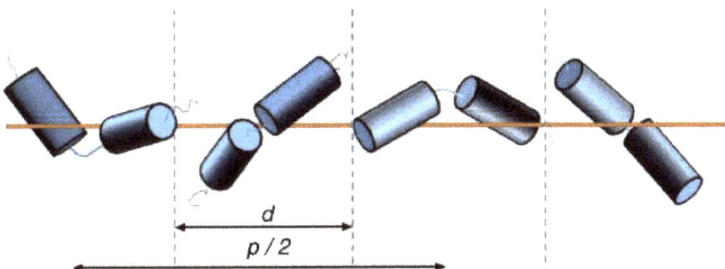

Figure 16. A schematic representation of the molecular arrangement within the SmC$_{TB}$ phase (taken from [3]).

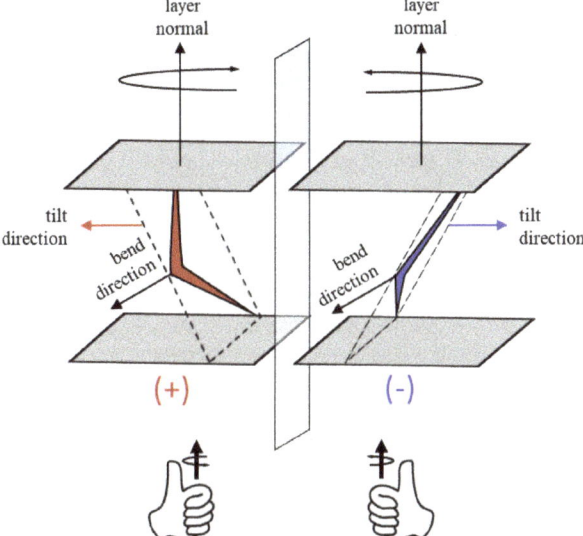

Figure 17. A sketch showing layer chirality in a tilted phase consisting of bent molecules.

The heliconical SmC$_{TB}$ phase was the first unambiguous example of a short pitch helical structure consisting of achiral dimers arranged into a lamellar phase. The driving force for the spontaneous formation of this short-pitch length heliconical structure is presumably the anomalously low-bend elastic constant which may be attributed to the bent molecules. Such a view is supported by the striped optical texture observed in the SmC$_{TB}$ phase which, by analogy to the interpretation of the striped texture observed for the N$_{TB}$ phase [95], suggests that the bend elastic constant is very low. In chiral systems, the SmCα* phase, found between the SmA* and SmC* phases, has a similarly short pitch length incommensurate with the layer spacing, typically between 5 and 8 smectic layers, and has a small tilt angle [96,97]. In this structure, the helix formation relieves the frustration in systems in which next nearest interactions favour antiparallel tilt. Given the similarity in structure between the SmC$_{TB}$ phase shown by CB6OIBeOm series and the SmCα* phase, we designate this SmC$_{TB}$ variant as the SmC$_{TB-\alpha}$ phase. Polar bent core molecules had also been reported to show a short pitch heliconical smectic phase designated Sm(CP)^{hel} and this is thought to be associated with the growth of polar order at the anticlinic-synclinic transition in smectic phases with a small tilt angle [98–100]. Again, the pitch length was incommensurate with the layer spacing, and typically around three-layer thicknesses. Earlier, liquid crystal phases containing helices of both handedness had been reported for bent-core liquid crystals by Sekine et al. [101], although this was later assigned as a B_4 phase, and technically crystalline in nature.

We saw earlier that CB6O.10 was the shortest member of the CB6O.m series to exhibit a smectic phase (Figure 13) and this was subsequently studied in detail and assigned as a SmC_{TB} phase [94]. The SmC_{TB}-N_{TB} transition was clearly observed using DSC, but was not associated with changes in optical textures. This suggested that in both phases there was a similar averaging of molecular orientation due to the formation of a helix. Only a small change in birefringence was seen at the SmC_{TB}-N_{TB} transition, evidencing that the conical angle in both phases is similar, and estimated to be 15°. Non-resonant XRD measurements revealed the layer spacing in the smectic phase to be almost two molecular lengths suggesting a bilayer arrangement with the molecules arranged head-to-head within a layer. A resonant signal was evident in the RSoXS pattern of the N_{TB} phase associated with the helix, and showed that the pitch length decreased with decreasing temperature (Figure 18). At the N_{TB}-SmC_{TB} transition, the resonant signal (q_4 in Figure 18) locks at approximately four molecular lengths and this persists for several degrees below the phase transition. This signal was purely resonant indicating it was associated with the helical structure. A signal associated with the bilayer structure was also observed but off-scale in Figure 18. The continuous evolution of structure at the N_{TB}-SmC_{TB} transition suggested that in the smectic phase just below the transition, the molecules in adjacent layers are azimuthally rotated by exactly 90° on the tilt cone and this may be described as an ideal clock structure (see Figure 19a). A model used to estimate the intensities of the RSoXS signals suggested that deeper in the phase, the structure may be described as a distorted clock with δ between 60–70° (Figure 19b). In such an arrangement the helical pitch is commensurate with two smectic bilayer spacings; in other words, approximately four molecular lengths. On cooling the SmC_{TB} phase, the resonant signal (q_4) split symmetrically into two resonant signals, ($q_4 + q_m$) and ($q_4 - q_m$), with the former much more intense. This splitting increased with decreasing temperature. A very weak signal at $2q_m$ was also observed.

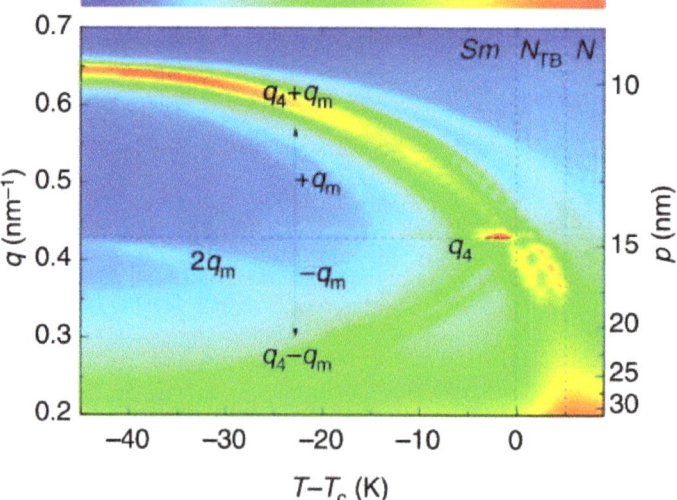

Figure 18. The resonant soft-X-ray scattering pattern for CB6O.10 (taken from [94]).

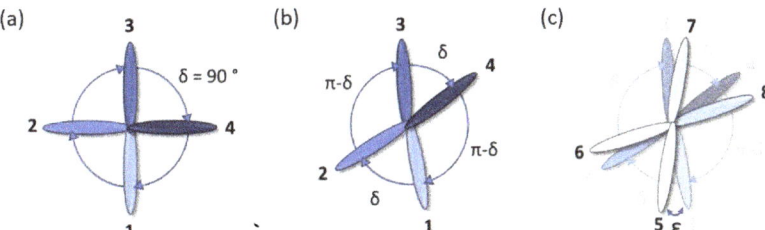

Figure 19. Schematic representations of the SmC$_{TB}$ phases using the clock model in which the ellipse represents the direction of the tilt projected onto the layer plane. (**a**) The ideal clock in which molecules rotate on the tilt cone by exactly 90° between consecutive layers. (**b**) The distorted clock in which the azimuthal angle between the projections of the long molecular axis in adjacent layers onto the smectic plane is no longer 90°. (**c**) A change in azimuthal angle to the adjacent layer causing the four-layer unit cell to rotate giving an additional helical structure.

The RSoXS pattern of either a smectic phase in which the basic four-layer unit repeats with a regular distribution of the molecules on the tilt cone (Figure 19a,b), or the all-in-one-plane bilayer structure with alternating synclinic and anticlinic interfaces between the constituent molecular layers, should contain only the signal q_4 and its harmonics. The splitting of the q_4 signal in Figure 18 indicated that an additional modulation, with a longer periodicity, is superimposed over the basic four-layer unit. It was noted that the pattern shown in Figure 18 was very similar to that seen for the SmC$_{FI2}^*$ phase observed over narrow temperature ranges between anticlinic and synclinic SmC* phases in chiral rod-like materials [102]. The SmC$_{FI2}^*$ has also a four-layer unit cell in which the molecules in consecutive layers form a distorted helix such that the azimuthal angle, δ, between the projections of the long molecular axis in adjacent layers on to the smectic plane is not 90° (Figure 19b). In addition, the azimuthal angle increases by some amount, $\mathcal{E}$, with respect to the adjacent layer causing the four-layer unit cell to rotate giving an additional helical structure (Figure 19c). The smectic phase shown by CB6O.10 is thought to be similar to this structure, but it should be stressed that given the molecules are achiral, the driving force for its formation must be different than in case of the SmC$_{FI2}^*$ phase. We termed this phase SmC$_{TB-DH}$ where DH indicates double helix and show the structure as a distorted clock in Figure 19c and schematically in Figure 20. An important difference to the structure of the SmC$_{FI2}^*$ phase is that in the SmC$_{TB-DH}$ phase, the modulation superimposed on the basic four-layer unit is much stronger, and similar to the basic modulation, is on the nanometre scale. A model used to predict the RSoXS signal intensities has shown that the handedness of the short four-layer structure determines that of the longer helical modulation. The angle δ is essentially temperature independent whereas $\mathcal{E}$ is strongly temperature dependent giving rise to the evolution of the helical structure shown in Figure 20. Close to the N$_{TB}$-SmC$_{TB-DH}$ transition, we see an almost ideal clock four-layer structure (Figure 19a) but as $\mathcal{E}$ increases on decreasing temperature, the structure evolves towards an anticlinic arrangement. We should also note that the bent nature of the molecules introduces another level of structural chirality, the sign of which is defined by the relative orientation of three vectors: the layer normal, the tilt projection onto the smectic layer, and the molecular bend direction. This structural chirality is referred to as layer chirality and was described earlier (Figure 17) [92]. Presumably the sign of the layer chirality is coupled to the handedness of the four-layer basic unit and hence to the longer, superimposed helix. We have seen earlier that in the N$_{TB}$ phase steric interactions drive the formation of the helical structure, and the pitch length decreases on reducing the temperature. On cooling into the SmC$_{TB-DH}$ phase an ideal four-layer clock helical structure forms, in which the azimuthal tilt angle changes by 90° on passing from layer to layer. As the temperature is reduced, the competition increases between the tendency to twist and the entropy-driven tendency for molecules to remain in one tilt plane which allows for easier molecular movement along the layer normal, and thus, the structure continuously evolves from the four-layer clock to an almost anticlinic bilayer arrangement. By comparison, in chiral materials a series of discontinuous phase transitions have been observed rather than a continuous evolution of structure [96].

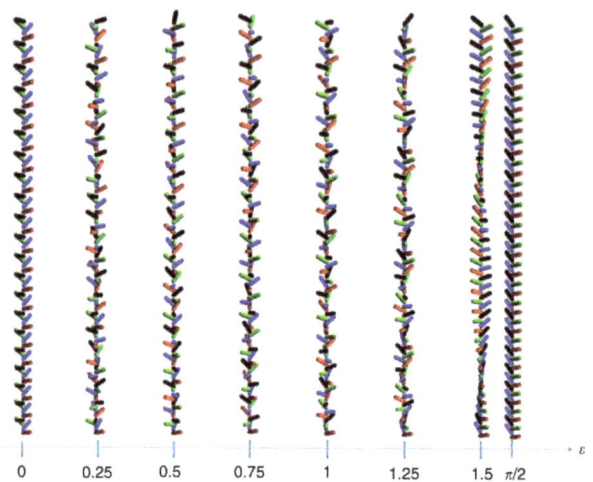

Figure 20. A schematic representation of the evolution of the four-layer structure in the SmC$_{TB-DH}$ phase on increasing $\mathcal{E}$, the difference between the azimuthal angle between adjacent layers. For clarity the tilt angle has been increased. (Taken from [94]).

A study of the longer homologues of the CB6O.*m* series, (*m* = 12, 14, 16, 18) revealed a sequence of four smectic phases: SmA—SmC$_{TB-SH}$—SmC$_{TB-DH}$—SmX below the I or N phase (Figure 13) [91]. The lowest temperature phase, SmX, was not studied in detail, but its X-ray diffraction pattern indicated that it is a hexatic-type smectic phase. The SmA phase was optically uniaxial (Figure 21) and the layer spacing corresponded to two molecular lengths indicating a bilayer structure in which the cyano groups are concentrated at alternating interfaces, as seen for CB6O.10. X-ray diffraction revealed that the SmA, SmC$_{TB-SH}$, and SmC$_{TB-DH}$ phases are liquid-like in terms of in-plane ordering. The SmC$_{TB-SH}$ phase was optically biaxial whereas the SmC$_{TB-DH}$ is optically uniaxial (Figure 21). The RSoXS pattern of the SmC$_{TB-SH}$ phase contained a resonant peak corresponding to four molecular layers and its position was temperature independent, and using non-resonant XRD, a bilayer structure was observed. SH is used to indicate that the SmC$_{TB-SH}$ phase is characterised by having a single helix. On entering the SmC$_{TB-DH}$ phase, symmetric satellites of the resonant peak developed indicating an additional modulation superimposed on the basic four-layer helix. Unlike the continuous development of this peak splitting seen in the RSoXS pattern for CB6O.10, the satellites' peaks change discontinuously with temperature, and this is thought to be associated with surface interactions. The SmC$_{TB-SH}$–SmC$_{TB-DH}$ phase transition is not associated with a change in layer spacing as measured using non-resonant XRD, implying that the tilt angle does not change. The optical biaxiality of the SmC$_{TB-SH}$ phase suggests a distorted clock structure for its four-layer unit cell (Figure 19b). In the optically uniaxial SmC$_{TB-DH}$ phase an additional shift in azimuthal angle, $\mathcal{E}$, between consecutive layers gives rise to a longer pitch-length helical modulation superimposed on the basic 4-layer helical structure. This additional helix has a pitch length corresponding to around 16 layers deep into the SmC$_{TB-DH}$ phase but as the temperature approaches the SmC$_{TB-SH}$–SmC$_{TB-DH}$ transition, the pitch length jumps to around 46-layer distances. As we saw for CB6O.10, this change in the pitch length reflects the evolution of the structure from a bilayer towards a four-layer arrangement with alternating synclinic and anticlinic interfaces. The equal intensities of the satellite peaks in the RSoXS pattern for the SmC$_{TB-DH}$ phase imply that the angle δ must be small such that the basic four-layer unit is a strongly distorted clock in which alternating interfaces approach synclinic and anticlinic arrangements. A study of the SmC_{TB-DH}–SmC_{TB-SH} transition using optical microscopy revealed selective reflection colours covering the whole optical spectrum at temperatures very close to the transition, which indicated unwinding of the additional helix of the SmC_{TB-DH} phase (Figure 22) [91]. The selective reflection of light occurs when the pitch of the helix, as it unwinds, becomes comparable to the wavelength of visible light, and this was the first observation of such behaviour for achiral liquid crystals.

Figure 21. Optical textures of CB6O.14 observed between crossed polarizers in a 3 µm thick cell with homeotropic anchoring (taken from [91]).

Figure 22. The optical texture shown by CB6O.12 taken in a 3 µm thick cell with homeotropic anchoring at the SmC_{TB-DH}–SmC_{TB-SH} transition showing selective reflection. The colours appear simultaneously because of a small temperature gradient in the sample. (Taken from [91]).

The corresponding dimers containing a thioether-linked terminated chain, CB6O.Sm, [103]

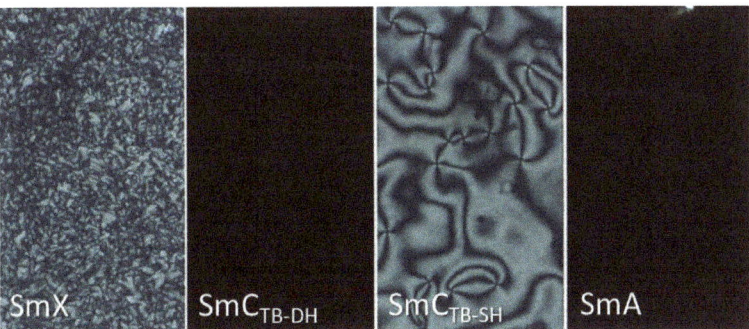

also show the SmC_{TB-DH}–SmC_{TB-SH} transition for $m > 13$, but the monotropic nature of these phases prevented their detailed study. This series of bent dimers also showed the helical filament B_4 phase described earlier.

To investigate the effects of increasing the spacer length on the formation of the heliconical SmC_{TB} phases, we studied the properties of the 1-(4-cyanobiphenyl-4′-yl)-10-(4-alkylaniline-benzylidene-4′-oxy)decanes (CB10O.m) [104]:

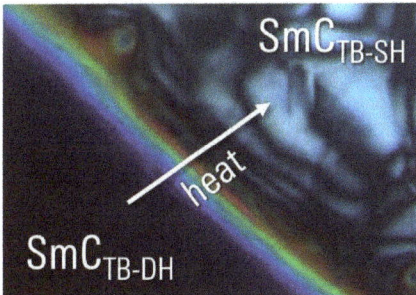

with the terminal chain length, m = 1–10, 12, 14, 16 and 18. The CB10O.m series shows a rich phase polymorphism including the N and N_{TB} phases, and six different smectic phases (Figure 23). All the homologues showed the conventional N phase. For the shortest members, m = 1–3, the N_{TB} phase was seen. An intercalated SmC$_A$ phase emerged for CB10O.3, and for m = 4–6 the N_{TB} phase was extinguished and a direct SmC$_A$-N transition observed. The N_{TB} phase remerged with CB10O.7, and the SmC$_A$ phase extinguished after m = 8. The homologues with m = 9 and 10 were exclusively nematogenic, showing both N and N_{TB} phases. Smectic behaviour re-emerged at m = 12, and the longer homologues exhibited heliconical SmC$_{TB}$ phases. This pattern of behaviour has clear similarities to that seen for the CBO4O.m series (Figure 5) for which smectic phases were observed for short and long terminal chain lengths and solely nematic behaviour for intermediate chain lengths [18]. For the CBO4O.m series, this was interpreted in terms of the change in the structure of the smectic phases on increasing chain length from being interdigitated to intercalated (Figure 6). A similar explanation accounts for the behaviour seen for the CB10O.m series except that with increasing m we now see a switch from intercalated to interdigitated bilayer smectic phases. Three homologues (m = 12, 14, 16) show SmC$_{TB-SH}$ and SmC$_{TB-DH}$ phases whereas for the longest homologue only the SmC$_{TB-SH}$ phase is seen, presumably the transition to the SmC$_{TB-DH}$ phase is precluded by the formation of the underlying SmY phase. As described for the CB6O.m series, the SmC$_{TB-SH}$ phase is optically biaxial implying a strongly distorted clock arrangement (Figure 19b) whereas the SmC$_{TB-DH}$ phase is optically uniaxial, given the additional modulation superimposed on the basic four-layer structure leading to space-averaging of the azimuthal positions of the molecules along the layer normal (Figure 20). It is interesting to note that there is no apparent change in layer spacing at the SmC$_{TB-SH}$-SmC$_{TB-DH}$ transition implying the tilt angle is similar in both. The striking difference in the nature of the smectic phases shown by m = 3–8, and $m \geq 12$ is revealed in the behaviour of an approximately equimolar mixture of homologues with m = 6 and 16 that exhibited the N_{TB} phase over a broad temperature range although neither individual component does. This shows that the intercalated and interdigitated smectic phases are incompatible and destabilised in the mixture, revealing the underlying N_{TB} phase.

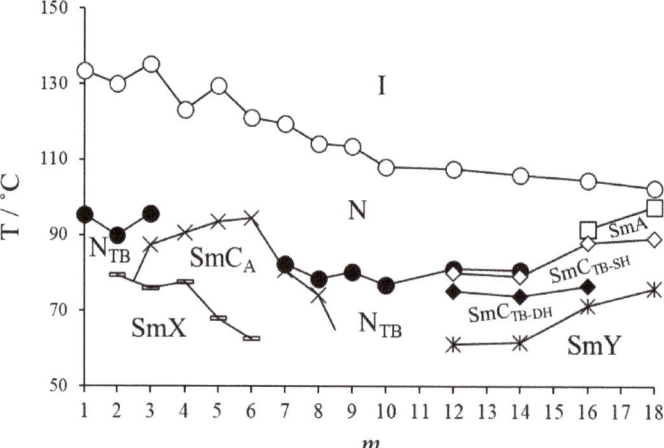

Figure 23. The dependence of the transition temperatures on the length of the terminal chain, m, for the CB10O.m series [104]. The melting points have been omitted for the sake of clarity. The dependence of the transition temperatures on the length of the terminal chain, m, for the CB10O.m series [104]. The melting points have been omitted for the sake of clarity. Unfilled circles denoted T_{NI}; filled circles T_{NTBN}; crosses $T_{SmCN/NTB}$; dashes T_{SmX}-; squares T_{SmAN}; unfilled diamonds $T_{SmCTB-SH}$-; filled diamonds $T_{SmCTB-DHSmCTB-SH}$; stars T_{SmY}-.

7. Summary and Outlook

In this review we have discussed the relationships between the smectic behaviour of liquid crystal dimers and their molecular structures. These are dominated by the average shape of the molecule and markedly different behaviour is seen for bent-odd membered dimers than for linear even-membered dimers. For symmetric dimers there is a strong preference for the formation of monolayer smectic phases driven apparently by the incompatibility between the mixing of the spacers and terminal chains. This may be overcome for nonsymmetric dimers for which intercalated and interdigitated smectic phases are observed depending on the relative lengths of the spacers and terminal chains. The stronger tendency of odd-membered dimers to exhibit tilted phases is attributed to the difficulty associated with the packing of their bent shape into orthogonal phases. By adjusting the molecular curvature, it is possible to drive the formation of spontaneously chiral twist-bend smectic phases, and to date, we have reported three variants of the SmC_{TB} phase. In the SmC_{TB}-α phase, the helical pitch length is incommensurate with the layer spacing and the phase is optically uniaxial. In the SmC_{TB-SH} phase the helical pitch corresponds to approximately four molecular lengths, and may be described by the distorted clock model and the phase is optically biaxial. The SmC_{TB-DH} phase has an additional helical modulation described by the rotation of the basic four-layer units by a temperature dependent angle, and thus is optically uniaxial. To date, the twist-bend arrangement has only been reported for either monolayer or bilayer smectic phases, and not for intercalated structures. We anticipate, however, the discovery of further variants of the SmC_{TB} phase as a wider expanse of molecular structure space is explored. More widely, the spontaneous formation of chiral structures with multiple different levels of chirality by achiral molecules is an area of intense global interest and of fundamental importance in the physical and biological sciences. The study of liquid crystal dimers has an exciting and vibrant future!

Author Contributions: All the authors have made a substantial, and intellectual contribution to the work and approved it for publication. All authors have read and agreed to the published version of the manuscript.

Funding: This research received no external funding.

Acknowledgments: The authors gratefully acknowledge Professor Nataša Vaupotič for her central role in the interpretation of the resonant soft X-ray scattering (RSoXS) data and the preparation of the original figures. The authors also wish to thank Ewan Cruickshank for many helpful discussions.

Conflicts of Interest: The authors declare no conflict of interest.

References

1. Cestari, M.; Diez-Berart, S.; Dunmur, D.A.; Ferrarini, A.; de la Fuente, M.R.; Jackson, D.J.B.; Lopez, D.O.; Luckhurst, G.R.; Perez-Jubindo, M.A.; Richardson, R.M.; et al. Phase behavior and properties of the liquid-crystal dimer 1″,7″-bis(4-cyanobiphenyl-4′-yl) heptane: A twist-bend nematic liquid crystal. *Phys. Rev. E* **2011**, *84*, 031704. [CrossRef] [PubMed]
2. Dozov, I. On the spontaneous symmetry breaking in the mesophases of achiral banana-shaped molecules. *Europhys. Lett.* **2001**, *56*, 247–253. [CrossRef]
3. Abberley, J.P.; Killah, R.; Walker, R.; Storey, J.M.D.; Imrie, C.T.; Salamonczyk, M.; Zhu, C.H.; Gorecka, E.; Pociecha, D. Heliconical smectic phases formed by achiral molecules. *Nat. Commun.* **2018**, *9*, 228. [CrossRef] [PubMed]
4. Imrie, C.T.; Henderson, P.A. Liquid crystal dimers and oligomers. *Curr. Opin. Colloid Interface Sci.* **2002**, *7*, 298–311. [CrossRef]
5. Imrie, C.T.; Henderson, P.A. Liquid crystal dimers and higher oligomers: Between monomers and polymers. *Chem. Soc. Rev.* **2007**, *36*, 2096–2124. [CrossRef]
6. Imrie, C.T.; Henderson, P.A.; Yeap, G.Y. Liquid crystal oligomers: Going beyond dimers. *Liq. Cryst.* **2009**, *36*, 755–777. [CrossRef]
7. Attard, G.S.; Garnett, S.; Hickman, C.G.; Imrie, C.T.; Taylor, L. Asymmetric dimeric liquid-crystals with charge-transfer groups. *Liq. Cryst.* **1990**, *7*, 495–508. [CrossRef]
8. Griffin, A.C.; Britt, T.R. Effect of molecular-structure on mesomorphism.12. Flexible-center siamese-twin liquid-crystalline diesters—A prepolymer model. *J. Am. Chem. Soc.* **1981**, *103*, 4957–4959. [CrossRef]
9. Vorlander, D. On the nature of carbon chains in crystalline-fluid substances. *Z. Phys. Chem.* **1927**, *126*, 449–472.
10. Date, R.W.; Imrie, C.T.; Luckhurst, G.R.; Seddon, J.M. Smectogenic dimeric liquid-crystals—The preparation and properties of the alpha,ω-bis(4-n-alkylanilinebenzylidine-4′-oxy)alkane. *Liq. Cryst.* **1992**, *12*, 203–238. [CrossRef]

11. Smith, G.W.; Gardlund, Z.G.; Curtis, R.J. Phase-transitions in mesomorphic benzylideneanilines. *Mol. Cryst. Liq. Cryst.* **1973**, *19*, 327–330. [CrossRef]
12. Date, R.W.; Luckhurst, G.R.; Shuman, M.; Seddon, J.M. Novel modulated hexatic phases in symmetrical liquid-crystal dimers. *J. Phys II Fr.* **1995**, *5*, 587–605.
13. Imrie, C.T. Non-symmetric liquid crystal dimers: How to make molecules intercalate. *Liq. Cryst.* **2006**, *33*, 1449–1454. [CrossRef]
14. Henderson, P.A.; Niemeyer, O.; Imrie, C.T. Methylene-linked liquid crystal dimers. *Liq. Cryst.* **2001**, *28*, 463–472. [CrossRef]
15. Henderson, P.A.; Seddon, J.M.; Imrie, C.T. Methylene- and ether-linked liquid crystal dimers, I.I. Effects of mesogenic linking unit and terminal chain length. *Liq. Cryst.* **2005**, *32*, 1499–1513. [CrossRef]
16. Henderson, P.A.; Imrie, C.T. Methylene-linked liquid crystal dimers and the twist-bend nematic phase. *Liq. Cryst.* **2011**, *38*, 1407–1414. [CrossRef]
17. Hogan, J.L.; Imrie, C.T.; Luckhurst, G.R. Asymmetric dimeric liquid-crystals—The preparation and properties of the α-(4-cyanobiphenyl-4′-oxy)-ω-(4-n-alkylanilinebenzylidene-4′-oxy)hexanes. *Liq. Cryst.* **1988**, *3*, 645–650. [CrossRef]
18. Attard, G.S.; Date, R.W.; Imrie, C.T.; Luckhurst, G.R.; Roskilly, S.J.; Seddon, J.M.; Taylor, L. Nonsymmetrical dimeric liquid-crystals—The preparation and properties of the alpha-(4-cyanobiphenyl-4′-yloxy)-omega-(4-n-alkylanilinebenzylidene-4′-oxy)alkanes. *Liq. Cryst.* **1994**, *16*, 529–581. [CrossRef]
19. Park, J.W.; Bak, C.S.; Labes, M.M. Effects of molecular complexing on properties of binary nematic liquid-crystal mixtures. *J. Am. Chem. Soc.* **1975**, *97*, 4398–4400. [CrossRef]
20. Cladis, P.E. The re-entrant nematic, enhanced smectic-a phases and molecular composition. *Mol. Cryst. Liq. Cryst.* **1981**, *67*, 833–847. [CrossRef]
21. Blatch, A.E.; Fletcher, I.D.; Luckhurst, G.R. The intercalated smectic A phase—The liquid-crystal properties of the alpha(4-cyanobiphenyl-4′-yloxy)-omega-(4-alkyloxycinnamoate)alkanes. *Liq. Cryst.* **1995**, *18*, 801–809. [CrossRef]
22. Takanishi, Y.; Takezoe, H.; Fukuda, A.; Komura, H.; Watanabe, J. Simple method for confirming the antiferroelectric structure of smectic liquid-crystals. *J. Mater. Chem.* **1992**, *2*, 71–73. [CrossRef]
23. Le Masurier, P.J.; Luckhurst, G.R. Structural studies of the intercalated smectic C phases formed by the non-symmetric alpha-(4-cyanobiphenyl-4′-yloxy)-omega-(4-alkylaniline-benzylidene-4′-oxy) alkane dimers using EPR spectroscopy. *J. Chem. Soc. Faraday Trans.* **1998**, *94*, 1593–1601. [CrossRef]
24. Walker, R.; Pociecha, D.; Faidutti, C.; Perkovic, E.; Storey, J.M.D.; Gorecka, E.; Imrie, C.T. Remarkable stabilisation of the intercalated smectic phases of nonsymmetric dimers by tert-butyl groups. *Liq. Cryst.* **2022**, *49*, 969–981. [CrossRef]
25. Watanabe, J.; Komura, H.; Niiori, T. Thermotropic liquid-crystals of polyesters having a mesogenic 4,4-bibenzoate unit—Smectic mesophase properties and structures in dimeric model compounds. *Liq. Cryst.* **1993**, *13*, 455–465. [CrossRef]
26. Reddy, R.A.; Tschierske, C. Bent-core liquid crystals: Polar order, superstructural chirality and spontaneous desymmetrisation in soft matter systems. *J. Mater. Chem.* **2006**, *16*, 907–961. [CrossRef]
27. Bialecka-Florjanczyk, E.; Sledzinska, I.; Gorecka, E.; Przedmojski, J. Odd-even effect in biphenyl-based symmetrical dimers with methylene spacer—Evidence of the B4 phase. *Liq. Cryst.* **2008**, *35*, 401–406. [CrossRef]
28. Hough, L.E.; Jung, H.T.; Kruerke, D.; Heberling, M.S.; Nakata, M.; Jones, C.D.; Chen, D.; Link, D.R.; Zasadzinski, J.; Heppke, G.; et al. Helical Nanofilament Phases. *Science* **2009**, *325*, 456–460. [CrossRef] [PubMed]
29. Le, K.V.; Takezoe, H.; Araoka, F. Chiral Superstructure Mesophases of Achiral Bent-Shaped Molecules—Hierarchical Chirality Amplification and Physical Properties. *Adv. Mater. Interfaces* **2017**, *29*, 1602727. [CrossRef]
30. Borshch, V.; Kim, Y.K.; Xiang, J.; Gao, M.; Jakli, A.; Panov, V.P.; Vij, J.K.; Imrie, C.T.; Tamba, M.G.; Mehl, G.H.; et al. Nematic twist-bend phase with nanoscale modulation of molecular orientation. *Nat. Commun.* **2013**, *4*, 2635. [CrossRef] [PubMed]
31. Zhu, C.H.; Tuchband, M.R.; Young, A.; Shuai, M.; Scarbrough, A.; Walba, D.M.; Maclennan, J.E.; Wang, C.; Hexemer, A.; Clark, N.A. Resonant Carbon K-Edge Soft X-ray Scattering from Lattice-Free Heliconical Molecular Ordering: Soft Dilative Elasticity of the Twist-Bend Liquid Crystal Phase. *Phys. Rev. Lett.* **2016**, *116*, 147803. [CrossRef] [PubMed]
32. Dunmur, D.A. Anatomy of a Discovery: The Twist-Bend Nematic Phase. *Crystals* **2022**, *12*, 309. [CrossRef]
33. Sepelj, M.; Lesac, A.; Baumeister, U.; Diele, S.; Nguyen, H.L.; Bruce, D.W. Intercalated liquid-crystalline phases formed by symmetric dimers with an alpha,omega-diiminoalkylene spacer. *J. Mater. Chem.* **2007**, *17*, 1154–1165. [CrossRef]
34. Panov, V.P.; Nagaraj, M.; Vij, J.K.; Panarin, Y.P.; Kohlmeier, A.; Tamba, M.G.; Lewis, R.A.; Mehl, G.H. Spontaneous Periodic Deformations in Nonchiral Planar-Aligned Bimesogens with a Nematic-Nematic Transition and a Negative Elastic Constant. *Phys. Rev. Lett.* **2010**, *105*, 167801. [CrossRef] [PubMed]
35. Ferrarini, A.; Luckhurst, G.R.; Nordio, P.L.; Roskilly, S.J. Understanding the unusual transitional behavior of liquid-crystal dimers. *Chem. Phys. Lett.* **1993**, *214*, 409–417. [CrossRef]
36. Ferrarini, A.; Luckhurst, G.R.; Nordio, P.L.; Roskilly, S.J. Understanding the dependence of the transitional properties of liquid crystal dimers on their molecular geometry. *Liq. Cryst.* **1996**, *21*, 373–382. [CrossRef]
37. Emerson, A.; Luckhurst, G.R.; Phippen, R.W. The average shapes of flexible mesogenic molecules—On the choice of reference frame. *Liq. Cryst.* **1991**, *10*, 1–14. [CrossRef]
38. Ferrarini, A.; Luckhurst, G.R.; Nordio, P.L.; Roskilly, S.J. Prediction of the transitional properties of liquid-crystal dimers—A molecular-field calculation based on the surface tensor parametrization. *J. Chem. Phys.* **1994**, *100*, 1460–1469. [CrossRef]
39. Paterson, D.A.; Abberley, J.P.; Harrison, W.T.; Storey, J.M.; Imrie, C.T. Cyanobiphenyl-based liquid crystal dimers and the twist-bend nematic phase. *Liq. Cryst.* **2017**, *44*, 127–146. [CrossRef]

40. Walker, R.; Pociecha, D.; Strachan, G.J.; Storey, J.M.D.; Gorecka, E.; Imrie, C.T. Molecular curvature, specific intermolecular interactions and the twist-bend nematic phase: The synthesis and characterisation of the 1-(4-cyanobiphenyl-4-yl)-6-(4-alkylanilinebenzylidene-4-oxy)hexanes (CB6O.m). *Soft Matter.* **2019**, *15*, 3188–3197. [CrossRef]
41. Lu, Z.B.; Henderson, P.A.; Paterson, B.J.A.; Imrie, C.T. Liquid crystal dimers and the twist-bend nematic phase. The preparation and characterisation of the alpha,omega-bis(4-cyanobiphenyl-4′-yl) alkanedioates. *Liq. Cryst.* **2014**, *41*, 471–483. [CrossRef]
42. Paterson, D.A.; Gao, M.; Kim, Y.K.; Jamali, A.; Finley, K.L.; Robles-Hernandez, B.; Diez-Berart, S.; Salud, J.; de la Fuente, M.R.; Timimi, B.A.; et al. Understanding the twist-bend nematic phase: The characterisation of 1-(4-cyanobiphenyl-4′-yloxy)-6-(4-cyanobiphenyl-4′-yl)hexane (CB6OCB) and comparison with CB7CB. *Soft Matter.* **2016**, *12*, 6827–6840. [CrossRef] [PubMed]
43. Cruickshank, E.; Salamonczyk, M.; Pociecha, D.; Strachan, G.J.; Storey, J.M.D.; Wang, C.; Feng, J.; Zhu, C.H.; Gorecka, E.; Imrie, C.T. Sulfur-linked cyanobiphenyl-based liquid crystal dimers and the twist-bend nematic phase. *Liq. Cryst.* **2019**, *46*, 1595–1609. [CrossRef]
44. Arakawa, Y.; Komatsu, K.; Feng, J.; Zhu, C.H.; Tsuji, H. Distinct twist-bend nematic phase behaviors associated with the ester-linkage direction of thioether-linked liquid crystal dimers. *Mater. Adv.* **2021**, *2*, 261–272. [CrossRef]
45. Arakawa, Y.; Komatsu, K.; Ishida, Y.; Igawa, K.; Tsuji, H. Carbonyl- and thioether-linked cyanobiphenyl-based liquid crystal dimers exhibiting twist-bend nematic phases. *Tetrahedron* **2021**, *81*, 131870. [CrossRef]
46. Arakawa, Y.; Tsuji, H. Selenium-linked liquid crystal dimers for twist-bend nematogens. *J. Mol. Liq.* **2019**, *289*, 111097. [CrossRef]
47. Lesac, A.; Baumeister, U.; Dokli, I.; Hamersak, Z.; Ivsic, T.; Kontrec, D.; Viskic, M.; Knezevic, A.; Mandle, R.J. Geometric aspects influencing N-N-TB transition—Implication of intramolecular torsion. *Liq. Cryst.* **2018**, *45*, 1101–1110. [CrossRef]
48. Archbold, C.T.; Andrews, J.L.; Mandle, R.J.; Cowling, S.J.; Goodby, J.W. Effect of the linking unit on the twist-bend nematic phase in liquid crystal dimers: A comparative study of two homologous series of methylene- and ether-linked dimers. *Liq. Cryst.* **2017**, *44*, 84–92. [CrossRef]
49. Mandle, R.J.; Voll, C.C.A.; Lewis, D.J.; Goodby, J.W. Etheric bimesogens and the twist-bend nematic phase. *Liq. Cryst.* **2016**, *43*, 13–21. [CrossRef]
50. Forsyth, E.; Paterson, D.A.; Cruickshank, E.; Strachan, G.J.; Gorecka, E.; Walker, R.; Storey, J.M.D.; Imrie, C.T. Liquid crystal dimers and the twist-bend nematic phase: On the role of spacers and terminal alkyl chains. *J. Mol. Liq.* **2020**, *320*, 114391. [CrossRef]
51. Stevenson, W.D.; Zou, H.X.; Zeng, X.B.; Welch, C.; Ungar, G.; Mehl, G.H. Dynamic calorimetry and XRD studies of the nematic and twist-bend nematic phase transitions in a series of dimers with increasing spacer length. *Phys. Chem. Chem. Phys.* **2018**, *20*, 25268–25274. [CrossRef] [PubMed]
52. Merkel, K.; Loska, B.; Welch, C.; Mehl, G.H.; Kocot, A. Molecular biaxiality determines the helical structure—Infrared measurements of the molecular order in the nematic twist-bend phase of difluoro terphenyl dimer. *Phys. Chem. Chem. Phys.* **2021**, *23*, 4151–4160. [CrossRef] [PubMed]
53. Arakawa, Y.; Komatsu, K.; Shiba, T.; Tsuji, H. Methylene- and thioether-linked cyanobiphenyl-based liquid crystal dimers CBnSCB exhibiting room temperature twist-bend nematic phases and glasses. *Mater. Adv.* **2021**, *2*, 1760–1773. [CrossRef]
54. Paterson, D.A.; Walker, R.; Abberley, J.P.; Forestier, J.; Harrison, W.T.A.; Storey, J.M.D.; Pociecha, D.; Gorecka, E.; Imrie, C.T. Azobenzene-based liquid crystal dimers and the twist-bend nematic phase. *Liq. Cryst.* **2017**, *44*, 2060–2078. [CrossRef]
55. Walker, R.; Majewska, M.; Pociecha, D.; Makal, A.; Storey, J.M.D.; Gorecka, E.; Imrie, C.T. Twist-Bend Nematic Glasses: The Synthesis and Characterisation of Pyrene-based Nonsymmetric Dimers. *Chemphyschem* **2021**, *22*, 461–470. [CrossRef] [PubMed]
56. Strachan, G.J.; Harrison, W.T.A.; Storey, J.M.D.; Imrie, C.T. Understanding the remarkable difference in liquid crystal behaviour between secondary and tertiary amides: The synthesis and characterisation of new benzanilide-based liquid crystal dimers. *Phys. Chem. Chem. Phys.* **2021**, *23*, 12600–12611. [CrossRef]
57. Abberley, J.P.; Storey, J.M.D.; Imrie, C.T. Structure-property relationships in azobenzene-based twist-bend nematogens. *Liq. Cryst.* **2019**, *46*, 2102–2114. [CrossRef]
58. Sebastian, N.; Tamba, M.G.; Stannarius, R.; de la Fuente, M.R.; Salamonczyk, M.; Cukrov, G.; Gleeson, J.; Sprunt, S.; Jakli, A.; Welch, C.; et al. Mesophase structure and behaviour in bulk and restricted geometry of a dimeric compound exhibiting a nematic-nematic transition. *Phys. Chem. Chem. Phys.* **2016**, *18*, 19299–19308. [CrossRef]
59. Ahmed, Z.; Welch, C.; Mehl, G.H. The design and investigation of the self-assembly of dimers with two nematic phases. *RSC Adv.* **2015**, *5*, 93513–93521. [CrossRef]
60. Arakawa, Y.; Ishida, Y.; Tsuji, H. Ether- and Thioether-Linked Naphthalene-Based Liquid-Crystal Dimers: Influence of Chalcogen Linkage and Mesogenic-Arm Symmetry on the Incidence and Stability of the Twist-Bend Nematic Phase. *Chem. Eur. J.* **2020**, *26*, 3767–3775. [CrossRef]
61. Arakawa, Y.; Komatsu, K.; Ishida, Y.; Tsuji, H. Thioether-linked azobenzene-based liquid crystal dimers exhibiting the twist-bend nematic phase over a wide temperature range. *Liq. Cryst.* **2021**, *48*, 641–652. [CrossRef]
62. Knezevic, A.; Dokli, I.; Novak, J.; Kontrec, D.; Lesac, A. Fluorinated twist-bend nematogens: The role of intermolecular interaction. *Liq. Cryst.* **2021**, *48*, 756–766. [CrossRef]
63. Al-Janabi, A.; Mandle, R.J. Utilising Saturated Hydrocarbon Isosteres of para Benzene in the Design of Twist-Bend Nematic Liquid Crystals. *Chemphyschem* **2020**, *21*, 697–701. [CrossRef] [PubMed]
64. Mandle, R.J.; Goodby, J.W. Does Topology Dictate the Incidence of the Twist-Bend Phase? Insights Gained from Novel Unsymmetrical Bimesogens. *Chem. Eur. J.* **2016**, *22*, 18456–18464. [CrossRef]

65. Abberley, J.P.; Jansze, S.M.; Walker, R.; Paterson, D.A.; Henderson, P.A.; Marcelis, A.T.M.; Storey, J.M.D.; Imrie, C.T. Structure-property relationships in twist-bend nematogens: The influence of terminal groups. *Liq. Cryst.* **2017**, *44*, 68–83. [CrossRef]
66. Ivsic, T.; Baumeister, U.; Dokli, I.; Mikleusevic, A.; Lesac, A. Sensitivity of the N-TB phase formation to the molecular structure of imino-linked dimers. *Liq. Cryst.* **2017**, *44*, 93–105.
67. Mandle, R.J.; Davis, E.J.; Archbold, C.T.; Voll, C.C.A.; Andrews, J.L.; Cowling, S.J.; Goodby, J.W. Apolar Bimesogens and the Incidence of the Twist-Bend Nematic Phase. *Chem. Eur. J.* **2015**, *21*, 8158–8167. [CrossRef]
68. Mandle, R.J.; Goodby, J.W. Dependence of Mesomorphic Behaviour of Methylene-Linked Dimers and the Stability of the N-TB/N-X Phase upon Choice of Mesogenic Units and Terminal Chain Length. *Chem. Eur. J.* **2016**, *22*, 9366–9374. [CrossRef]
69. Mandle, R.J.; Goodby, J.W. A Liquid Crystalline Oligomer Exhibiting Nematic and Twist-Bend Nematic Mesophases. *Chemphyschem* **2016**, *17*, 967–970. [CrossRef]
70. Arakawa, Y.; Komatsu, K.; Inui, S.; Tsuji, H. Thioether-linked liquid crystal dimers and trimers: The twist-bend nematic phase. *J. Mol. Struct.* **2020**, *1199*, 126913. [CrossRef]
71. Mandle, R.J.; Goodby, J.W. A Nanohelicoidal Nematic Liquid Crystal Formed by a Non-Linear Duplexed Hexamer. *Angew Chem. Int. Ed.* **2018**, *57*, 7096–7100. [CrossRef] [PubMed]
72. Tuchband, M.R.; Paterson, D.A.; Salamonczykc, M.; Norman, V.A.; Scarbrough, A.N.; Forsyth, E.; Garcia, E.; Wang, C.; Storey, J.M.D.; Walba, D.M.; et al. Distinct differences in the nanoscale behaviors of the twist-bend liquid crystal phase of a flexible linear trimer and homologous dimer. *Proc. Nat. Acad. Sci. USA* **2019**, *116*, 10698–10704. [CrossRef] [PubMed]
73. Arakawa, Y.; Komatsu, K.; Shiba, T.; Tsuji, H. Phase behaviors of classic liquid crystal dimers and trimers: Alternate induction of smectic and twist-bend nematic phases depending on spacer parity for liquid crystal trimers. *J. Mol. Liq.* **2021**, *326*, 115319. [CrossRef]
74. Majewska, M.M.; Forsyth, E.; Pociecha, D.; Wang, C.; Storey, J.M.D.; Imrie, C.T.; Gorecka, E. Controlling spontaneous chirality in achiral materials: Liquid crystal oligomers and the heliconical twist-bend nematic phase. *Chem. Commun.* **2022**, *58*, 5285–5288. [CrossRef] [PubMed]
75. Jansze, S.M.; Martinez-Felipe, A.; Storey, J.M.D.; Marcelis, A.T.M.; Imrie, C.T. A Twist-Bend Nematic Phase Driven by Hydrogen Bonding. *Angew Chem. Int. Ed.* **2015**, *54*, 643–646. [CrossRef]
76. Walker, R.; Pociecha, D.; Abberley, J.P.; Martinez-Felipe, A.; Paterson, D.A.; Forsyth, E.; Lawrence, G.B.; Henderson, P.A.; Storey, J.M.D.; Gorecka, E.; et al. Spontaneous chirality through mixing achiral components: A twist-bend nematic phase driven by hydrogen-bonding between unlike components. *Chem. Commun.* **2018**, *54*, 3383–3386. [CrossRef]
77. Walker, R.; Pociecha, D.; Martinez-Felipe, A.; Storey, J.M.D.; Gorecka, E.; Imrie, C.T. Twist-Bend Nematogenic Supramolecular Dimers and Trimers Formed by Hydrogen Bonding. *Crystals* **2020**, *10*, 175. [CrossRef]
78. Walker, R.; Pociecha, D.; Crawford, C.A.; Storey, J.M.D.; Gorecka, E.; Imrie, C.T. Hydrogen bonding and the design of twist-bend nematogens. *J. Mol. Liq.* **2020**, *303*, 112630. [CrossRef]
79. Walker, R.; Pociecha, D.; Salamonczyk, M.; Storey, J.M.D.; Gorecka, E.; Imrie, C.T. Supramolecular liquid crystals exhibiting a chiral twist-bend nematic phase. *Mater. Adv.* **2020**, *1*, 1622–1630. [CrossRef]
80. Chen, D.; Nakata, M.; Shao, R.; Tuchband, M.R.; Shuai, M.; Baumeister, U.; Weissflog, W.; Walba, D.M.; Glaser, M.A.; Maclennan, J.E.; et al. Twist-bend heliconical chiral nematic liquid crystal phase of an achiral rigid bent-core mesogen. *Phys. Rev. E.* **2014**, *89*, 022506. [CrossRef]
81. Sreenilayam, S.P.; Panov, V.P.; Vij, J.K.; Shanker, G. The N-TB phase in an achiral asymmetrical bent-core liquid crystal terminated with symmetric alkyl chains. *Liq. Cryst.* **2017**, *44*, 244–253.
82. Stevenson, W.D.; An, J.G.; Zeng, X.B.; Xue, M.; Zou, H.X.; Liu, Y.S.; Ungar, G. Twist-bend nematic phase in biphenylethane-based copolyethers. *Soft Matter.* **2018**, *14*, 3003–3011. [CrossRef] [PubMed]
83. Mandle, R.J. A Ten-Year Perspective on Twist-Bend Nematic Materials. *Molecules* **2022**, *27*, 2689. [CrossRef] [PubMed]
84. Greco, C.; Luckhurst, G.R.; Ferrarini, A. Molecular geometry, twist-bend nematic phase and unconventional elasticity: A generalised Maier-Saupe theory. *Soft Matter.* **2014**, *10*, 9318–9323. [CrossRef]
85. Paterson, D.A.; Xiang, J.; Singh, G.; Walker, R.; Agra-Kooijman, D.M.; Martinez-Felipe, A.; Gan, M.; Storey, J.M.D.; Kumar, S.; Lavrentovich, O.D.; et al. Reversible Isothermal Twist-Bend Nematic-Nematic Phase Transition Driven by the Photoisomerization of an Azobenzene-Based Nonsymmetric Liquid Crystal Dinner. *J. Am. Chem. Soc.* **2016**, *138*, 5283–5289. [CrossRef]
86. Zaton, D.; Karamoula, A.; Strachan, G.J.; Storey, J.M.D.; Imrie, C.T.; Martinez-Felipe, A. Photo-driven effects in twist-bend nematic phases: Dynamic and memory response of liquid crystalline dimers. *J. Mol. Liq.* **2021**, *344*, 117680. [CrossRef]
87. Aya, S.; Salamon, P.; Paterson, D.A.; Storey, J.M.D.; Imrie, C.T.; Araoka, F.; Jakli, A.; Buka, A. Fast-and-Giant Photorheological Effect in a Liquid Crystal Dimer. *Adv. Mater. Interfaces* **2019**, *6*, 1802032. [CrossRef]
88. Dawood, A.A.; Grossel, M.C.; Luckhurst, G.R.; Richardson, R.M.; Timimi, B.A.; Wells, N.J.; Yousif, Y.Z. Twist-bend nematics, liquid crystal dimers, structure-property relations. *Liq. Cryst.* **2017**, *44*, 106–126.
89. Mandle, R.J.; Goodby, J.W. Intercalated soft-crystalline mesophase exhibited by an unsymmetrical twist-bend nematogen. *Crystengcomm* **2016**, *18*, 8794–8802. [CrossRef]
90. Mandle, R.J.; Goodby, J.W. A twist-bend nematic to an intercalated, anticlinic, biaxial phase transition in liquid crystal bimesogens. *Soft Matter.* **2016**, *12*, 1436–1443. [CrossRef]
91. Pociecha, D.; Vaupotic, N.; Majewska, M.; Cruickshank, E.; Walker, R.; Storey, J.M.D.; Imrie, C.T.; Wang, C.; Gorecka, E. Photonic Bandgap in Achiral Liquid Crystals-A Twist on a Twist. *Adv. Mater.* **2021**, *33*, 2103288. [CrossRef] [PubMed]

92. Link, D.R.; Natale, G.; Shao, R.; Maclennan, J.E.; Clark, N.A.; Korblova, E.; Walba, D.M. Spontaneous formation of macroscopic chiral domains in a fluid smectic phase of achiral molecules. *Science* **1997**, *278*, 1924–1927. [CrossRef] [PubMed]
93. Matraszek, J.; Topnani, N.; Vaupotic, N.; Takezoe, H.; Mieczkowski, J.; Pociecha, D.; Gorecka, E. Monolayer Filaments versus Multilayer Stacking of Bent-Core Molecules. *Angew Chem. Int. Ed.* **2016**, *55*, 3468–3472. [CrossRef] [PubMed]
94. Salamonczyk, M.; Vaupotic, N.; Pociecha, D.; Walker, R.; Storey, J.M.D.; Imrie, C.T.; Wang, C.; Zhu, C.H.; Gorecka, E. Multi-level chirality in liquid crystals formed by achiral molecules. *Nat. Commun.* **2019**, *10*, 1922. [CrossRef] [PubMed]
95. Salili, S.M.; Almeida, R.R.R.; Challa, P.K.; Sprunt, S.N.; Gleeson, J.T.; Jaklia, A. Spontaneously modulated chiral nematic structures of flexible bent-core liquid crystal dimers. *Liq. Cryst.* **2017**, *44*, 160–167. [CrossRef]
96. Takezoe, H.; Gorecka, E.; Cepic, M. Antiferroelectric liquid crystals: Interplay of simplicity and complexity. *Rev. Mod. Phys.* **2010**, *82*, 897–937. [CrossRef]
97. Mach, P.; Pindak, R.; Levelut, A.M.; Barois, P.; Nguyen, H.T.; Huang, C.C.; Furenlid, L. Structural characterization of various chiral smectic-C phases by resonant X-ray scattering. *Phys. Rev. Lett.* **1998**, *81*, 1015–1018. [CrossRef]
98. Tschierske, C. The Magic 4-Cyanoresocinols-Their Role in the Understanding of Phenomena at the Rod-Banana Cross-Over and Relations to Twist-Bend Phases and Other Newly Emerging LC Phase Types. *Liq. Cryst.* **2022**, *49*, 1043–1077. [CrossRef]
99. Sreenilayam, S.P.; Panarin, Y.P.; Vij, J.K.; Panov, V.P.; Lehmann, A.; Poppe, M.; Prehm, M.; Tschierske, C. Spontaneous helix formation in non-chiral bent-core liquid crystals with fast linear electro-optic effect. *Nat. Commun.* **2016**, *7*, 11369. [CrossRef]
100. Sreenilayam, S.P.; Panarin, Y.P.; Vij, J.K.; Lehmann, A.; Poppe, M.; Tschierske, C. Development of ferroelectricity in the smectic phases of 4-cyanoresorcinol derived achiral bent-core liquid crystals with long terminal alkyl chains. *Phys. Rev. Mater.* **2017**, *1*, 035604. [CrossRef]
101. Sekine, T.; Niori, T.; Watanabe, J.; Furukawa, T.; Choi, S.W.; Takezoe, H. Spontaneous helix formation in smectic liquid crystals comprising achiral molecules. *J. Mater. Chem.* **1997**, *7*, 1307–1309. [CrossRef]
102. Cady, A.; Pitney, J.A.; Pindak, R.; Matkin, L.S.; Watson, S.J.; Gleeson, H.F.; Cluzeau, P.; Barois, P.; Levelut, A.M.; Caliebe, W.; et al. Orientational ordering in the chiral smectic-C-F12* liquid crystal phase determined by resonant polarized x-ray diffraction. *Phys. Rev. E* **2001**, *64*, 050702. [CrossRef] [PubMed]
103. Cruickshank, E.; Anderson, K.; Storey, J.M.D.; Imrie, C.T.; Gorecka, E.; Pociecha, D.; Makal, A.; Majewska, M.M. Helical phases assembled from achiral molecules: Twist-bend nematic and helical filamentary B-4 phases formed by mesogenic dimers. *J. Mol. Liq.* **2022**, *346*, 118180. [CrossRef]
104. Alshammari, A.F.; Pociecha, D.; Walker, R.; Storey, J.M.D.; Gorecka, E.; Imrie, C.T. New patterns of twist-bend liquid crystal phase behaviour: The synthesis and characterisation of the 1-(4-cyanobiphenyl-4′-yl)-10-(4-alkylaniline-benzylidene-4′-oxy)decanes (CB10O.m). *Soft Matter.* **2022**, *18*, 4679–4688. [CrossRef]

Review

Steroid-Based Liquid Crystalline Polymers: Responsive and Biocompatible Materials of the Future

Bartlomiej Czubak [1], Nicholas J. Warren [2] and Mamatha Nagaraj [1,*]

[1] School of Physics and Astronomy, University of Leeds, Leeds LS2 9JT, UK; pybcz@leeds.ac.uk
[2] School of Chemical and Process Engineering, University of Leeds, Leeds LS2 9JT, UK; n.warren@leeds.ac.uk
* Correspondence: M.Nagaraj@leeds.ac.uk; Tel.: +44-(0)-113-343-8475

Abstract: Steroid-based liquid crystal polymers and co-polymers have come a long way, with new and significant advances being made every year. This paper reviews some of the recent key developments in steroid-based liquid crystal polymers and co-polymers. It covers the structure–property relationship between cholesterol and sterol-based compounds and their corresponding polymers, and the influence of chemical structure and synthesis conditions on the liquid crystalline behaviour. An overview of the nature of self-assembly of these materials in solvents and through polymerisation is given. The role of liquid crystalline properties in the applications of these materials, in the creation of nano-objects, drug delivery and biomedicine and photonic and electronic devices, is discussed.

Keywords: liquid crystal; polymer; cholesterol; block copolymer; self-assembly; polymerisation-induced self-assembly

Citation: Czubak, B.; Warren, N.J.; Nagaraj, M. Steroid-Based Liquid Crystalline Polymers: Responsive and Biocompatible Materials of the Future. *Crystals* **2022**, *12*, 1000. https://doi.org/10.3390/cryst12071000

Academic Editor: Ingo Dierking

Received: 1 June 2022
Accepted: 12 July 2022
Published: 19 July 2022

Publisher's Note: MDPI stays neutral with regard to jurisdictional claims in published maps and institutional affiliations.

Copyright: © 2022 by the authors. Licensee MDPI, Basel, Switzerland. This article is an open access article distributed under the terms and conditions of the Creative Commons Attribution (CC BY) license (https://creativecommons.org/licenses/by/4.0/).

1. Introduction

In 1888, botanist Freidrich Reinitzer observed an unexpected double melting point in a crystal, cholesteryl benzoate, collected from the root of a carrot [1]. The material, cholesteryl benzoate, was the first liquid crystal and its discovery marks the start of the modern liquid crystal research. Over the last century, liquid crystals (LCs) have contributed to some of the greatest technological leaps, such as flat-screen televisions, mobile phones and sensors [2–9]. Whilst cholesterol by itself is not liquid crystalline, a number of its derivatives are; most notably, esters of cholesterol [10–14]. These compounds, due to the inherent chiral nature of sterols, form a chiral nematic (N*) phase. Cholesterol plays an important role within biological systems. It is an essential component of all animal cells. It controls the permeability of the cell membrane and plays a key part in the formation of lipid rafts [15].

Liquid crystal polymers are higher-molecular-weight liquid crystals that exhibit mesomorphic behaviour. They have been extensively investigated for their fundamental properties and as thermoplastics, photochromic, semiconducting, etc., materials [16–21]. Even though the N* phase observed back in 1888 was in low-molecular-weight liquid crystals (LMWLC), it was hypothesised that this phase could be observed in polymers or other higher-molecular weight liquid crystals (HMWLCs). Early work on liquid crystalline polymers forming the N* phase was performed with cholesterol and its derivatives [22]. The resulting polymers, however, did not form the N* phase, unlike their corresponding monomers; instead, they formed higher ordered smectic phases [23,24]. In 1978, Finkelmann et al., prepared the first enantiotropic cholesteric polymer by polymerising mixtures of cholesterol-based monomers with short and long spacers [25]. Expanding on their earlier work on liquid crystalline polymer synthesis, Fineklmann then proceeded to develop a polymer series with an N* phase, moving on to describe the first polymer with a biaxial N* phase [26]. The key step in this process was decoupling the motion of the polymer backbone from the mesogenic group via a 'spacer' group between the two.

There is a wide arsenal of synthetic approaches for the preparation of monomers based on cholesterol and other steroids (Figure 1). However, the liquid crystalline order in their corresponding polymers is highly influenced by the properties of the polymer, such as its molar mass dispersity, degree of polymerisation and, in the case of block copolymers (Figure 1A), the weight ratio of each of the blocks. Work by Shibaev in 1979 showed an ambiguous phase formation in methacrylic copolymers bearing cholesterol and butylmethacrylate [27]. While they displayed textures and a selective reflection similar to those observed in an N* phase in LMWLCs in bulk, these textures were not observed in thin films. In thin film they instead showed textures similar to those of standard smectics. Another example of a polymer forming higher-order structures compared to the corresponding monomers was observed by Xu et al. [28]. They found that the N* phase was only observed in monomers, but not in polymers bearing the same mesogen. The polymers instead formed a smectic A (SmA) phase due to the limited flexibility of the backbone and spacer units hindering the formation of the helix of the N* phase.

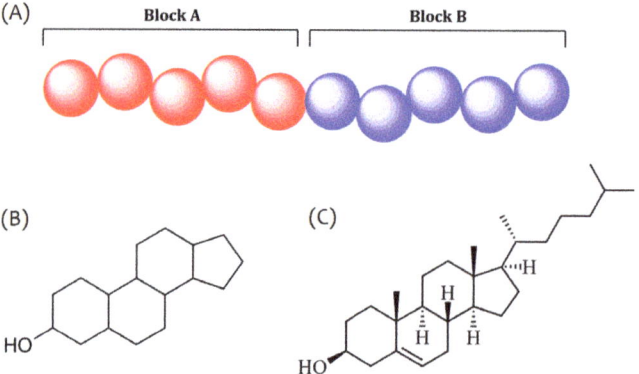

Figure 1. (**A**) Cartoon representation of a block copolymer. The repeating units are bound together by a covalent bond resulting in an ampiphilic molecule. (**B**) Chemical structure of generic sterol and (**C**) cholesterol.

2. Liquid Crystalline Polymers Bearing Sterol Side Groups

Synthesis and Structure–Property Relationships

Homopolymers bearing side-chain cholesteryl mesogens tend to form smectic phases (predominantely SmA) over a wide range of temperatures, with only a selected few showing both chiral nematic and smectic phases [29]. This is due to the choice of the chemical structures of the spacer groups and backbone. As the cholesteryl molecule is intrinsically chiral, it only forms chiral phases in polymers when the mesogen is free to do so. Initial preparation of the homopolymers bearing side-chain cholesteryl mesogens was achieved using free radical polymerisation. However, this led to the poor dispersity of the polymer, which hindered the formation of the mesophases. Nowadays, reversible-deactivation radical polymerisation (RDRP) reactions are employed, which offer a much more sophisticated approach to the preparation of the polymers. Methods such as reversible addition-fragmentation chain-transfer polymerisation (RAFT) [30,31] and atom transfer radical polymerisation (ATRP) [32,33] are commonly used as they offer tighter control over the degree of polymerisation and narrow polydispersity over the resulting polymer.

The nature and, hence, the influence of the flexible spacer on the polymers is a key parameter that controls the phases and transition temperatures. It does so by decoupling the mesogenic units and allowing them to freely form the mesophases. Both the type of the spacer and the length of the spacer are important [34]. Additionally, the stability and formation of the N* phase in polymers can be affected even by a small change in the chemical structure of the polymer.

Finkelmann et al. also prepared an enantiotropic liquid crystalline polymer with an N* phase by mixing different equal parts monomers with short spacers and longer spacers based on alkylbenzene ester [35]. Whilst this was not a homopolymer, it paved the way for a better understanding of the influence of the spacer group on the resulting mesophases. The influence of the spacer on the properties of the polymer was further demonstrated by Hu et al. [36], where they showed that longer spacer length lead to lower phase transition temperatures and wider LC temperature ranges in polymers bearing cholesteryl side groups. The selective reflection of the polymers was also shown to blue shift as the spacer lengths were increased or the rigidity of the mesogens decreased. This was further explored by Yang et al., who showed that the degree of polymerisation of the polymer does not have a significant effect on the polymers without flexible spacers and with a stiffer methacrylic backbone [37]. However, it does have an effect on the more flexible acrylic backbone where the LC behaviour is observed only upon passing a critical threshold of molecular weight of 12×10^3 gmol^{-1}. By switching the spacer from an ether (Figure 2A) to an equivalent ester, (Figure 2B) Yang et al. demonstrated that the transition temperature to the N* mesophase shifted from $-27\,°C$ in the ether linker to $72\,°C$ in the ester case whilst having little impact on the clearing temperature [34]. This switching of the spacer from an ether to an equivalent ester also caused the polymer to display crystallisation rather than a glass transition. A much more dramatic change was observed when the second ether in the linker was replaced with another ester group (Figure 2C). In this example, the polymer only displayed a monotropic smectic phase and no N* phase was observed. For the N* phase to form, the spacer length had to be increased to n = 10. This arises from the relative stiffness of the C=O bond compared to C-O, which hinders the freedom of the mesogen to order itself. Klok et al. showed that the covalent incorporation of cholesterol moiety to a low-molecular-weight L-lactic acid oligomers resulted in the formation of thermotropic liquid crystal smectic phases [38]. While the mesogen was a small part of the overall molecule, its strong tendency to form the mesophase was expressed in the larger molecule.

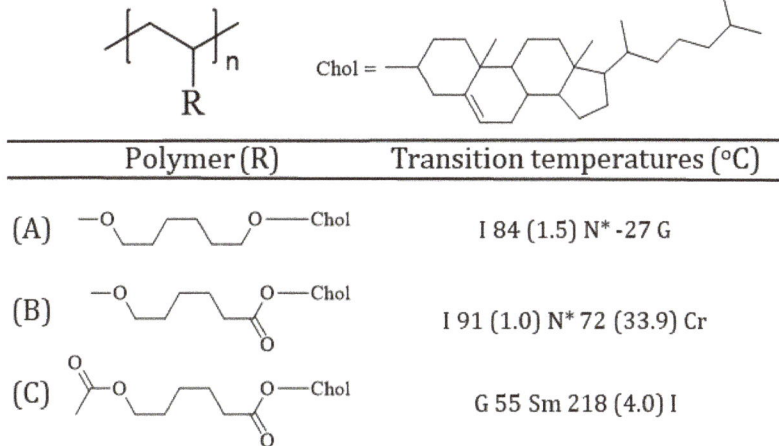

Figure 2. An example of the influence of chemical differences in the spacer connecting the backbone and the cholesteryl mesogen, on the liquid crystallinity of the material. The introduction of each of the ester groups (A–C) gradually increases the relative stiffness of the backbone. N*—chiral nematic, I—isotropic, G—glass, Cr—crystal and Sm—smectic.

A number of studies have been carried out on block-copolymers (BCPs) bearing side-chain cholesteryl mesogens. A BCP bearing a chiral but non-mesogenic menthol group and a cholesterol mesogen was prepared by Wang et al. [39] The introduction of the non-mesogenic unit resulted in the lowering of the glass transition temperature, T_g but led to a narrower mesophase temperature range. The corresponding cholesterol homopolymer did not form the N* phase, and instead showed a chiral smectic phase. Only upon the addition of chiral menthol, an N* phase was observed in the BCPs. This, however, lead to a significant decrease in the temperature range at which the mesophase was observed, from 145 °C to 24 °C, resulting in limited applications of this polymer system. The self-assembly behaviour of block copolymers has been a topic of great interest due to their application potential. Therefore, in the next section, the structure–property relationship of BCPs and their synthesis methods are covered while discussing their applications.

3. Applications of Sterol-Based Liquid Crystalline Block Copolymers

3.1. Self-Assembly Behaviour: Creation of Micelles and Nano-Objects

Block copolymers (BCPs) are well known to self-assemble in bulk and in solution due to their intrinsic amphiphilic properties. Hamley et al. studied a series of block copolymers based on polystyrene and a methacrylic cholesterol monomer [32]. They reported the formation of a stable SmA phase within all of the polymers with varied LC content. The copolymers were probed using small-angle X-ray scattering (SAXS) where multiple scattering peaks were observed. The smectic A monolayer had a distinct and sharp peak at approx. d = 2.1 nm with a more diffuse peak observed at temperatures below the smectic transition. These lower-intensity peaks corresponded to smectic-like ordering with interdigitating groups where the sample was unable to align as a result of the partial crystallisation of the material. Venkataraman et al. prepared a series of disk-like micelles and stacked columns of the micelles from mPEG$_{113}$-b-P(MTC-Chol)$_{11}$ block copolymer [40]. Figure 3 shows the transmission electron microscopy (TEM) images of aqueous self-assembled nanostructures formed by mPEG$_{113}$-b-P(MTC-Chol)$_{11}$. It shows the various orientations of the stacked structures, including face-on, intermediate ellipsoidal, edge-on and stacked-disk. This unique formation of micelles was attributed to the self-association of the cholesterol moieties in smectic order. Depending on the percentage of the LC content, the disks displayed a unique ability to stack perpendicular to the major axis in a process driven by the narrow polydispersity (PD) of the copolymer chains, combined with the liquid crystalline order as a driving force.

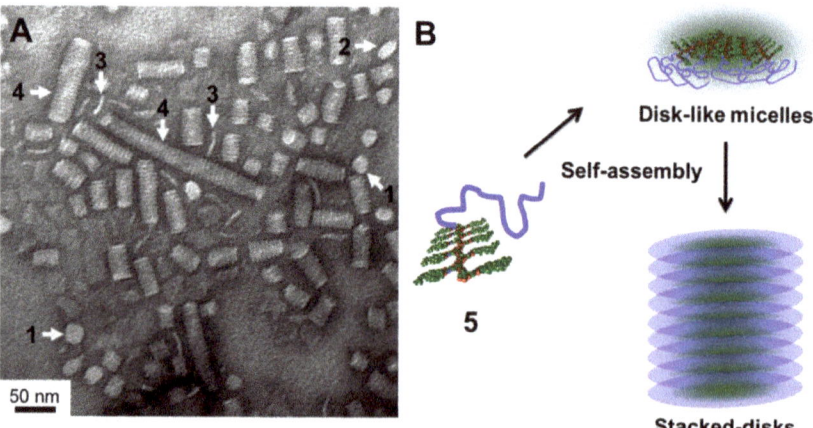

Figure 3. (**A**) TEM micrographs of stained aqueous self-assembled nanostructures formed by mPEG$_{113}$-b-P(MTC-Chol)$_{11}$. Arrows 1–3 show the face-on, intermediate ellipsoidal, and edge-on orientations of the structures, respectively; arrow 4 shows the stacked-disk orientation (**B**) Schematic representation of the mechanism driving the formation of stacked-disk micelles. Reprinted with permission from Venkataraman et al., *Macromolecules*, **2013**, *46*, 4839–4846 [40].

The morphology of BCP nano-objects as a function of pH was investigated by Guo et al. [41]. for a diosgenyl functionalised block copolymer, mPEG$_{43}$-PMCC$_{25}$-P(MCC-DHO)$_1$. Under basic pH conditions, the polymer formed nanospheres. However, a transition to nanofibres was facilitated by acidifying the solution (Figure 4). This transition was due to the ionization of the carboxylic groups, which subsequently repelled each other to promote an increase in the preferred curvature of the interface. Interestingly, the size of the nanospheres was directly correlated with the pH. Particles prepared at pH = 2 had a radius of 324 nm and those prepared at pH = 6 had a radius of 232 nm.

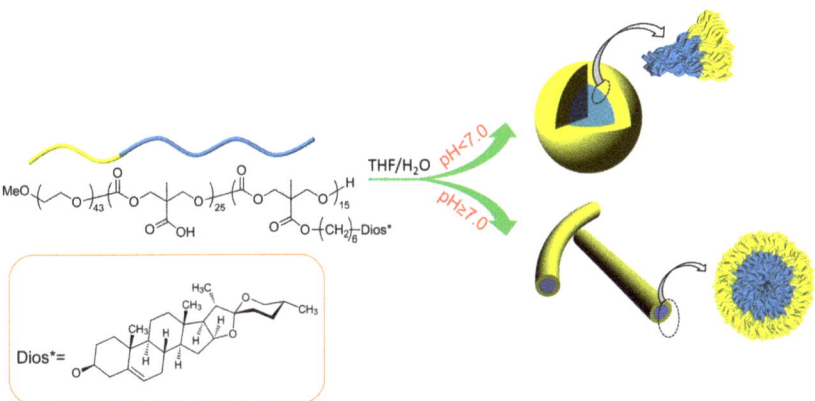

Figure 4. Self-assembly behaviour of mPEG$_{43}$-PMCC$_{25}$-P(MCC-DHO)$_1$ as a function of pH in tetrahydrofuran/water. Reprinted with permission from Guo et al., *Nanomaterials*, **2017**, *7*, 169 [41].

A comprehensive study on the self-assembly of a cholesterol-functionalised liquid crystalline block copolymer of poly(cholesteryl methacryloyloxy ethyl carbonate)-b-(poly (ethylene glycol)methyl ether methacrylate), (PChEMA$_m$-b-POEGMA$_n$), in different solvents was carried out by Li et al. [42]. By controlling the weight fraction of the LC block, a series of structures was observed (Figure 5). The two factors that dictated the self-assembly, the smectic order within the LC block and the ampiphilc nature of the copolymer were expressed differently depending on the solvent. The LC order was found to be absent in a tetrahydrofuran(THF)/water system, which indicates that the ampiphilicity plays an important role in the self-assembly process. In the 1,4-dioxane/water system, a smectic order was observed within micelles, indicating that the LC order plays a more important role in the growth and morphology of the micelles. This change in behaviour was explained by the difference in the affinity between the cholesteryl block and the solvents. The Flory–Huggins interaction parameter (X) between the PChol and THF is 0.63, whereas between PChol and dioxane it is 0.92. This implies that THF is a much better solvent for the polymer, which results in a reduced driving force for self-assembly and an absence of LC order. A transition between smectic nanofibers and smectic ellipsoidal vesicles was observed for a cholesteryl-containing polymer with a biodegradable polytrimethylenecarbonate backbone (Figure 5). These structures assembled into rods under dioxane, but reoriented from lamellar to ellipsoidal vesicles upon the addition of water.

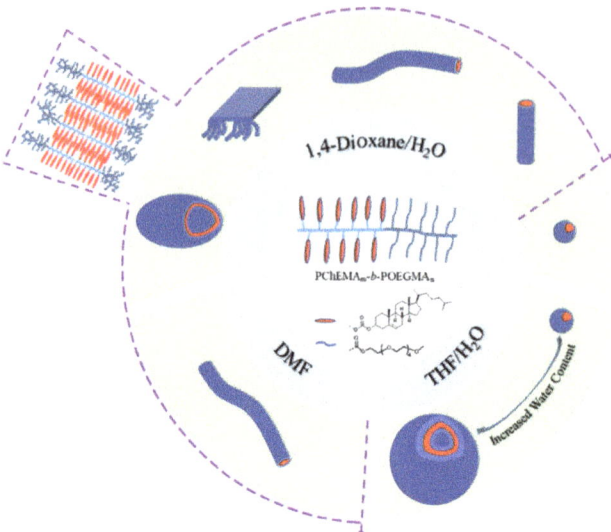

Figure 5. The adaptable nature of the cholesterol-bearing diblock BCP allows it to self-assemble into a plethora of structures in different solvents and concentrations. Reprinted with permission from Li et al., *Langmuir*, **2018**, *34*, 11034–11041 [42].

Krishnasam et al. investigated the reversible inversion of chirality in block copolymers of poly(methyl methacrylate) and poly(cholesteryloxyhexyl methacrylate) [43]. Chirality switching is potentially a useful method for the application of block copolymer in novel electronic and photonic materials, especially when it happens in response to a thermal change [44–46]. This system of copolymers of poly(methyl methacrylate) and poly(cholesteryloxyhexyl methacrylate) was the first reported system capable of switching chirality without chiral dopants, making it additive-free. The inversion of chirality was also observed in polymers bearing shorter poly(cholesteryloxyhexyl methacrylate) (PChMA) chains, which posses syndiotacticity (repeating units have alternating stereochemical configurations). The combination seem to lead to tightly packed side chains where the inversion of chirality overcomes the energy penalty at lower temperature. Figure 6 shows the inversion of chirality observed in PChMA-b-PMMA BCP. The graphs show the plot of ellipticity measured as a function of wavelength in chiral SmA and SmC phases, measured through circular dichroic (CD) spectroscopy.

The self-assembly behaviour in aqueous solutions of a cholesterol-bearing block copolymer PP_2-g-LC_4 with a flexible hydrophobic backbone was studied by Yang et al. [47]. The BCP consisted of poly(ethylene oxide) block and a mesogenic cholesteryl pendant group. One of the structures observed was a hollow concentric spherical vesicle with SmA order in the shells. These structures exhibited an interdigitating smectic A phase, with LC mesogens grafted onto a backbone of poly-(ethylene oxide)-block-polybutadiene via the "thiol-ene" radical addition reaction. This also marks the first time a spherical arrangement was observed for copolymers comprising a polybutadiene backbone and an LC group. Previously, ellipsoidal or faceted structures were observed [48,49].

A biodegradable BCP bearing cholesteryl mesogen was prepared by Zhou et al. via ring-opening polymerisation [50]. After solvent switching from dioxane to water, nanofibres were obtained (Figure 7). The vesicular wall possessed a smectic order, leading to uniform wall thickness. The nanofibres then, upon contact with each other, formed lamellaes that further transformed into ellipsoidal vesicles.

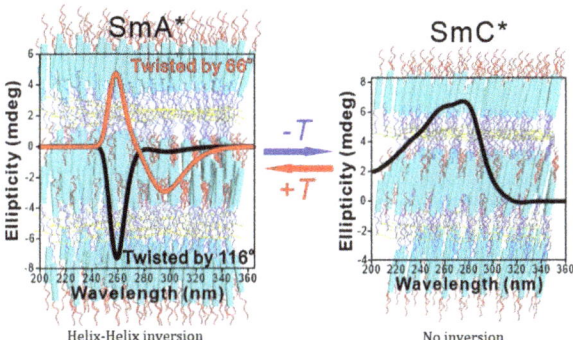

Figure 6. Inversion of chirality observed in PChMA-b-PMMA BCP measured through circular dichroism (CD) experiments. The graphs show the plots of ellipticity, measured as a function of wavelength in chiral SmA and SmC phases. The blue rods represent the LC mesogens arranged in layer structures. T represents the temperature. Reprinted with permission from Krishnasamy et al., *Macromolecules*, **2020**, *53*, 4193–4203 [43].

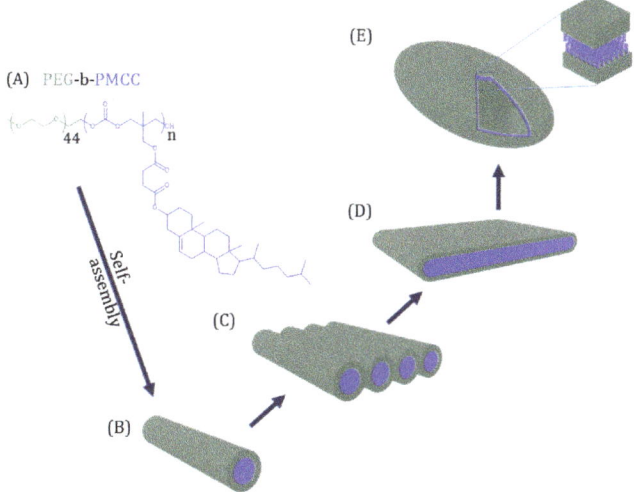

Figure 7. The morphological evolution of self-assembled nanostructures observed for PEG-b-PMCC BCP. (**A**) Chemical structure of PEG-b-PMCC BCP. The schematic representation of (**B**) the initial nanofibres formed, (**C**) the fusion of nanofibres, (**D**) the formation of lamella structure from fused nanofibres and (**E**) the rearrangement of lamellar structures into elliposidal vesicles with LC order within the membrane.

A unique morphology of self-assembled nanostructures was reported by Jia et al. [51]. Nanoribbons bearing smectic stripes were formed upon nanoprecipitation into dioxane (Figure 8). These ribbons were flexible enough to form small creases. Increasing the initial concentration of the BCP during the nanoprecipitation lead to the formation of wider ribbons with more folds. Figure 8 shows TEM images of these nanoribbons for different polymer concentrations.

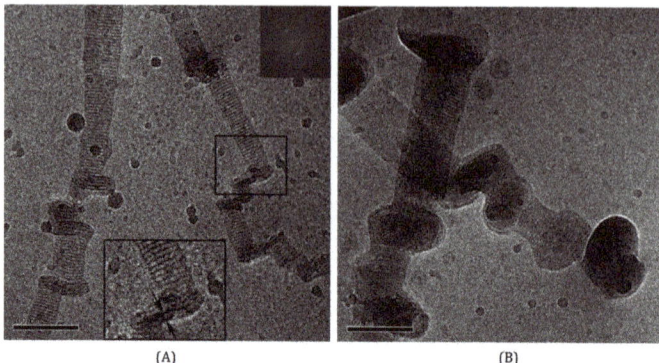

Figure 8. Nanoribbons formed by PEG$_{45}$-b-PCpEChol$_3$. (**A**) Starting polymer concentration = 0.1 wt%; (**B**) starting concentration = 0.5 wt%. Note that the smectic stripes formed at the sides of the ribbons. Reprinted with permission from Jia et al., *Langmuir*, **2012**, *28*, 11215–11224 [51].

A metal-mediated self-assembly of hydrophilic BCPs was developed as a method for the facile preparation of organic–inorganic hybrid nanomaterials under aqueous conditions. Jeong et al. demonstrated that the liquid crystalline order that a cholesterol mesogen contributes to a chelating double-hydrophilic BCP offers a way of controlling chelating sites in nanoparticles [52]. This was a novel method for creating the density modulation of nanoparticles which can be exploited for controlled drug delivery applications.

Jia et al. studied the the self-assembly of a PEG-PAChol diblock BCP system in THF/water and obtained monodisperse spherical aggregates with no obvious features [53]. However, the same BCP in a dioxane/water system formed higher-ordered structures, where the smectic order of the cholesterol block was characterised by a striped pattern observed in TEM images (Figure 9). Due to the highly hydrophobic nature of cholesterol mesogen, the self-assembly of BCPs bearing it did not tend to result in complex structures, instead forming spherical structures. On the contrary, the reduced polarity of dioxane/water mixtures had a tendency to plasticise the particle core, providing enhanced mobility to facilitate reorganisation into more complex structures, including ellipsoids and nanocylinders.

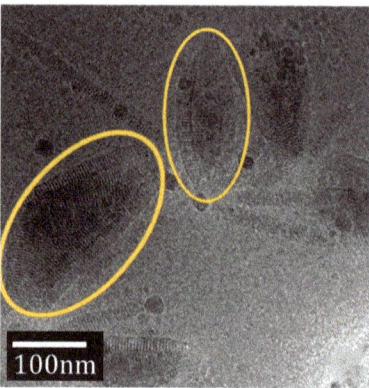

Figure 9. Self-assembled structures of PEG$_{114}$-b-PAChol$_{60}$ prepared in dioxane/water (14:86) showing the notable smectic stripes decorated on nanocylinders and ellipsoids (circled). Reprinted with permission from Jia et al., *Polymer*, **2011**, *52*, 2565–2575 [53].

Boisse et al. [30] showed different possible structures formed by a cholesterol-bearing ampiphilic BCP with a poly(N,N-diethylacrylamide) hydrophilic block. The polymers predominantly formed 1D long fibres (Figure 10A–C). A comparative polymer without LC side groups was shown to instead assemble into vesicles and short cylindrical micelles (Figure 10D), demonstrating the important influence of LC order on the self-assembly of BCPs.

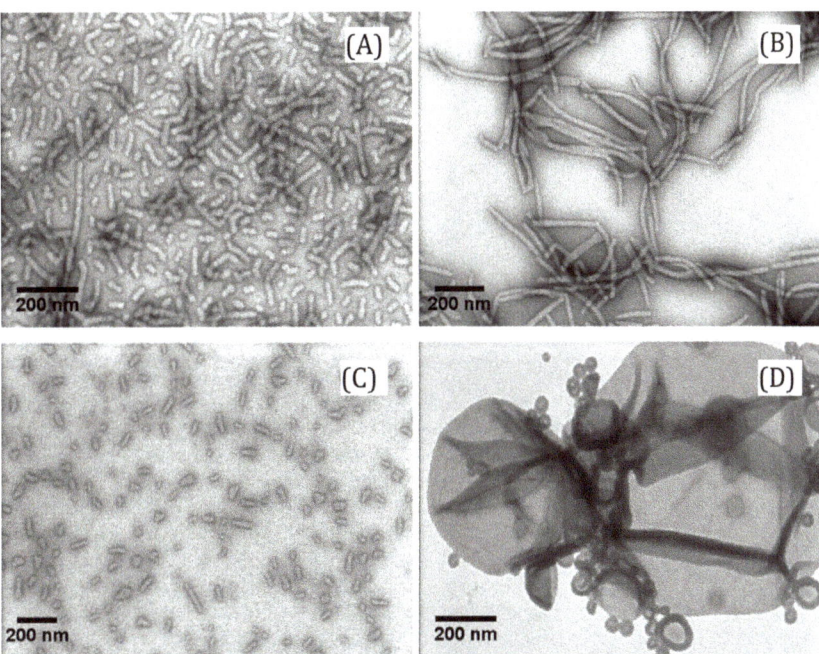

Figure 10. TEM micrographs of self-assemblies of (**A**–**C**) PMAChol-bPDEAAm at different degrees of polymerisation (DP) of the hydrophilic block. (**A**) DP = 103, (**B**) = 30, (**C**) = 21. (**D**) PS$_{136}$-b-PDEAAm$_{32}$. Reprinted with permission from Boissé et al., Macromolecules 2009, 42, 22, 8688–8696 [30].

3.1.1. Polymerisation-Induced Self Assembly

Polymerisation-induced self-assembly (PISA) provides a convenient route for preparing block copolymer nanostructures with high fidelity and tunable functionality. During the PISA process, a soluble precursor polymer is chain extended with a monomer that forms a solvophobic block, resulting in an amphiphilic polymer, which self-assembles to avoid unfavourable interaction with the solvent (Figure 11). It has been widely used to prepare nanoparticles of various morphologies, from simple spherical micelles, worms and vesicles, through to more exotic structures such as ellipsoidal micelles and faceted cuboids. These sorts of structures also frequently exhibit light-, pH- or temperature-responsive transitions between morphologies. While PISA with groups bearing LC mesogens is a relatively under-explored area of research as a whole, it was in fact reported in one of the earliest known PISA papers by Zhang et al. [54], who polymerised CholA monomer ion in the presence of a P(AA-co-PEGA) macromolecular chain transfer agent in an ethanol/water (95:5) solvent to produce poly(acrylic acid-co-poly(ethylene glycol) monomethyl ether acrylate)-b-poly(cholesteryl acryloyloxyethyl carbonate).

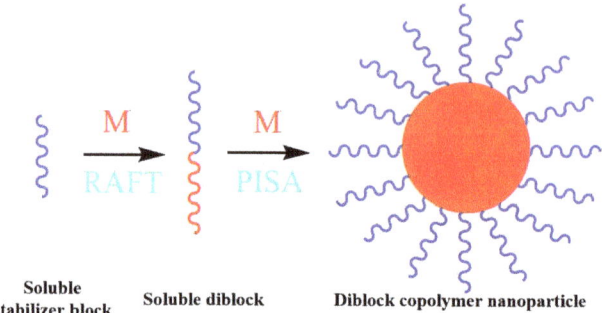

Figure 11. Schematic representation of the self-assembly process in polymerisation-induced self-assembly (PISA).

The result was a dispersion of nanocylinders (see Figure 3 in reference [54]) with internal smectic-type order over a broad range of solid concentrations and relative block ratios. The radius of the nanocylinders increased as a function of the chain length of the hydrophobic block. In this case, the LC order was assumed to play a significant role in the assembly formation alongside the traditional driving forces. The study showed that the LC order was preserved and the introduction of it to the system led to the preferential formation of nanofibers over vesicles, observed in an equivalent polystyrene system. An advantage observed upon the introduction of a LC group was the reduction in the polymer dispersity (Đ) with higher LC content. While few PISA studies have been performed with steroid-based monomers, the field is starting to gain traction with recent publications focused on the introduction of perfluorinated LC, monomers [55–60], biphenyls [61] and azobenzenes [62–64]. The structures seen within these systems show unique properties such as physical rearrangements upon thermal stimuli. Analogous to systems mentioned earlier, the linker length plays an important role in determining the resulting PISA morphology, with a biphenyl mesogen providing insights into the mechanism underpinning the impact of LC groups within PISA [61]. Longer linkers lead to slower polymerisation kinetics and only the longest (C = 11) displayed direct liquid crystalline behaviour with smectic order in the nanorods. The linker also plays a key role in the resulting morphology of the PISA structures. The shortest linker (C = 0) formed spheres and sphere aggregates, the intermediate linker (C = 6) led to spheres and worms and the longest (C = 11) led to the formation of rigid nanorods/nanowires [63].

3.1.2. Cholesteryl Polymer-Based Nano-Objects Showing Aggregation-Induced Emission

Triblock BCP bearing a PEG group, a cholesteryl acrylate block and a tetraphenylethene AIEgen (aggregation-induced emission material) side group were synthesised by Zham et al. [65]. The BCP was then self-assembled in a mixture of either dioxane/water or THF/water. Each of the systems self-assembled into a range of nanostructures. The dioxane/water particles formed ellipsodial vesicles with a constant membrane thickness despite heterogeneity in the overall vesicle diameter. In standard, non-LC PISA systems, the membrane thickness of vesicles is known to be dictated by the solvophobic chain length, where the chains reside in an intermediate state between collapsed globules and fully stretched chains [66,67]. In LC polymers, it has been shown that the membrane formed is a bilayer of fully stretched PAChol groups laying along the normal edge of the membrane. The nanoparticles formed in THF/water were instead spherical and more monodisperse compared to dioxane/water vesicles. The unique combination of LC and AIE properties was probed to see their effect on each other. The LC order arising from the cholesteryl mesogen was undisturbed by either TPE (tetraphenylethylene) or PEG as similar LC behaviour was observed in homopolymers of the cholesteryl acrylate. Both nanospheres and vesicles presented AIE properties with a relatively low quantum yield of 0.74% and 0.76%, respectively. Another triblock BCP bearing tert-Butyl methacrylate (tBA), tetraphenylethylene (TPE) and methacrylic cholesteryl mesogen was studied by Wang et al., [68] in relation to the AIE properties. In this system, the AIE group was introduced to the corona-forming block. While the TPE was still partially dissolved, strong AIE behaviour was observed after micellisation.

3.2. Biomedical Applications

Artificial tissue regeneration requires a biocompatible 3D scaffold. The incorporation of liquid crystal order within the scaffold can have the beneficial effect of guiding the direction of cell growth [69]. Particularly, the incorporation of biocompatible components such as cholesterol expands the potential applications for medical devices both in vivo and in vitro. A cholesteryl-bearing BCP has been shown to improve the adhesion and spreading of fibroblast tissue while remaining biocompatible. While the definitive mechanism is not fully understood, the improvement was accredited to the smectic liquid crystal order which helped direct the cells [70]. A triblock BCP was investigated as a potential candidate to use as implant material for cellular scaffolding for dynamic organs such as heart or blood vessels.

A cholesterol-based elastomer was prepared for use in artificial tissue regeneration. The material was made porous by the salt-leaching method in order to promote cell infiltration and promote growth [71]. The SmA order found in the material acted as a template for the pore preparation. These properties made it a suitable candidate for the in vitro growth of neuronal tissue. Xie et al. [72] studied the cytotoxicity of cholesterol-based elastomers. In particular, they focused on the possibility of using the elastomer as a suitable substrate for cell cultures. The results, with glass substrates acting as a control, show that unpolymerised samples are toxic to the cell cultures; however, polymers display high biocompatibility with the cultures. Additionally, due to biocompatibility of the substrate, the cells had better spread and morphology and larger diffusion areas, making these polymers highly suitable for tissue growth and regeneration. Figure 12 shows a representation of the biocompatibilty of the cholesterol-based elastomers with the tissue cultures and the resulting tissues samples. The LC content on the surface of the substrate helps the cells adhere, resulting in higher density of pseudopods.

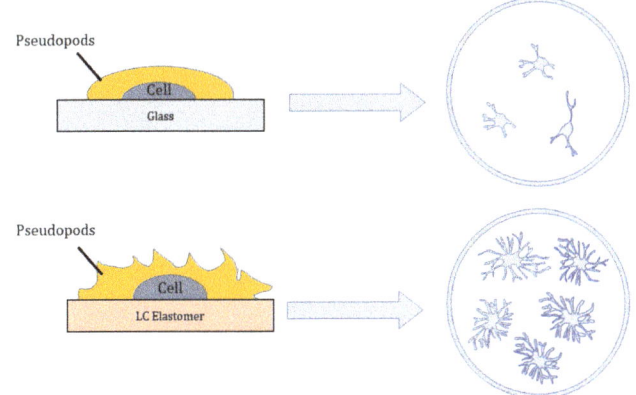

Figure 12. An illustration demonstrating the biocompatibilty of the cholesterol-based elastomers with the tissue cultures and the resulting tissue samples. The LC content on the surface of the substrate helps the cells adhere, which results in the higher density of pseudopods [72].

A room-temperature smectic elastomer bearing cholesterol mesogen was used for cell culture [73]. The liquid crystalline order rendered the elastomer softer relative to the liquid-crystal-free equivalent elastomer. The porosity, an important factor for cell growth, can be controlled by tuning the amount of cross-linker, and the liquid crystalline properties were controlled by attaching the cholesteryl to either the alpha or gamma position relative to the carboxyl group.

Liu et al. [74] investigated the micelles of copolymer bearing poly(ascrobyl acrylate) and sidechain cholesteryl mesogen for potential applications in the drug delivery of ibuprofen and Nile red dye. Polymers were prepared using RAFT polymerisation followed by desulfurisation and hydrogenolysis. The micelles were then prepared using dialysis over a course of two days. The drug delivery applications were studied as a function of loading the drug molecule and the entrapment efficiency. Nile red was studied as an example of a hydrophobic drug. It was found that having a longer hydrophobic (i.e., cholesterol)-based block helps increase the loading efficiency of the Nile red (Figure 13A) and ibuprofen, as well as increasing the entrapment efficiency of ibuprofen (Figure 13B). The increase in entrapment efficiency was significant (up to 40%) when compared to no polymer at all (5%) and another polymer studied for this application poly(ethylene oxide)-block-poly(lactic acid) (8–9%).

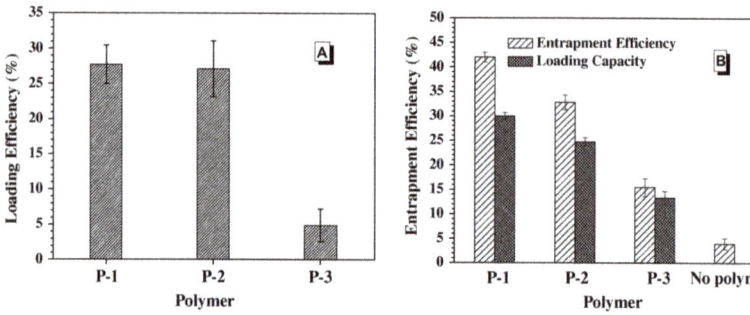

Figure 13. (A) Loading efficiency of Nile red. (B) Entrapment efficiency and loading efficiency of ibuprofen poly(lactide-co-glycolide)-cholesterol nanoparticles have also been tested for tumor-targeted delivery of curcumin through intravenous routes, which resulted in low cytotoxicity and high serum stability. P-1, P-2 and P-3 are polymers with gradually decreasing ratios of cholesterol content. Reprinted with permission from Liu et al., *Journal of Colloid and Interface Science*, **2012**, 377, 197–206 [74].

A series of BCPs bearing cholesteryl and PEG were prepared via metal-free organocatalyic ring opening polymerisation [75]. These were then self-assembled into micelles and loaded with paclitaxel, an anticancer drug notorious for its difficulty of delivery within aqueous systems. The loaded micelles showed high kinetic stability and demonstrated higher cytotoxicity compared with other delivery methods. Jia et al. [76] reported the development of robust polymersomes capable of controlled cargo release in response to the intercellular reducing environment, which led to the breakage of the disulfide bridge linking the hydrophobic PAChol block and hydrophilic PEG blocks. The system showed good loading capacity and efficiency of calcein, and in vitro studies demonstrated high (up to 80%) drug release upon trigger.

Aliphatic BCPs consisting of PEG blocks and a methacrylic cholesteryl group with a C_6 spacer have been shown to form both smectic A and N* phases [77]. These polymers also possessed a mesophase range much wider then their corresponding monomers. They also further demonstrated the influence of the spacer on the mesomorphic properties with longer spacers showing wider mesophase range. The mesophase range extended below body temperature, making them suitable for applications in site-specific drug delivery. The incorporation of a cholesterol-bearing sidegroup to a guandium-rich ampihipatic carbonate BCP was studied for its influence on the complexation to siRNA (small-interfering rinucleic acid). The cholesteryl group made it insoluble in phosphate-buffered saline (PBS), rendering it impractical to use for siRNA drug delivery [78].

PEG- and cholesterol-based triblock BCPs were prepared via attaching the cholesteryl pendant group to a polymer of fixed length [79]. This led to the formation of SmA phase, spreading to below body temperature. The BCP was then self-assembled by solvent switching into THF/water, leading to the formation of spherical nanoparticles, similar to those described in Section 3.1 above. The nanoparticles were then probed for their potential applications for the targeted delivery of doxorubicin (DOX) for use in vivo. The material displayed acceptable loading content and efficiency. Switching the pH from neutral to low promoted drug release [80].

Another promising BCP bearing cholesterol was studied for applications in the delivery of doxorubicin composed of polynorbonene-cholesterol/PEG [81]. It demonstrated no cytotoxicity by itself, but showed the steady release of DOX at 2% per day in PBS. It had additional positive effects, as it showed decreased cardiac accumulation of DOX compared to free DOX in vivo mice studies.

3.3. Photonic and Electronic Applications

Cholesterol-containing polymers, on top of their excellent biocompatibility, have shown promise as future materials in photonic and electronic applications. Table 1 summarises some of the recent developments within these fields.

Table 1. List of various cholesterol-containing polymers and their applications in photonic and electronic devices.

Polymer Type	Components	Synthesis Method	Application	References
Elastomers	Poly(methylsilylene), cholesterylcarbonate and a cross linker	Hydrosilylation	Lasing	[82–84]
Block copolymers	Photochromic spirooxazine and cholesterol mesogen	Free radical polymerisation	Data storage	[85–87]
Homopolymers and copolymers	Methacrylates bearing cholesterol pendant group and methly methacrylate	Atom transfer radical polymerisation (ATRP)	Organic field-effect transistor (OFET)	[88,89]
Cross-linked copolymer	Polybenzoxazine functionalised with cholesterol	Ring-opening polymerisation	High-thermal conductivity and heat-resistance	[90–92]
Block-copolymers	Azobenzene and cholesteryl mesogens	Ring-opening polymerisation	Photo-optical switching	[93,94]
Block-copolymers	Polyoxyethylene and cholesterol methacrylate	Reversible addition–fragmentation chain transfer (RAFT) polymerisation	Energy storage	[95]
Homopolymers	Polyphenylallenes with various spacers between backbone and cholesterol mesogen	Coordination polymerisation	Fluorescent sensors	[96–99]

As discussed in the Introduction, cholesterol-based liquid crystal polymers are well-known for their chiral nematic phase. Figure 14 shows the colour shift observed for a cholesteric liquid crystal elastomer as a function of strain. The colour change was observed due to the change in pitch of the elastomer. This polymer was then used for a low-threshold, mirrorless lasing [82] The lasing effect was observed after doping the elastomer network with a 4-(dicyanomethylene)-2-methyl-6-(4-dimethyl-amino styryl)-4-H-pyran (DCM), a well-known laser dye. It was shown that the laser line of narrow line width, approx. 3 Å, could be shifted by applying mechanical strain to the elastomer (Figure 14). Another potential application of chiral nematic-based cholesterol polymers is data storage. Hattori et al. demonstrated the controlled pitch change of an N* film upon exposure to UV irradiation [85]. Ndaya et al. [93] prepared a series of responsive copolymers based on azobenzene and cholesterol mesogens, which displayed a reversible change in their selective reflection, from near IR to visible light, as a function of temperature. This allows these materials to be employed as photo-optical switches.

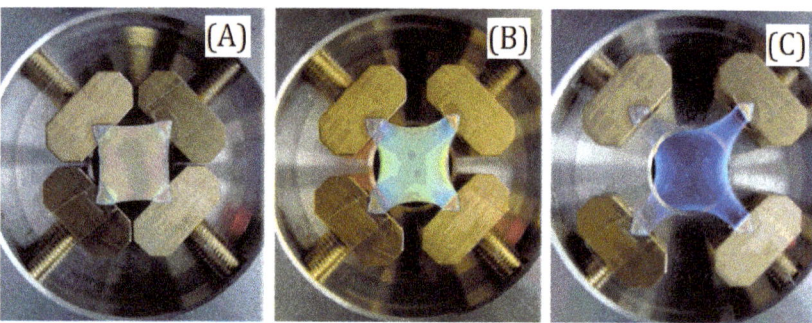

Figure 14. Colour shift observed for cholesteric liquid crystal elastomer as a function of strain. The colour change is observed due to change in the position of the reflection band of the elastomer as a function of mechanical deformation with unstrained sample (**A**) being green through to (**B**) light green, with blue in the corners, to strained sample (**C**), where the entire sample is blue. Reprinted with permission from Finkelmann et al., *Advanced Materials*, **2001**, *13*, 1069–1072.

Dognaci et al. [88] prepared a series of methacrylic homopolymers and copolymers bearing a cholesterol pendant group and methyl methacrylate. The Cole–Cole plot of most of the polymers prepared showed non-Debye relaxation behaviours, with the exception of two individual polymers. A homopolymer with a spacer length of 10 showed promise as a potential material for applications within organic photovoltaics, as it demonstrated DC conductivity behaviour at high frequencies.

A series of heat-resistant cross-linked polybenzoxazine-based polymers containing cholesterol mesogens was prepared by Liu et al. [90]. They demonstrated high thermal conductivity and showed thermal stability up to 300 °C, with a smectic range between 100 °C and 220 °C. Another potential application for a cholesterol-containing polymer is within solid-state energy storage and lithium ion batteries. Zhou et al. [95] prepared a series of copolymers of polyoxyethylene (PEO) and a methacrylic cholesteryl mesogen, which formed amorphous nanostructures of PEO at room temperature but maintained their mechanical properties at elevated temperatures. Shape memory polymers (SMPs) have a wide variety of applications, including withing the field of photonics in the design of functional and responsive gratings. Mahajan et al. [100] studied SMP based on side chain cholesterol mesogens. They observed reversible shape transformation after stretching the polymer at elevated temperature and quenching to room temperature (Figure 15A–C). Ahn et al. [101] demonstrated the memory effect within terpolymer and explored it within the context of the smectic order director in relation to the long axis of the film formed (Figure 15D).

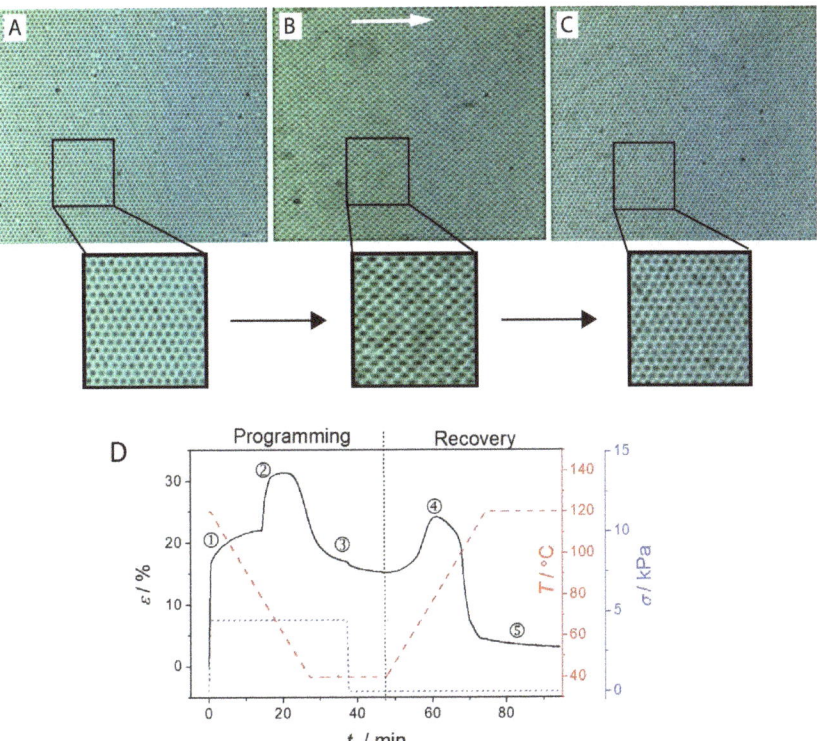

Figure 15. Shape memory effect (SMP) displayed by a cholesteric terpolymer film (**A**) stretched at 110 °C (**B**) quenched to room temperature and (**C**) recovering its original shape upon reheating back to 110 °C. (**D**) Shape memory effect of a film of a terpolymer and its subsequent recovery on heating shown as function of time. Red dashed line represents the response as a function of temperature, blue dotted line shows the strain applied and solid black line the strain response. Numbers 1 and 5 correspond to different isotropic states and 2 to 4 correspond to various SmA polymorphism change during shape memory effect. Reprinted with permission from Mahajan et al., *Macromolecules*, 2017, **50**, 5929–5939 and Ahn et al., *ACS Nano*, **2011**, *5*, 3085–3095 [100,101].

As mentioned in Section 3.1.2, cholesterol-containing polymers can be made to display aggregation-induced emission (AIE) behaviour, which can act as a fluorescent probe. Zhang et al. [96] showed that a series of hydrogen-bonding polymers, depending on the linker between the backbone and cholesteryl mesogen, displayed both aggregation-induced emission and aggregation-induced quenching (AIQ).

4. Conclusions and Outlook

Steroid-based liquid crystal polymers and co-polymers is a rapidly developing area of research with applications in a variety of fields. There are few reviews available on the topic of steroid-based polymers discussing the relevant chemical aspects and self-assembly capabilities [69,102–104]. In this review, cholesterol and related sterol-based polymers are described and their structure–property relationship is discussed, with particular emphasis on their liquid crystalline properties and their potential benefit in applications. These underlying principles, in combination with sterols' natural biocompatibility, make them excellent materials for use in drug delivery devices for anticancer, antimicrobial and anti-inflammatory drugs, as well as for imaging dyes. Their responsiveness has been utilised for controlled payload release in vivo with mammalian cells and in vitro. Their intrinsic liquid crystalline behaviour has demonstrated a positive impact on artificial tissue growth. Applications of LC cholesterol-based elastomers within biomedicine as scaffolds for tissue growth have only recently been explored, but show tremendous promise due to low cytotoxicity and good biocompatibility. This is a feature not commonly found in standard liquid crystalline polymers, which makes the cholesterol-based polymers more relevant as materials of the future.

The versatile self-assembly behaviour and responsiveness to changes in the system, including solvent exposure, light and temperature, make steroid-based liquid crystal polymers and copolymers prime candidates for sensing and imaging applications. Their uniform, 1D structures have been shown to adapt, making them suitable for applications within nanoelectronics as charge carriers for micro- and nano-sized devices. Their liquid crystalline properties allow them to self-assemble into stable structures otherwise unavailable with non liquid crystalline blocks. Future directions of research in this area promise novel medical and technological developments. Further expanding and optimising the array of biocompatible polymers with cholesterol is being explored, particularly the potential applications in artificial tissue growth, not just for medical applications but also for lab-grown meat. The materials are being further developed for applications in the next generation of photovoltaics and fuel cells. The natural abundance of cholesterol and their ease of acquisition makes them suitable for scaling up on an industrial scale for the fabrication of such devices.

Funding: This research received no external funding.

Institutional Review Board Statement: Not applicable.

Informed Consent Statement: Not applicable.

Data Availability Statement: Not applicable.

Conflicts of Interest: The authors declare no conflict of interest.

References

1. Reinitzer, F. Contributions to the knowledge of cholesterol. *Liq. Cryst.* **1989**, *5*, 7–18. [CrossRef]
2. Bullard, B.; Stumpf, C.H.; Zhao, W.; Kuzenko, S.; Niehaus, G.D. Crystal Diagnostics Xpress S Kit for the rapid detection of Salmonella spp. in selected food matrixes. *J. AOAC Int.* **2017**, *100*, 1038–1050. [CrossRef] [PubMed]
3. Schadt, M.; Helfrich, W. Voltage-dependent optical activity of a twisted nematic liquid crystal. *Appl. Phys. Lett.* **1971**, *18*, 127–128. [CrossRef]
4. Chang, C.; Bang, K.; Wetzstein, G.; Lee, B.; Gao, L. Toward the next-generation VR/AR optics: A review of holographic near-eye displays from a human-centric perspective. *Optica* **2020**, *7*, 1563. [CrossRef]
5. Milton, H.E.; Morgan, P.B.; Clamp, J.H.; Gleeson, H.F. Electronic liquid crystal contact lenses for the correction of presbyopia. *Opt. Express* **2014**, *22*, 8035. [CrossRef]
6. Heilmeier, G.H.; Zanoni, L.A.; Barton, L.A. Dynamic scattering in nematic liquid crystals. *Appl. Phys. Lett.* **1968**, *13*, 46–47. [CrossRef]
7. Kowerdziej, R.; Garbat, K.; Walczakowski, M. Nematic liquid crystal mixtures dedicated to thermally tunable terahertz devices. *Liq. Cryst.* **2018**, *45*, 1040–1046. [CrossRef]
8. Caputo, R.; Palermo, G.; Infusino, M.; De Sio, L. Liquid Crystals as an Active Medium: Novel Possibilities in Plasmonics. *Nanospectroscopy* **2015**, *1*, 40–53. [CrossRef]
9. Wang, Z.; Xu, T.; Noel, A.; Chen, Y.C.; Liu, T. Applications of liquid crystals in biosensing. *Soft Matter* **2021**, *17*, 4675–4702. [CrossRef]
10. Hata, Y.; Insull, W. Significance of Cholesterol Esters as Liquid Crystal in Human Atherosclerosis. *Jpn. Circ. J.* **1973**, *37*, 269–275. [CrossRef]
11. Edward Barrall, b.M.; Porter, R.S.; Johnson, J.F.; Barrall, E.M.; Porter, R.S.; Johnson, J.F.; Phys, J.; Martin, H.; Müller, F.H. Molecular Structure and the Properties of Liquid Crystals. *Z. Physik Chem.* **1964**, *68*, 51.
12. Usoltseva, V.A.; Chistyakov, I.G. Chemical characteristics, structure, and properties of liquid crystal. *Russ. Chem. Rev.* **1963**, *32*, 495–509. [CrossRef]
13. Barrall, E.M.; Porter, R.S.; Johnson, J.F.; S Andrews, J.T. Heats of Transition for Some Cholesteryl Esters by Differential Scanning Calorimetry. *J. Phys. Chem.* **1967**, *71*, 1224–1228. [CrossRef]
14. Baessler, H.; Labes, M.M. Helical twisting power of steroidal solutes in cholesteric mesophases. *J. Chem. Phys.* **1970**, *52*, 631–637. [CrossRef] [PubMed]

15. Yu, X.H.; Zhang, D.W.; Zheng, X.L.; Tang, C.K. Cholesterol transport system: An integrated cholesterol transport model involved in atherosclerosis. *Prog. Lipid Res.* **2019**, *73*, 65–91. [CrossRef] [PubMed]
16. Thakur, V.K.; Kessler, M.R. *Liquid Crystalline Polymers*, 1st ed.; Springer International Publishing: Cham, Switzerland, 2016. [CrossRef]
17. Wang, X.J.; Zhou, Q.F. *Liquid Crystalline Polymers*; World Scientific: Singapur, 2004. [CrossRef]
18. Donald, A.M.; Windle, A.H.; Hanna, S. *Liquid Crystallie Polymers*; Cambridge University Press: Cambridge, UK, 2006. [CrossRef]
19. Wen, Z.; Yang, K.; Raquez, J.M. A review on liquid crystal polymers in free-standing reversible shape memory materials. *Molecules* **2020**, *25*, 1241. [CrossRef]
20. Trimmel, G.; Riegler, S.; Fuchs, G.; Slugovc, C.; Stelzer, F. Liquid Crystalline Polymers by Metathesis Polymerization. In *Metathesis Polymerization*; Buchmeiser, M.R., Ed.; Springer: Berlin/Heidelberg, Germany, 2005; pp. 43–87. [CrossRef]
21. Lyu, X.; Xiao, A.; Shi, D.; Li, Y.; Shen, Z.; Chen, E.Q.; Zheng, S.; Fan, X.H.; Zhou, Q.F. Liquid crystalline polymers: Discovery, development, and the future. *Polymer* **2020**, *202*, 122740. [CrossRef]
22. Platé, N.A.; Shibaev, V.P. Thermotropic liquid crystalline polymers-problems and trends. *Die Makromol. Chem.* **1984**, *6*, 3–27. [CrossRef]
23. Xie, Y.; Hu, J.; Dou, Q.; Xiao, L.; Yang, L. Synthesis and mesomorphism of new aliphatic polycarbonates containing side cholesteryl groups. *Liq. Cryst.* **2016**, *43*, 1486–1494. [CrossRef]
24. Doganci, E.; Davarci, D. Synthesized and mesomorphic properties of cholesterol end-capped poly(ϵ-caprolactone) polymers. *J. Polym. Res.* **2019**, *26*, 1–10. [CrossRef]
25. Finkelmann, H.; Ringsdorf, H.; Sol, W.; Wendofl, J.H. Synthesis of Cholesteric Liquid Crystalline Polymers Polyreactions in Ordered Systems, 15′). *Makromol. Chem* **1978**, *179*, 829–832. [CrossRef]
26. Hessel, F.; Herr, R.P.; Finkelmann, H. Synthesis and characterization of biaxial nematic side chain polymers with laterally attached mesogenic groups. *Makromol. Chem* **1987**, *188*, 1597–1611. [CrossRef]
27. Shibaev, V.P.; Plate, N.A.; Freidzon, Y.S. Thermotropic Liquid Crystalline Polymers. I. Cholesterol-Containing Polymers and Copolymers. *J. Polym. Sci. Polym. Chem. Ed.* **1979**, *17*, 1655–670. [CrossRef]
28. Xu, X.; Liu, X.; Li, Q.; Hu, J.; Chen, Q.; Yang, L.; Lu, Y. New amphiphilic polycarbonates with side functionalized cholesteryl groups as biomesogenic units: Synthesis, structure and liquid crystal behavior. *RSC Adv.* **2017**, *7*, 14176–14185. [CrossRef]
29. Cowie, J.M.G.; Arrighi, V. *Polymers: Chemistry and Physics of Modern Materials*, 3rd ed.; CRC Press: Boca Raton, FL, USA, 2007.
30. Boissé, S.; Rieger, J.; Di-Cicco, A.; Albouy, P.A.; Bui, C.; Li, M.H.; Charleux, B. Synthesis via RAFT of amphiphilic block copolymers with liquid-crystalline hydrophobic block and their self-assembly in water. *Macromolecules* **2009**, *42*, 8688–8696. [CrossRef]
31. Huang, Z.H.; Zhou, Y.Y.; Wang, Z.M.; Li, Y.; Zhang, W.; Zhou, N.C.; Zhang, Z.B.; Zhu, X.L. Recent advances of CuAAC click reaction in building cyclic polymer. *Chin. J. Polym. Sci.* **2017**, *35*, 317–341. [CrossRef]
32. Hamley, I.W.; Castelletto, V.; Parras, P.; Lu, Z.B.; Imrie, C.T.; Itoh, T. Ordering on multiple lengthscales in a series of side group liquid crystal block copolymers containing a cholesteryl-based mesogen. *Soft Matter* **2005**, *1*, 355–363. [CrossRef]
33. Xu, J.P.; Ji, J.; Chen, W.D.; Shen, J.C. Novel biomimetic surfactant: Synthesis and micellar characteristics. *Macromol. Biosci.* **2005**, *5*, 164–171. [CrossRef]
34. Yang, S.Y.; Kang, J.H.; Jeong, S.Y.; Choi, K.H.; Lee, S.; Shin, G.J. Influence of chemical structure on the optical properties of new side-chain polymers containing cholesterol. *Liq. Cryst.* **2014**, *41*, 1286–1292. [CrossRef]
35. Finkelmann, H.; Kock, H.J.; Rehage, G. Investigations on Liquid Crystalline Polysiloxanes 3a) Liquid Crystalline Elastomers—A New Type of Liquid. *Makromol. Chem. Rapid Commun.* **1981**, *2*, 317–322. [CrossRef]
36. Hu, J.S.; Zhang, B.Y.; Jia, Y.G.; Wang, Y. Structures and Properties of Side-Chain Cholesteric Liquid Crystalline Polyacrylates. *Polym. J.* **2003**, *35*, 160–166. [CrossRef]
37. Yang, X.; Chen, S.; Chen, S.; Xu, H. Influencing factors on liquid crystalline properties of cholesterol side-chain liquid crystalline polymers without spacer: Molecular weight and copolymerisation. *Liq. Cryst.* **2019**, *46*, 1827–1842. [CrossRef]
38. Klok, H.A.; Hwang, J.J.; Iyer, S.N.; Stupp, S.I. Cholesteryl-(L-lactic acid) building blocks for self-assembling biomaterials. *Macromolecules* **2002**, *35*, 746–759. [CrossRef]
39. Wang, Y.; Zhang, B.Y.; He, X.Z.; Wang, J.W. Side-chain cholesteric liquid crystalline polymers containing menthol and cholesterol—Synthesis and characterization. *Colloid Polym. Sci.* **2007**, *285*, 1077–1084. [CrossRef]
40. Venkataraman, S.; Lee, A.L.; Maune, H.T.; Hedrick, J.L.; Prabhu, V.M.; Yang, Y.Y. Formation of disk- and stacked-disk-like self-assembled morphologies from cholesterol-functionalized amphiphilic polycarbonate diblock copolymers. *Macromolecules* **2013**, *46*, 4839–4846. [CrossRef]
41. Guo, Z.H.; Liu, X.F.; Hu, J.S.; Yang, L.Q.; Chen, Z.P. Synthesis and self-assembled behavior of pH-responsive chiral liquid crystal amphiphilic copolymers based on diosgenyl-functionalized aliphatic polycarbonate. *Nanomaterials* **2017**, *7*, 169. [CrossRef]
42. Li, L.; Zhou, F.; Li, Y.; Chen, X.; Zhang, Z.; Zhou, N.; Zhu, X. Cooperation of Amphiphilicity and Smectic Order in Regulating the Self-Assembly of Cholesterol-Functionalized Brush-Like Block Copolymers. *Langmuir* **2018**, *34*, 11034–11041. [CrossRef]
43. Krishnasamy, V.; Qu, W.; Chen, C.; Huo, H.; Ramanagul, K.; Gothandapani, V.; Mehl, G.H.; Mehl, G.H.; Zhang, Q.; Liu, F. Self-Assembly and Temperature-Driven Chirality Inversion of Cholesteryl-Based Block Copolymers. *Macromolecules* **2020**, *53*, 4193–4203. [CrossRef]
44. Hembury, G.A.; Borovkov, V.V.; Inoue, Y. Chirality-sensing supramolecular systems. *Chem. Rev.* **2008**, *108*, 1–73. [CrossRef]
45. Aviram, A.; Ratner, M.A. Molecular rectifiers. *Chem. Phys. Lett.* **1974**, *29*, 277–283. [CrossRef]

46. Morgenstern, K. Switching individual molecules by light and electrons: From isomerisation to chirality flip. *Prog. Surf. Sci.* **2011**, *86*, 115–161. [CrossRef]
47. Yang, H.; Jia, L.; Zhu, C.; Di-Cicco, A.; Levy, D.; Albouy, P.A.; Li, M.H.; Keller, P. Amphiphilic poly(ethylene oxide)-block-poly(butadiene-graft-liquid crystal) copolymers: Synthesis and self-assembly in water. *Macromolecules* **2010**, *43*, 10442–10451. [CrossRef]
48. Jia, L.; Cao, A.; Lévy, D.; Xu, B.; Albouy, P.A.; Xing, X.; Bowick, M.J.; Li, M.H. Smectic polymer vesicles. *Soft Matter* **2009**, *5*, 3446–3451. [CrossRef]
49. Jones, R.A. Challenges in soft nanotechnology. *Faraday Discuss.* **2009**, *143*, 9–14. [CrossRef] [PubMed]
50. Zhou, L.; Zhang, D.; Hocine, S.; Pilone, A.; Trépout, S.; Marco, S.; Thomas, C.; Guo, J.; Li, M.H. Transition from smectic nanofibers to smectic vesicles in the self-assemblies of PEG-: B -liquid crystal polycarbonates. *Polym. Chem.* **2017**, *8*, 4776–4780. [CrossRef]
51. Jia, L.; Liu, M.; Di Cicco, A.; Albouy, P.A.; Brissault, B.; Penelle, J.; Boileau, S.; Barbier, V.; Li, M.H. Self-assembly of amphiphilic liquid crystal polymers obtained from a cyclopropane-1,1-dicarboxylate bearing a cholesteryl mesogen. *Langmuir* **2012**, *28*, 11215–11224. [CrossRef]
52. Jeong, Y.H.; Ahn, T.; Yu, W.; Lee, S.M. Cholesterol-Functionalized Linear/Brush Block Copolymers for Metal-Incorporated Nanostructures with Modulated Core Density and Enhanced Self-Assembly Efficiency. *ACS Macro Lett.* **2021**, *10*, 492–497. [CrossRef]
53. Jia, L.; Albouy, P.A.; Di Cicco, A.; Cao, A.; Li, M.H. Self-assembly of amphiphilic liquid crystal block copolymers containing a cholesteryl mesogen: Effects of block ratio and solvent. *Polymer* **2011**, *52*, 2565–2575. [CrossRef]
54. Zhang, X.; Boissé, S.; Bui, C.; Albouy, P.A.; Brûlet, A.; Li, M.H.; Rieger, J.; Charleux, B. Amphiphilic liquid-crystal block copolymer nanofibers via RAFT-mediated dispersion polymerization. *Soft Matter* **2012**, *8*, 1130–1141. [CrossRef]
55. Huo, M.; Zhang, Y.; Zeng, M.; Liu, L.; Wei, Y.; Yuan, J. Morphology Evolution of Polymeric Assemblies Regulated with Fluoro-Containing Mesogen in Polymerization-Induced Self-Assembly. *Macromolecules* **2017**, *50*, 8192–8201. [CrossRef]
56. Li, Y.; Lu, Q.; Chen, Q.; Wu, X.; Shen, J.; Shen, L. Directional effect on the fusion of ellipsoidal morphologies into nanorods and nanotubes. *RSC Adv.* **2021**, *11*, 1729–1735. [CrossRef] [PubMed]
57. Huo, M.; Song, G.; Zhang, J.; Wei, Y.; Yuan, J. Nonspherical Liquid Crystalline Assemblies with Programmable Shape Transformation. *ACS Macro Lett.* **2018**, *7*, 956–961. [CrossRef] [PubMed]
58. Huo, M.; Li, D.; Song, G.; Zhang, J.; Wu, D.; Wei, Y.; Yuan, J. Semi-Fluorinated Methacrylates: A Class of Versatile Monomers for Polymerization-Induced Self-Assembly. *Macromol. Rapid Commun.* **2018**, *39*, 1700840–1700846. [CrossRef] [PubMed]
59. Shen, L.; Guo, H.; Zheng, J.; Wang, X.; Yang, Y.; An, Z. RAFT Polymerization-Induced Self-Assembly as a Strategy for Versatile Synthesis of Semifluorinated Liquid-Crystalline Block Copolymer Nanoobjects. *ACS Macro Lett.* **2018**, *7*, 287–292. [CrossRef] [PubMed]
60. Li, X.; Jin, B.; Gao, Y.; Hayward, D.W.; Winnik, M.A.; Luo, Y.; Manners, I. Monodisperse Cylindrical Micelles of Controlled Length with a Liquid-Crystalline Perfluorinated Core by 1D "Self-Seeding". *Angew. Chem.* **2016**, *128*, 11564–11568. [CrossRef]
61. Guan, S.; Chen, A. Influence of Spacer Lengths on the Morphology of Biphenyl-Containing Liquid Crystalline Block Copolymer Nanoparticles via Polymerization-Induced Self-Assembly. *Macromolecules* **2020**, *53*, 6235–6245. [CrossRef]
62. Guan, S.; Zhang, C.; Wen, W.; Qu, T.; Zheng, X.; Zhao, Y.; Chen, A. Formation of Anisotropic Liquid Crystalline Nanoparticles via Polymerization-Induced Hierarchical Self-Assembly. *ACS Macro Lett.* **2018**, *7*, 358–363. [CrossRef]
63. Wen, W.; Ouyang, W.; Guan, S.; Chen, A. Synthesis of azobenzene-containing liquid crystalline block copolymer nanoparticles: Via polymerization induced hierarchical self-assembly. *Polym. Chem.* **2021**, *12*, 458–465. [CrossRef]
64. Wen, W.; Chen, A. The self-assembly of single chain Janus nanoparticles from azobenzene-containing block copolymers and reversible photoinduced morphology transitions. *Polym. Chem.* **2021**, *12*, 2447–2456. [CrossRef]
65. Zhang, N.; Fan, Y.; Chen, H.; Trépout, S.; Brûlet, A.; Li, M.H. Polymersomes with a smectic liquid crystal structure and AIE fluorescence. *Polym. Chem.* **2022**, *13*, 1107–1115. [CrossRef]
66. Derry, M.J.; Fielding, L.A.; Warren, N.J.; Mable, C.J.; Smith, A.J.; Mykhaylyk, O.O.; Armes, S.P. In situ small-angle X-ray scattering studies of sterically-stabilized diblock copolymer nanoparticles formed during polymerization-induced self-assembly in non-polar media. *Chem. Sci.* **2016**, *7*, 5078–5090. [CrossRef]
67. Warren, N.J.; Mykhaylyk, O.O.; Ryan, A.J.; Williams, M.; Doussineau, T.; Dugourd, P.; Antoine, R.; Portale, G.; Armes, S.P. Testing the vesicular morphology to destruction: Birth and death of diblock copolymer vesicles prepared via polymerization-induced self-assembly. *J. Am. Chem. Soc.* **2015**, *137*, 1929–1937. [CrossRef] [PubMed]
68. Wang, S.; Luo, Y.; Li, X. Liquid crystal block copolymer micelles with aggregation-induced emission from corona chains. *J. Phys. Conf. Ser.* **2020**, *1605*, 012173. [CrossRef]
69. Albuquerque, H.M.; Santos, C.M.; Silva, A.M. Cholesterol-based compounds: Recent advances in synthesis and applications. *Molecules* **2019**, *24*, 116. [CrossRef] [PubMed]
70. Hwang, J.J.; Iyer, S.N.; Li, L.S.; Claussen, R.; Harrington, D.A.; Stupp, S.I. Self-assembling biomaterials: Liquid crystal phases of cholesteryl oligo(L-lactic acid) and their interactions with cells. *Proc. Natl. Acad. Sci. USA* **2002**, *99*, 9662–9667. [CrossRef]
71. Prévôt, M.E.; Andro, H.; Alexander, S.L.; Ustunel, S.; Zhu, C.; Nikolov, Z.; Rafferty, S.T.; Brannum, M.T.; Kinsel, B.; Korley, L.T.; et al. Liquid crystal elastomer foams with elastic properties specifically engineered as biodegradable brain tissue scaffolds. *Soft Matter* **2018**, *14*, 354–360. [CrossRef]

72. Xie, W.; Ouyang, R.; Wang, H.; Li, N.; Zhou, C. Synthesis and cytotoxicity of novel elastomers based on cholesteric liquid crystals. *Liq. Cryst.* **2020**, *47*, 449–464. [CrossRef]
73. Sharma, A.; Neshat, A.; Mahnen, C.J.; Nielsen, A.D.; Snyder, J.; Stankovich, T.L.; Daum, B.G.; Laspina, E.M.; Beltrano, G.; Gao, Y.; et al. Biocompatible, biodegradable and porous liquid crystal elastomer scaffolds for spatial cell cultures. *Macromol. Biosci.* **2015**, *15*, 200–214. [CrossRef]
74. Liu, Y.; Wang, Y.; Zhuang, D.; Yang, J.; Yang, J. Bionanoparticles of amphiphilic copolymers polyacrylate bearing cholesterol and ascorbate for drug delivery. *J. Colloid Interface Sci.* **2012**, *377*, 197–206. [CrossRef]
75. Lee, A.L.; Venkataraman, S.; Sirat, S.B.; Gao, S.; Hedrick, J.L.; Yang, Y.Y. The use of cholesterol-containing biodegradable block copolymers to exploit hydrophobic interactions for the delivery of anticancer drugs. *Biomaterials* **2012**, *33*, 1921–1928. [CrossRef]
76. Jia, L.; Cui, D.; Bignon, J.; Di Cicco, A.; Wdzieczak-Bakala, J.; Liu, J.; Li, M.H. Reduction-responsive cholesterol-based block copolymer vesicles for drug delivery. *Biomacromolecules* **2014**, *15*, 2206–2217. [CrossRef] [PubMed]
77. Liu, X.; Guo, Z.; Xie, Y.; Chen, Z.; Hu, J.; Yang, L. Synthesis and liquid crystal behavior of new side chain aliphatic polycarbonates based on cholesterol. *J. Mol. Liq.* **2018**, *259*, 350–358. [CrossRef]
78. Geihe, E.I.; Cooley, C.B.; Simon, J.R.; Kiesewetter, M.K.; Edward, J.A.; Hickerson, R.P.; Kaspar, R.L.; Hedrick, J.L.; Waymouth, R.M.; Wender, P.A.; et al. Designed guanidinium-rich amphipathic oligocarbonate molecular transporters complex, deliver and release siRNA in cells. *Proc. Natl. Acad. Sci. USA* **2012**, *109*, 13171–13176. [CrossRef] [PubMed]
79. Liu, X.; Xie, Y.; Hu, Z.; Chen, Z.; Hu, J.; Yang, L. pH responsive self-assembly and drug release behavior of aliphatic liquid crystal block polycarbonate with pendant cholesteryl groups. *J. Mol. Liq.* **2018**, *266*, 405–412. [CrossRef]
80. Zhang, C.Y.; Xiong, D.; Sun, Y.; Zhao, B.; Lin, W.J.; Zhang, L.J. Self-assembled micelles based on pH-sensitive PAE-g-MPEG-cholesterol block copolymer for anticancer drug delivery. *Int. J. Nanomed.* **2014**, *9*, 4923–4933. [CrossRef] [PubMed]
81. Tran, T.H.; Nguyen, C.T.; Gonzalez-Fajardo, L.; Hargrove, D.; Song, D.; Deshmukh, P.; Mahajan, L.; Ndaya, D.; Lai, L.; Kasi, R.M.; et al. Long circulating self-assembled nanoparticles from cholesterol-containing brush-like block copolymers for improved drug delivery to tumors. *Biomacromolecules* **2014**, *15*, 4363–4375. [CrossRef] [PubMed]
82. Finkelmann, H.; Kim, S.T.; Muñoz, A.; Palffy-Muhoray, P.; Taheri, B. Tunable Mirrorless Lasing in Cholesteric Liquid Crystalline Elastomers. *Adv. Mater.* **2001**, *13*, 1069–1072. [CrossRef]
83. Kim, S.T.; Finkelmann, H. Cholesteric Liquid Single-Crystal Elastomers (LSCE) Obtained by the Anisotropic Deswelling Method. *Macromol. Rapid Commun* **2001**, *22*, 429–433. [CrossRef]
84. Varanytsia, A.; Nagai, H.; Urayama, K.; Palffy-Muhoray, P. Tunable lasing in cholesteric liquid crystal elastomers with accurate measurements of strain. *Sci. Rep.* **2015**, *5*, 1–8. [CrossRef]
85. Hattori, H.; Uryu, T. Synthesis and Properties of Photochromic Liquid-Crystalline Copolymers Containing Both Spironaphthoxazine and Cholesteryl Groups. *J. Polym. Sci. Part A Polym. Chem.* **2000**, *38*, 887–894. [CrossRef]
86. Petri, A.; Kummer, S.; Anneser, H.; Feiner, F.; Briiuchles, C. Photoinduced Reorientation of Cholesteric Liquid Crystalline Polysiloxanes Photoinduced Reorientation of Cholesteric Liquid Crystalline Polysiloxanes and Applications in Optical Information Storage and Second Harmonic Generation. *Berichte Bunseges. Phys. Chem.* **1993**, *97*, 1281–1286. [CrossRef]
87. He, X.Z.; Gao, Y.F.; Zheng, J.J.; Li, X.Y.; Meng, F.B.; Hu, J.S. Chiral photosensitive side-chain liquid crystalline polymers—Synthesis and characterization. *Colloid Polym. Sci.* **2016**, *294*, 1823–1832. [CrossRef]
88. Doganci, E.; Cakirlar, C.; Bayir, S.; Yilmaz, F.; Yasin, M. Methacrylate based side chain liquid crystalline polymers with pendant cholesterol group: Preparation and investigation of electrical conductivity mechanism. *J. Appl. Polym. Sci.* **2017**, *134*, 45207. [CrossRef]
89. Doganci, E.; Kayabasi, F.; Davarcı, D.; Demir, A.; Gürek, A.G. Synthesis of liquid crystal polymers containing cholesterol side groups and investigation of their usability potential as an insulator in organic field effect transistor (OFET) applications. *J. Polym. Res.* **2022**, *29*, 147. [CrossRef]
90. Liu, Y.; Chen, J.; Qi, Y.; Gao, S.; Balaji, K.; Zhang, Y.; Xue, Q.; Lu, Z. Cross-linked liquid crystalline polybenzoxazines bearing cholesterol-based mesogen side groups. *Polymer* **2018**, *145*, 252–260. [CrossRef]
91. Suk-Kyun, A.H.; Le, L.T.; Kasi, R.M. Synthesis and characterization of side-chain liquid crystalline polymers bearing cholesterol mesogen. *J. Polym. Sci. Part A Polym. Chem.* **2009**, *47*, 2690–2701. [CrossRef]
92. Zhou, F.; Li, Y.; Jiang, G.; Zhang, Z.; Tu, Y.; Chen, X.; Zhou, N.; Zhu, X. Biomacrocyclic side-chain liquid crystalline polymers bearing cholesterol mesogens: Facile synthesis and topological effect study. *Polym. Chem.* **2015**, *6*, 6885–6893. [CrossRef]
93. Ndaya, D.; Bosire, R.; Kasi, R.M. Cholesteric-azobenzene liquid crystalline copolymers: Design, structure and thermally responsive optical properties. *Polym. Chem.* **2019**, *10*, 3868–3878. [CrossRef]
94. Ndaya, D.; Bosire, R.; Vaidya, S.; Kasi, R.M. Molecular engineering of stimuli-responsive, functional, side-chain liquid crystalline copolymers: Synthesis, properties and applications. *Polym. Chem.* **2020**, *11*, 5937–5954. [CrossRef]
95. Zhou, Y.; Ahn, S.K.; Lakhman, R.K.; Gopinadhan, M.; Osuji, C.O.; Kasi, R.M. Tailoring crystallization behavior of PEO-based liquid crystalline block copolymers through variation in liquid crystalline content. *Macromolecules* **2011**, *44*, 3924–3934. [CrossRef]
96. Zhang, X.J.; Gao, R.T.; Kang, S.M.; Wang, X.J.; Jiang, R.J.; Li, G.W.; Zhou, L.; Liu, N.; Wu, Z.Q. Hydrogen-bonding dependent nontraditional fluorescence polyphenylallenes: Controlled synthesis and aggregation-induced emission behaviors. *Polymer* **2022**, *245*. [CrossRef]

97. Luo, Z.W.; Tao, L.; Zhong, C.L.; Li, Z.X.; Lan, K.; Feng, Y.; Wang, P.; Xie, H.L. High-efficiency circularly polarized luminescence from chiral luminescent liquid crystalline polymers with aggregation-induced emission properties. *Macromolecules* **2020**, *53*, 9758–9768. [CrossRef]
98. Wu, Y.; You, L.H.; Yu, Z.Q.; Wang, J.H.; Meng, Z.; Liu, Y.; Li, X.S.; Fu, K.; Ren, X.K.; Tang, B.Z. Rational Design of Circularly Polarized Luminescent Aggregation-Induced Emission Luminogens (AIEgens): Promoting the Dissymmetry Factor and Emission Efficiency Synchronously. *ACS Mater. Lett.* **2020**, *2*, 505–510. [CrossRef]
99. Song, F.; Cheng, Y.; Liu, Q.; Qiu, Z.; Lam, J.W.; Lin, L.; Yang, F.; Tang, B.Z. Tunable circularly polarized luminescence from molecular assemblies of chiral AIEgens. *Mater. Chem. Front.* **2019**, *3*, 1768–1778. [CrossRef]
100. Mahajan, L.H.; Ndaya, D.; Deshmukh, P.; Peng, X.; Gopinadhan, M.; Osuji, C.O.; Kasi, R.M. Optically Active Elastomers from Liquid Crystalline Comb Copolymers with Dual Physical and Chemical Cross-Links. *Macromolecules* **2017**, *50*, 5929–5939. [CrossRef]
101. Ahn, S.K.; Deshmukh, P.; Gopinadhan, M.; Osuji, C.O.; Kasi, R.M. Side-chain liquid crystalline polymer networks: Exploiting nanoscale smectic polymorphism to design shape-memory polymers. *ACS Nano* **2011**, *5*, 3085–3095. [CrossRef]
102. Zhou, Y.; Briand, V.A.; Sharma, N.; Ahn, S.k.; Kasi, R.M. Polymers comprising cholesterol: Synthesis, self-assembly, and applications. *Materials* **2009**, *2*, 636–660. [CrossRef]
103. Hosta-Rigau, L.; Zhang, Y.; Teo, B.M.; Postma, A.; Städler, B. Cholesterol-A biological compound as a building block in bionanotechnology.T *Nanoscale* **2013**, *5*, 89–109. [CrossRef]
104. Misiak, P.; Markiewicz, K.H.; Szymczuk, D.; Wilczewska, A.Z. Polymeric drug delivery systems bearing cholesterol moieties: A review. *Polymers* **2020**, *12*, 2620. [CrossRef]

Review

Recent Progresses on Experimental Investigations of Topological and Dissipative Solitons in Liquid Crystals

Yuan Shen and Ingo Dierking *

Department of Physics and Astronomy, School of Natural Sciences, University of Manchester, Oxford Road, Manchester M13 9PL, UK; yuan.shen@postgrad.manchester.ac.uk
* Correspondence: ingo.dierking@manchester.ac.uk

Abstract: Solitons in liquid crystals have received increasing attention due to their importance in fundamental physical science and potential applications in various fields. The study of solitons in liquid crystals has been carried out for over five decades with various kinds of solitons being reported. Recently, a number of new types of solitons have been observed, among which, many of them exhibit intriguing dynamic behaviors. In this paper, we briefly review the recent progresses on experimental investigations of solitons in liquid crystals.

Keywords: liquid crystal; soliton; toron; skyrmion; nematic; cholesteric; smectic; micro-cargo transport; dissipative dynamics

Citation: Shen, Y.; Dierking, I. Recent Progresses on Experimental Investigations of Topological and Dissipative Solitons in Liquid Crystals. *Crystals* **2022**, *12*, 94. https://doi.org/10.3390/cryst12010094

Academic Editor: Borislav Angelov

Received: 15 December 2021
Accepted: 8 January 2022
Published: 11 January 2022

Publisher's Note: MDPI stays neutral with regard to jurisdictional claims in published maps and institutional affiliations.

Copyright: © 2022 by the authors. Licensee MDPI, Basel, Switzerland. This article is an open access article distributed under the terms and conditions of the Creative Commons Attribution (CC BY) license (https://creativecommons.org/licenses/by/4.0/).

1. Introduction

Solitons are self-sustained localized packets of waves in nonlinear media that propagate without changing shape. They are found everywhere in our daily life from nerve pluses in our bodies to eyes of storms in the atmosphere and even density waves in galaxies. They were first observed as water waves in a shallow canal by a Scottish engineer John Scott Russell in 1834 [1], which initiated the theoretical work of Rayleigh and Boussinesq and eventually led to the well-known KdV (Korteweg, de Vries) equation which has been broadly used as an approximate description of solitary waves [2]. However, the significance of soliton was not widely appreciated until 1965 when the word "soliton" was coined by Zabusky and Kruskal [3]. Nowadays, solitons have appeared in every branch of physics, such as nonlinear photonics [4], Bose-Einstein condensates [5], superconductors [6], and magnetic materials [7], just to name a few. Generally, solitons appear as self-organized localized waves that preserve their identities after pairwise collisions [8]. This ideal nonlinear property of solitons may enable distortion-free long-distance transport of matter or information and thus makes them considerably attractive to both fundamental research and technological applications [9–11].

Liquid crystals (LCs) are self-organized anisotropic fluids that are thermodynamically intermediate between the isotropic liquid and the crystalline solid, exhibiting the fluidity of liquids as well as the order of crystals [12,13]. Generally, LCs consist of anisotropic building blocks with rod- or disc-like shapes, which spontaneously orient in a specific direction on average, called director, **n**. As a typical nonlinear material, LCs have been broadly used as an ideal testbed for studying solitons, in which different kinds of solitons have been generated in the past five decades.

In this review, we first give a brief overview of the early works of solitons in LCs (Section 2), which is followed by a short discussion of investigations on nematicons (Section 3) and a discussion of recent progress in studies of topological solitons in chiral nematics (Section 4). The article then continues by overviewing the investigations of dynamic particle-like dissipative solitons, "director bullets" or "directrons", which were first reported by Brand et al. in 1997 but did not receive much attention until recently (Section 5). This review is mainly focused on the recent experimental investigations on solitons in liquid crystals, readers who are interested in this topic can find more detailed early experimental and theoretical investigations in the excellent book edited by Lam and Prost [14].

2. Early Works

The study of solitons in LCs was started in 1968 by Wolfgang Helfrich [15]. He theoretically modelled alignment inversion walls as static solitons in an infinite sample of nematic order. By applying a magnetic field, **H**, depending on the assumed orientation of the director at infinity, there are three types of possible walls, i.e., twist wall, splay-bend wall parallel to the applied field, and the splay-bend wall perpendicular to the field, which are analogous to the Bloch and Neel walls in ferromagnetics (Figure 1). Such a model was later improved by de Gennes who studied the boundary effects of the substrate and the movement of the walls [16].

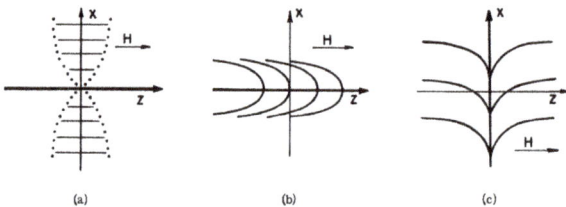

Figure 1. Schematic diagrams of different alignment inversion walls. (**a**) Twist wall. (**b**) Splay-bend wall parallel to the magnetic field. (**c**) Splay-bend wall vertical to the magnetic field. Reprinted with permission from Ref. [15]. Copyright 1968 American Physical Society.

The first experimental investigation of these inversion walls was probably reported by Leger in 1972 [17]. In this work, a 180° twist wall was generated at the top free surface of a nematic droplet by rotating the applied magnetic field by 180° (Figure 2a,b). The migration time of the twist walls was also measured. However, instead of working with a free surface geometry, the measurements were carried out in nematic droplets sandwiched between two rubbed glass plates due to the difficulty of measuring the sample thickness. In this case, the twist walls were generated near each glass plate and then moved toward the midplane of the sample. Because the pair had opposite twists, they annihilated each other once they met. The measured migration time was in great agreement with theoretical prediction [16]. However, because the observation was from the top of the sample, one could not really see the propagation process of the twist walls in the experiment.

In the same year, a different type of wall was investigated theoretically by Brochard [18] and experimentally by Leger [19,20]. These walls separate domains in which LC molecules rotate in two different directions, i.e., the so-called reverse tilt domains [21]. They can be generated by increasing the applied magnetic field above the Freedericksz transition (Figure 2c,d). The walls can either form in a straight line state where they are stable and static or in a closed loop state where they continuously shrink inward and eventually annihilate to minimize the free energy. The authors investigated the static structure and the dynamic behavior of the walls. Unlike the twist walls mentioned above, the walls discussed here move in the plane of the LC cell and can be observed directly.

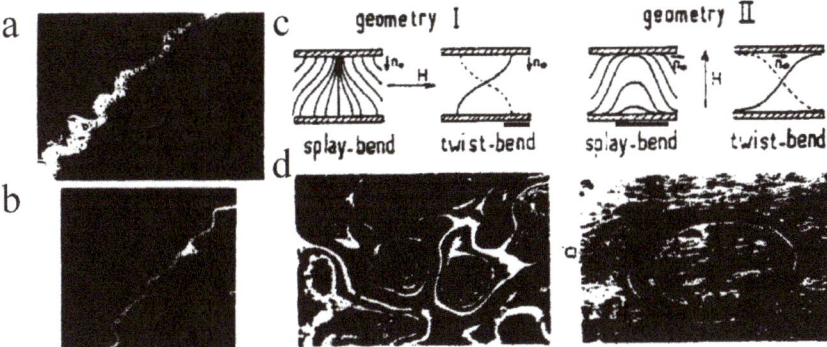

Figure 2. Typical image of the free surface of a MBBA droplet within which a twist wall was generated at (**a**) H = 3000 G and (**b**) H = 6000 G, where H represents the magnitude of the magnetic field. Reprinted with permission from Ref. [17]. Copyright 1972 Elsevier. (**c**) Schematic distortion of the director structure of different walls. (**d**) The microscopy of the walls corresponding to (**c**). Reprinted with permission from Ref. [19]. Copyright 1972 Elsevier.

The interactions between flows and director field in a nematic LC can introduce a nonlinear term in the director equation of motion. Thus, it is possible to induce solitons by shearing nematics without applying any external field [14]. In 1976, Cladis and Torza reported propagating solitary wave instabilities of a nematic in a Couette flow field [22]. They found that for small shears, a "tumbling" instability was observed, which was similar to the appearance of a solitary wave in a long torsion bar to which is attached a dense array of pendulums. By slightly increasing the shear rate, a cellular flow instability was induced (Figure 3a). Further increase of the shear rate, led to a dense mass of disclinations being formed, which aligned with their long axis parallel to the flow field. At very large shear rates, Taylor vortices were generated. In 1982, Zhu reported a soliton-like director wave in a nematic by mechanical shearing [23]. In his work, the nematic film was confined in a home-made cell with an exciter at the entrance. The nematic was homeotropically aligned. By moving the exciter, propagating director waves were observed through polarized white light as black lines, which travelled through the nematic bulk (Figure 3b). It should be noted that these solitons were first theoretically predicted and explained by Lin et al. [24–26]. Later, in 1987, C. Q. Shu et al. reported the generation of two-dimensional (2D) axisymmetric propagating solitons in a radial Poiseuille flow of homeotropic nematic LCs [27]. These solitons appear as dark rings in polarized white light and are large enough to be observed directly by the naked eye (Figure 3c). They are generated by periodically pressing the rim of the cell and can move through the nematic bulk at a constant speed with their shape remaining unchanged.

Convection as one of the simplest examples of hydrodynamic instability induces a rich variety of nonlinear phenomena in a fluid. In conventional Rayleigh-Benard and Taylor experiments, convective patterns emerge once the temperature gradient across the fluid layer, and the relative velocity field of the rotating planes, exceed some specific threshold (the Rayleigh number R_a and T_a). Generally, the convective pattern is composed of spatially periodic rolls (Williams rolls) with translational invariance, whose periodicity is of order of the thickness of the fluid layer. In LCs, the convective pattern exhibits similar characteristic features and can be generated by applying an electric field to a properly aligned nematic. It was reported that the nonlinear coupling between the convective flow and the director field may lead to the generation of a number of solitonic structures. In 1979, Ribotta measured the penetration length of a vortex into a subcritical region by using the electro-convective instability in a nematic LC [28]. In his experiment, the electrode on one plate is divided into two parts separated by a small gap of about 5 μm. Convective patterns are generated in one region (region 1) by applying an electric field. An electric field with the same frequency but a relatively low amplitude is applied in the other region (region 2), in which no convective pattern is generated. One then observes that small portions of individual rolls are "emitted" from the gap, and propagate into region 2 with a uniform group velocity (Figure 4a). The rolls behave like solitary waves with their shape and amplitude remaining constant during the motion. A different kind of soliton was later reported by Lowe and Gollub in 1985 in a convecting nematic subjected to spatially periodic forcing [29]. The solitons (also called domain walls or discommensurations) are regions of local compression of convective rolls (Figure 4b) and can be well described by the solutions to the Sine-Gordon equation. Generally, the convective pattern is stationary and spatially homogeneous with its velocity field and orientational field being time independent. However, in 1988, Joets and Ribotta reported a time-dependent localized state of electro-convection in a nematic [30]. These spatially localized domains exhibit elliptical shape and distribute randomly throughout the sample. Inside the domains, a periodic structure of Williams Rolls translates uniformly (Figure 4c). The velocity of the rolls has the same amplitude in different domains, but its sign changes randomly from one domain to another. The domains themselves do not show uniform translational motion, instead they fluctuate around an average location. Both the velocity of the rolls and the shape of the domains can be varied by tuning the applied electric field.

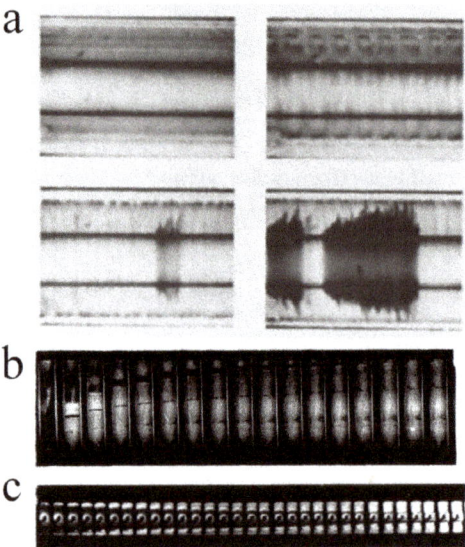

Figure 3. (**a**) Cellular flow in a nematic LC. Reprinted with permission from Ref. [22]. Copyright 1976 Elsevier. (**b**) Propagation process of the director wave. Reprinted with permission from Ref. [23]. Copyright 1982 American Physical Society. (**c**) Propagation of the 2D axisymmetric soliton. Reprinted with permission from Ref. [27]. Copyright 1987 Taylor & Francis.

Figure 4. (a) Portions of convective rolls "emitted" from region 1 to region 2. Reprinted with permission from Ref. [28]. Copyright 1979 American Physical Society. (b) Micrograph of a quasiperiodic convective structure. The solitons (indicated by arrows) are regions of compression of the rolls. Reprinted with permission from Ref. [29]. Copyright 1985 American Physical Society. (c) Micrograph of localized domains of travelling convective rolls. Reprinted with permission from Ref. [31]. Copyright 1988 American Physical Society.

3. Nematicons

Nematicons are self-focused light beams (spatial optical solitons) that propagate in nematic LCs. The beginning may date back to the early works by Braun et al. in which optical beams of complex structures, such as the formation of focal light spots, the onset of transverse beam undulations, and the development of multiple beam filaments, are realized by interacting a low-power laser beam with a nematic LC [32,33]. Compared to most materials, the nonlinear coefficient of nematic LC is extremely large (10^6 to 10^{10} times greater than that of typical optical materials such as CS_2), making it an ideal system for investigating spatial optical solitons [32]. As shown in Figure 5, a linearly polarized beam propagates along the z-axis and enters a nematic cell. The polarization of the beam is parallel to the y-axis. The nematic LC within the cell is homogeneously aligned with its director, **n**, being parallel to the cell substrates but making an angle θ with respect to the wave vector (**k**) of the beam. The extraordinary waves of the beam whose electric field **E** is in the **nk** plane propagate along the Poynting vector **S** which is deviated from **k** at the angle δ [34]. The light induces electric dipoles in the LC molecules which interact with the electric field and produce a torque $\Gamma = \varepsilon_0 \Delta\varepsilon (\mathbf{n}\cdot\mathbf{E})(\mathbf{n}\times\mathbf{E})$, where ε_0 is the dielectric susceptibility of vacuum, $\Delta\varepsilon = n_e^2 - n_o^2$ is the optical anisotropy, n_e and n_o are the extraordinary and ordinary refractive indices, respectively. For nematic LCs with $n_e > n_o$, the torque reorients the director and increases the angle θ, leading to the increase of the extraordinary refractive index $n_{e,\theta} = \frac{n_e n_o}{\sqrt{n_o^2\cdot\sin^2(\theta) + n_e^2\cdot\cos^2(\theta)}}$. Such an increase of the refractive index focuses the beam, and leads to the formation of a nematicon which propagates along **S**. The study of nematicons has attracted a great deal of interest since the beginning of the 21st century due to its promising applications in nonlinear optics and photonics [35]. Recently, different kinds of nematicons, such as vortex nematicons [36–38], have been reported. Since this review is mainly concentrated on topological and dissipative LC solitons, we refer the readers who are interested in nematicons, to a book and several reviews published by Assanto, et al., recommended here [35,39–41].

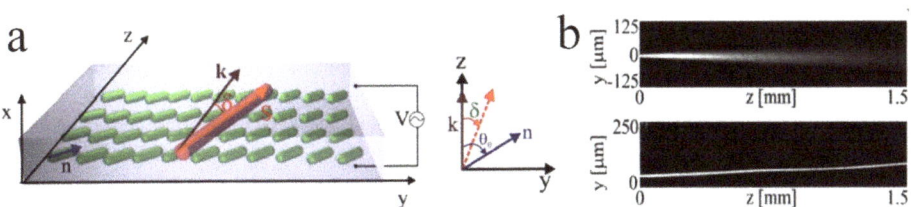

Figure 5. (a) Schematic of the formation of a nematicon in a planar nematic LC cell. (b) Photographs of a propagating ordinary light beam (top) and a nematicon (bottom). Reprinted with permission from Ref. [34]. Copyright 2019 Optical Society of America.

4. Topological Solitons in Chiral Nematics

Topological solitons are continuous but topologically nontrivial field configurations embedded in uniform physical fields that behave like particles and cannot be transformed into a uniform state through smooth deformations [42]. They were probably first proposed by the great mathematician Carl Friedrich Gauss, who envisaged that localized knots of physical fields, such as electric or magnetic fields, could behave like particles [43]. Kelvin and Tait noticed the importance of this concept in physics and proposed one of the early models of atoms, in which they tried to explain the diversity of chemical elements as different knotted vortices [43]. Based on these theories, Hopf proposed the celebrated mathematical Hopf fibration [44], which was later applied to three-dimensional physical fields by Finkelstein [45] and led to the increasing interest of topological solitons to mathematicians and physicists. Nowadays, topological solitons have been investigated in many branches of physics such as instantons in quantum theory [46,47], vortices in superconductors [48], rotons in Bose-Einstein condensates [49], and Skyrme solitons in particle physics [50], etc. The field of topological solitons in LCs started about 50 years ago with the discussion of static linear and planar solitons, which are actually the inversion walls discussed above. In this section, the attention will be mainly focused on the 3D particle-like topological solitons, i.e., the so-called "baby skyrmions", in chiral nematic LCs (CNLCs).

In CNLCs, topological solitons such as 2D merons and skyrmions (low-dimensional analogs of Skyrme solitons) can be generated and have recently received great attention. The molecules of a CNLC form a "layered" structure. In each molecular layer, the director, **n**, aligns in a specific direction. The director of different layers twists at a constant rate along a helical axis which is perpendicular to the layers. The distance over which **n** rotates by an angle of 2π is called the pitch, p. Generally, by applying an electric field to a CNLC or sandwiching it between surfaces of homeotropic anchoring, the helical superstructure of the CNLC will be deformed, leading to the formation of string-like cholesteric fingers [51] and/or nonsingular solitonic field configurations [52]. In 1974, Haas and Adams [52] reported the formation of densely packed particle-like director field configurations, called "spherulites" by the authors, which are now known as "skyrmions", following Skyrme who developed a 3D soliton model of nucleons [50]. In the experiment, the CNLC is confined in a cell with homeotropic anchoring. By applying an electric field to the sample, electro-hydrodynamic effects are induced, and the spherulites can be generated after removing the electric field. Almost at the same time, similar results were also reported by Kawachi et al. [53]. However, in their publication, the spherulites were called "bubble domains". These spherulites or bubble domains soon attracted a great deal of interest and fueled an explosive growth of studies in the next few years [54–62]. In 2009, with the help of laser tweezers, different kinds of spherulites or bubble domains were optically generated at will at a selected place of a homeotropically aligned CNLC by Smalyukh, et al. [63]. By characterizing and simulating the 3D director structure of these solitons, the authors recognized that these are low-dimensional analogs of Skyrme solitons. The solitons are composed of a double-twist cylinder closed on itself in the form of a torus and coupled to the surrounding uniform field by point or line topological defects and are called "torons" (Figure 6a). The authors successfully demonstrated the structure and stability of the torons by the basic field theory of elastic director deformations and obtained the equilibrium field configuration and elastic energy of torons through numerical simulations. Later, the same authors reported the generation of 2D reconfigurable photonic structures composed of ensembles of torons [64–66]. In 2013, Chen et al. reported the generation of Hopf fibration (Figure 6c) in a CNLC by manipulating the two point defects of torons [67]. They demonstrate the relationship between Hopf fibration and torons through a topological visualization technique derived from the Pontryagin-Thom construction. In the following years, a variety of different kinds of skyrmionic solitons, such as half-skyrmions, twistion, skyrmion bags, skyrmion spin ice, skyrmion-dressed colloidal particles, and more complicated structures composed of torons, hopfions, and various disclinations, were realized and reported by different groups [68–76]. The self-assembly of torons (Figure 6d) [77–79] and hopfions in ferromagnetic LCs were later realized by Ackerman et al. [80]. Furthermore, the continuous transformation of 3D Hopf solitons [81] and the generation of 3D knots dubbed "heliknotons" (Figure 6e,f) [82] in CNLCs were reported by Tai et al. Due to the continuous twist of the director field within topological solitons, they can be used as optical devices for controlling and modulating the propagation of light [83–86]. For instance, Varanytsia et al. reported that the surface-assisted assembly of a two-dimensional toron array could be utilized as a spatial light modulator to control the light transmission and scattering [83,87]. Recently, Hess et al. showed that the skyrmionic solitons can act as lenses to steer laser beams [84]. Papic et al. showed that the topological LC solitons inserted in a Fabry-Perot microcavity can be used as a tunable microlaser to generate structured laser beams. The structure of the emitted light could be easily controlled by tuning the topology and geometry of the solitons [88]. In addition, Mai et al. recently showed that topological solitons, such as heliknotons, can be used as micro-templates for spatial reorganization of nanoparticles [89].

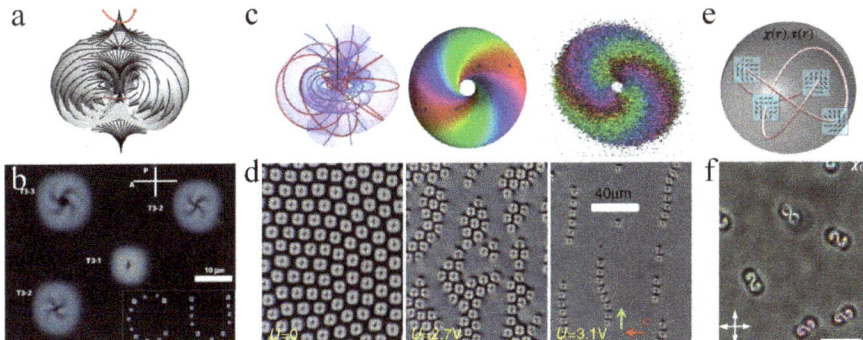

Figure 6. (a) Configuration of a toron. (b) Polarizing microscopy texture of different defect-proliferated torons. Reprinted with permission from Ref. [63]. Copyright 2010 Nature. (c) Flow lines and preimage surfaces of Hopf fibration. Reprinted with permission from Ref. [67]. Copyright 2013 American Physical Society. (d) Self-assembly of skyrmions. Reprinted with permission from Ref. [77]. Copyright 2015 Nature. (e) Knotted co-located half-integer vortex lines in a heliknoton. (f) Polarizing microscopy texture of heliknotons. Reprinted with permission from Ref. [82]. Copyright 2019 Science.

In most investigations, these topological solitons are viewed as static field configurations in LCs. However, it is found that they can be driven into motion by applying electric fields. In 2017, Ackerman et al. reported an electrically driven squirming motion of baby skyrmions in a chiral nematic [90]. By applying a modulated electric field, the skyrmions behave like defects in active matter and move in directions orthogonal to the electric field. Such a motion stems from the non-reciprocal rotational dynamics of LC director fields. During motion, the periodic relaxation and tightening of the twisted region make the skyrmions expand, contract, and morph, resembling squirming motion (Figure 7a). Both the direction and speed of the moving solitons can be controlled by tuning the applied electric field. Such a controllable motion of skyrmions may enable versatile applications such as micro-cargo transport [91]. In 2019, Sohn et al. reported an electrically driven collective motion of skyrmions, in which thousands to millions of skyrmions started from random orientations and motions, but then synchronized their motions and developed polar ordering within seconds (Figure 7b) [92]. They also showed that such a collective motion could even be enriched and guided by light (Figure 7c) [93] and could be used as a model for studying the dynamics of topological defects and grain boundaries in crystalline solid systems [94]. As we mentioned above, the topological defects and solitons are usually generated through local relief of geometric frustration of the helical structure of CNLCs.

As a result, in most works, the topological defects are generated in CNLCs confined by homeotropic anchoring conditions. Very recently, Shen and Dierking reported the creation of 3D topological solitons, i.e., the torons, in a CNLC which is confined in cells of homogeneous anchoring by applying electric fields [95]. In that work, the authors demonstrate the transformation between the cholesteric fingers and the solitons and the formation of "skyrmion bags" with a tunable topological degree. The solitons exhibit different static geometric textures and dynamic behaviors by changing the pitch of the CNLC system (Figure 7d). They undergo anomalous diffusion at equilibrium and performed directional motion driven by electric fields. The solitons could even form aggregates of tunable shape, anisotropy, and fractal dimension through inelastic collisions with each other.

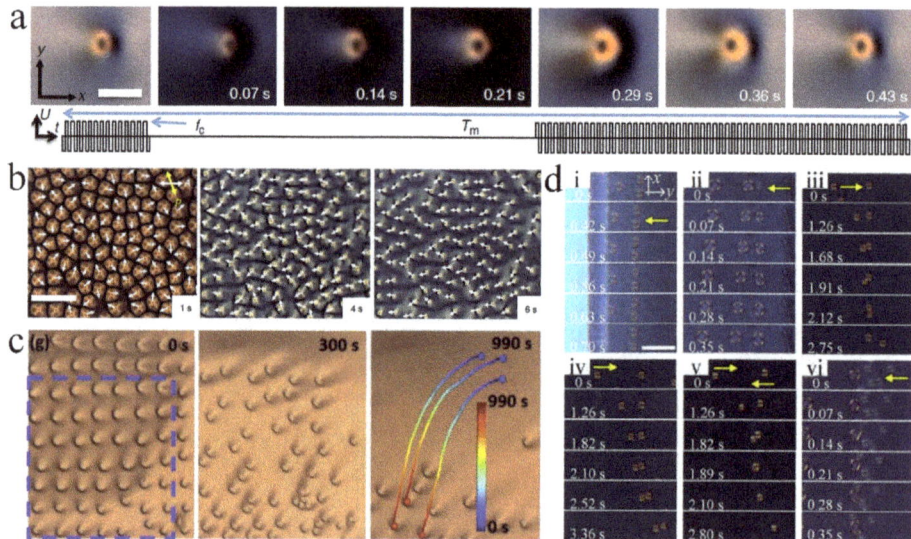

Figure 7. (**a**) Polarizing microscopy texture of a squirming skyrmion. Reprinted with permission from Ref. [90]. Copyright 2017 Nature. (**b**) Temporal evolution of skyrmion velocity order. Reprinted with permission from Ref. [92]. Copyright 2019 Nature. (**c**) Motion of skyrmions guided by light. Reprinted with permission from Ref. [93]. Copyright 2020 Optical Society of America. (**d**) Different dynamic behaviors of torons. Reprinted with permission from Ref. [95]. Copyright 2021 American Physical Society.

In this section, the recent experimental studies on 3D particle-like skyrmionic topological solitons were briefly introduced. For readers who are interested in more details concerning this topic, an excellent review written by Smalyukh is strongly recommended [96].

5. Dynamic Dissipative Solitons in Liquid Crystals

Dissipative solitons are stable localized solitary deviations of a state variable from an otherwise homogeneous stable stationary background distribution. They are generally powered by an external driver and vanish below a finite strength of the driver [97]. Experimentally, dissipative solitons were generated in the form of electric current filaments in a 2D planar gas-discharge system [98]. In LCs, different kinds of dissipative solitons have been generated and reported recently [99–105].

In 2018, Li et al. reported the formation of 3D dissipative solitons in an electrically driven nematic, which were called "director bullets" [99] or "directrons" [100] by the authors (Figure 8). These solitons were first reported by Brand et al. in 1997, and were called "butterflies", but did not receive much attention at that time. The directrons (we will refer to them as "directrons" to distinguish them from other solitons) are self-confined localized director deformations. While the nematic aligns homogeneously outside the directrons, the director field is distorted and oscillates with the frequency of the applied AC electric field within the directrons. Such an oscillation breaks the fore-aft symmetry of the structure of the directrons and leads to the rapid propagation perpendicular to the alignment direction. The directrons can move with speeds as large as 1000 µm s^{-1} through the homogeneous nematic bulk over a macroscopic distance thousands of times larger than their size. They survive collisions and pass through each other without losing their identities. Unlike the topological solitons, the directrons are topologically equivalent to a uniform state and disappear right after switching off the applied electric field. The nematic media in which the directrons were generated by Li et al., 4′-butyl-4-heptyl-bicyclohexyl-4-carbonitrile (CCN-47), is of the (−,−) type, which means that both the dielectric and conductivity anisotropies are negative, i.e., $\Delta\varepsilon = \varepsilon_\| - \varepsilon_\perp < 0$ and $\Delta\sigma = \sigma_\| - \sigma_\perp < 0$, respectively. The basic mechanism of many electro-hydrodynamic instabilities in nematics of the (−,+) type is now quite well understood and can be explained by the well-known "Carr-Helfrich" model, in which a subtle balance between the dielectric torque that stabilizes the initial planar director field and an anomalous conductive torque induced by space charges that breaks the planar state is reached [106,107]. However, the model cannot be used to explain the generation of the directrons observed in (−,−) nematics, in which case both the dielectric and conductivity torques can only stabilize the planar state. Instead, due to the equality of the frequency of the directron oscillation and that of the applied electric field, the main reason of the excitation of the directrons is attributed to the flexoelectric polarization [105].

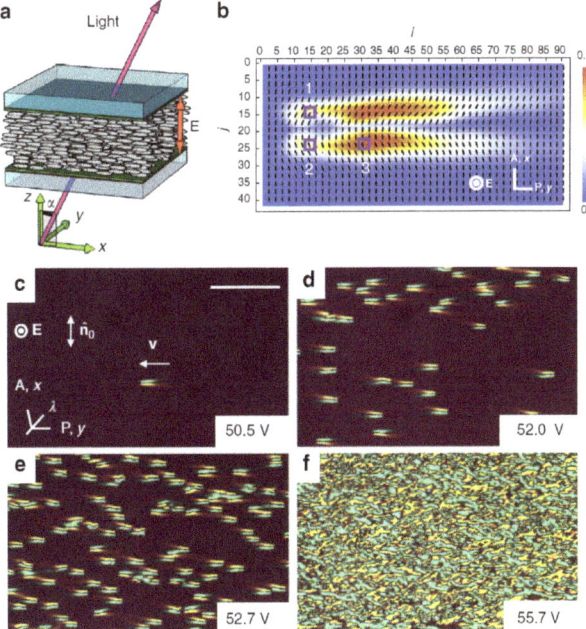

Figure 8. Director bullets in a planar nematic cell. (**a**) Cell scheme. (**b**) Transmitted light intensity map and director distortions in the xy plane within a single bullet. (**c–e**) Polarizing microscopy of the director bullets at varied voltages. (**f**) Polarizing microscopy of the electro-hydrodynamic pattern. Scale bar 200 µm. Reprinted with permission from Ref. [99]. Copyright 2018 Nature.

The study carried out by Aya and Araoka showed that similar directrons can also be generated in nematics of the (−,+) type [101]. In order to systemically examine the influences of the material parameters on the generation and dynamics of the directrons, the authors used a mixture of two different nematics which are of the (−,−) type and the (+,+) type, respectively. By altering the concentrations of the two nematics, a continuous transition of the dielectric and conductivity anisotropies of the mixture could be realized, leading to the formation of different kinds of electro-hydrodynamic patterns. The authors found that the conductivity is vital in determining the stability of the directrons. These directrons can only exist in the limited range of moderate conductivity, $0.8 \times 10^{-8} < \sigma < 4 \times 10^{-8}\ \Omega^{-1}\mathrm{m}^{-1}$. The unifying feature of the generation of the directrons in nematics of the (−,−) type and the (−,+) type is that the conductivity of the nematic host is relatively low compared to those typically used in exploring conventional Carr-Helfrich electro-hydrodynamic phenomena ($\sim 10^{-7}\ \Omega^{-1}\mathrm{m}^{-1}$) [108]. Li et al. reported that the conductivity of the nematic host (CCN-47) for producing dissipative solitons was about $(0.5–0.6) \times 10^{-8}\ \Omega^{-1}\mathrm{m}^{-1}$ [99], and the conductivity of the nematic (ZLI-2806) used in the experiments by Shen and Dierking was about $(0.6–1.9) \times 10^{-8}\ \Omega^{-1}\mathrm{m}^{-1}$ [102].

On the other hand, a recent study carried out by Shen and Dierking showed that similar directrons could even be produced in nematics of the (+,+) type (Figure 9) [103], which is unexpected by the "standard model" [109]. In their experiment, a nematic (5CB) with both positive dielectric and conductivity anisotropies was confined in a cell with homogeneous alignment. The alignment of the cells was induced through the photo-alignment technique instead of the conventional rubbing method, the former providing a relatively weak azimuthal anchoring. The directrons are generated after applying an AC electric field with a relatively low frequency to the sample. The dynamic behavior of the directrons is similar to that reported by Li et al. [87,88] and Aya et al. [89]. Directrons can propagate either parallel or perpendicular to the alignment direction which can be switched by tuning the frequency and amplitude of the applied electric field [99–101]. The directrons behave like waves when they pass through each other without losing their identities after collisions. Electro-hydrodynamic instabilities in LCs have been investigated for decades. Early studies mainly focus on the electro-convection effects in nematics with opposite signs of anisotropies, e.g., nematics of the (−,+) type, in which electro-convection rolls, such as the well-known "Williams domains" [110] and "chevrons" [28], are observed. In most of the cases, the destabilization can be explained by the charge separation mechanism introduced by Carr [106] and Helfrich [107], which was later extended to a 3D theory, i.e., the "standard model" [109]. On the other hand, most nonstandard electro-convection phenomena observed in nematics of the (−,−) type can be explained by adding flexoelectricity effects to the standard model. The standard model predicts no electro-hydrodynamic instability in nematics of the (+,+) type. However, complicated electro-convection patterns, such as fingerprint textures [111], Maltese crosses [111], and cellular patterns [112], were reported in nematic cells with homeotropic alignment. Different explanations, including isotropic ionic flows [111,113,114], charge injection (known as Felici-Benard mechanism) [115,116], flexoelectricity, and surface-polarization effects [117,118], etc., were proposed to account for the origin of the instabilities. However, a rigorous explanation is still to be found. In the case of homogeneous alignment, only stationary Williams domains were observed in nematics with small values of the dielectric anisotropy ($0 < \Delta\varepsilon < 0.4$). For nematics with large dielectric anisotropy (just as the situation in ref [103]), electro-hydrodynamic instabilities are usually suppressed by the Freedericksz transition and thus are not expected [119]. The formation of the directrons in 5CB reported by Shen and Dierking is attributed to the special conditions of their experimental setup, i.e., a relatively high ion concentration of the nematic host and a relatively weak azimuthal anchoring of the cells. Both of these factors lead to the strong nonlinear coupling between the isotropic ionic flows and the director field which induces the directrons [103].

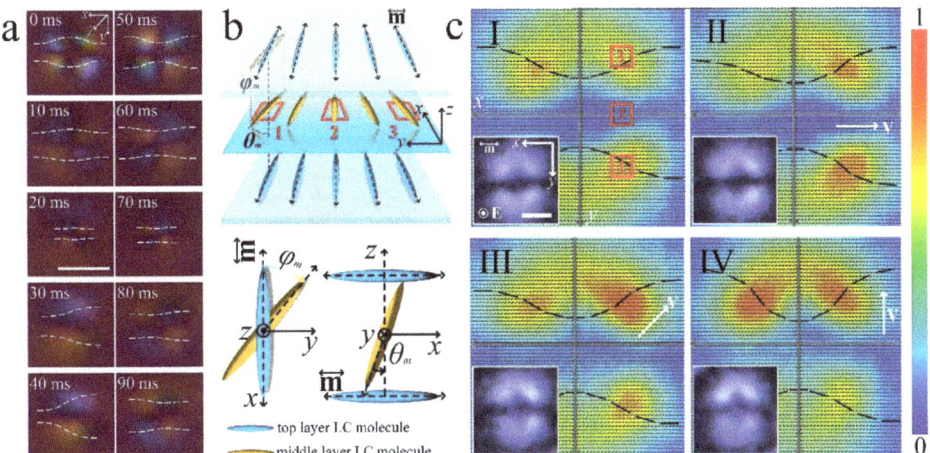

Figure 9. The structure of the dissipative solitons in 5CB. (**a**) Time series of polarizing micrographs of a soliton modulated by an AC electric field. Scale bar 20 µm. (**b**) The schematic structure of a soliton. **m** represents the alignment direction. φ_m and θ_m represent the azimuthal angle and the polar angle of the local mid-layer director. (**c**) Transmitted light intensity maps and the corresponding mid-layer director fields (black dashed lines) in the xy plane within solitons. **v** represents the velocity of the soliton. The color bar shows a linear scale of transmitted light intensity. Insets are the corresponding POM micrographs, scale bar 10 µm. Red squares 1, 2, and 3 are corresponding to the ones in (**b**). Reprinted with permission from Ref. [103]. Copyright 2020 Royal Society of Chemistry.

The formation of the dynamic directrons is not the privilege of nematics only. Shen and Dierking showed that the directrons can also form in chiral nematics of the (−,+) type and the (+,+) type, respectively [102,103]. In the experiments, the chiral nematic hosts are prepared by doping a chiral dopant into achiral nematics, and the mixtures are filled into LC cells with homogeneous alignment conditions. An AC electric field is applied parallel to the helical axis of the chiral nematics, and the directrons emerge as the amplitude of the electric field increases above some frequency-dependent thresholds. The dynamic behavior of the directrons was investigated and compared to those in achiral nematics. It was found that the directrons in the achiral nematics show a "butterfly-like" structure (Figure 10a), but the ones in chiral nematics exhibit a "bullet-like" structure (Figure 10b). In both cases (chiral and achiral nematics), the directrons move either parallel or perpendicular to the alignment direction, a behavior which is dependent on the applied electric field. Interestingly, the directrons in achiral nematics behave like waves in that they collide and pass through each other as reported by Li et al. [99]. The directrons in chiral nematics behave similar to a wave-particle dualism in so far that they either pass through each other without losing identity (solitary wave) (Figure 10c), or collide with each other and undergo reflection (hard particle) (Figure 10d). The authors also showed that the motion of the directrons can be controlled by the alignment. As shown in Figure 10e, the LC cell is divided into three regions with different alignment directions by using the photo-alignment technique. Tuning the applied electric field, the directrons either move parallel or perpendicular to the alignment direction in each individual region. However, once the directrons move across the boundaries of different regions, they change their directions continuously to fit the alignment condition. The directrons can therefore be used as vehicles for micro-cargo transport [102,103]. As shown in Figure 10f, a directron is induced around a dust particle once the electric field is applied. It then carries and translates the particle by moving it through the nematic bulk. Similar phenomena were also reported by Li et al. which was termed "directron-induced liquid crystal-enabled electrophoresis" [120].

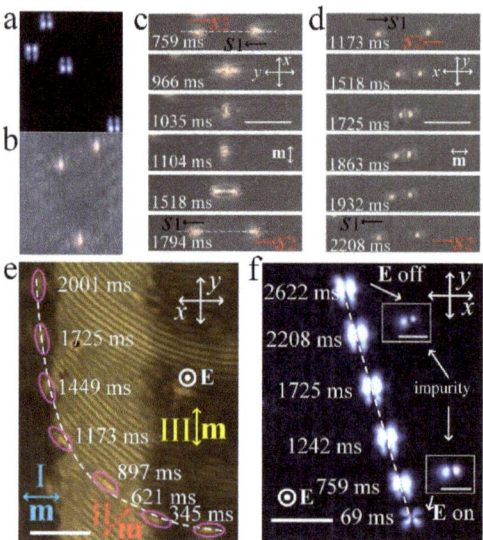

Figure 10. Dissipative solitons in chiral nematics. Solitons in achiral (**a**) and chiral (**b**) nematics. (**c**) Two solitons pass through each other. (**d**) Two solitons collide and reflect into opposite directions. (**e**) Motion of solitons in a cell divided into three regions with different alignment directions. (**f**) Micro-cargo transport by a soliton. Reprinted with permission from Ref. [102]. Copyright 2020 Nature.

So far, most dynamic solitons have been reported only in nematics. However, recently, it was shown that particle-like dynamic dissipative solitons can be formed in LCs of the fluid smectic A phase [104]. A smectic A phase is characterized by the formation of a layered structure of elongated molecules with orientational and 1D positional order. Within each layer, the LC molecules are orientated perpendicular to the layer, but their molecular centers of mass are distributed randomly without any further in-plane positional order [121]. The smectic phase is usually characterized by the remarkable patterns of singular ellipses, hyperbolas, and parabolas known as the focal conic domains (FCDs). In this work, a smectic LC (8CB) is confined in a LC cell with homogeneous alignment condition. The sample is kept at a temperature slightly below the nematic-smectic phase transition point. The solitons are formed by applying a low-frequency field with a mediate voltage to the so-called "scattering state" of the smectic A phase. The solitons exhibit a swallow-tail like texture under crossed polarizers with a static structure analogous to the parabolic focal conic domains (PFCDs) (Figure 11). They are characterized by an elliptical contour, which is generated as a result of the localization of stress. The contour is composed of the loci of the cusps of smectic layers. Outside the solitons, the equidistant smectic layers align homogeneously perpendicular to the alignment direction. Within the solitons, the transmitted light intensity increases, indicating azimuthal deviations of the director field from the alignment. As a result, the smectic layers continuously deform into curves within the solitons. The curvature of the curves exhibits a maximum at one of the foci of the elliptical contour, where a singular defect line is located and acts as the core of the solitons. The size of the solitons is dependent on the applied voltage, and decreases with increasing voltage. When driven by a low-frequency electric field, the director within the solitons tilts up and down due to the dielectric torque, which leads to a periodic shape transformation of the soliton. The solitons move bidirectionally along the alignment direction with constant speed, which is attributed to the permeation ion flow perpendicular to the smectic layers. The solitons behave like particles when they collide with each other. They can also interact with colloidal micro-particles.

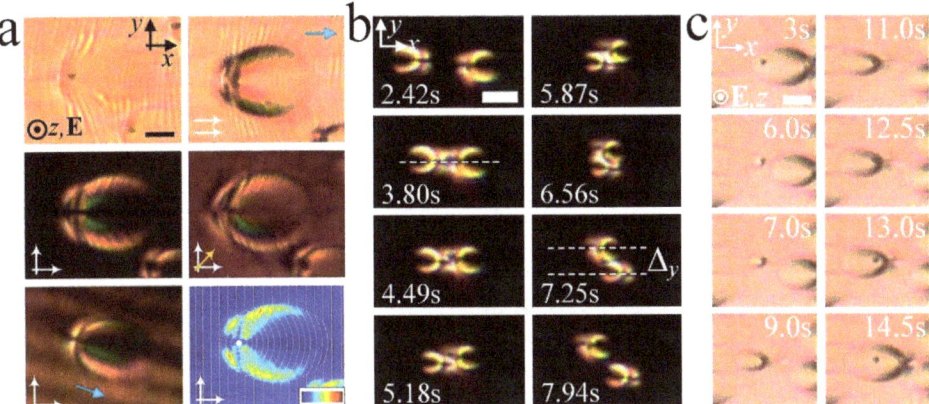

Figure 11. Swallow-tail solitons in smectics. (**a**) Polarizing micrographs of the swallow-tail soliton and its corresponding static structure. (**b**) Collision of two swallow-tail solitons. (**c**) Nucleation of swallow-tail solitons on a colloidal micro-particle. Reprinted with permission from Ref. [104]. Copyright 2021 Royal Society of Chemistry.

6. Conclusions

In summary, we have briefly discussed some important early works of solitons in LCs and the recent progresses made in the investigations of topological solitons and dissipative solitons in LCs. Although recent studies of topological solitons and dissipative solitons have received great attention, many fundamental questions remain unanswered. For instance, the existence of topological solitons with higher dimensions in biaxial liquid crystal systems, a systematic classification of the topological solitons, the stability of the topological and dissipative solitons, the transformation between different topological solitons, the influence of the topological structure on the dynamics and interactions of topological solitons, the formation mechanism of the directrons, the role of ions played in the formation and motion of dissipative solitons, the influence of surface anchoring on the stability, formation and dynamics of the solitons, the effect of chirality on the structure and dynamics of the solitons, the interactions between solitons and colloidal particles, the self-assembly and collective behavior of the solitons, the existence of topological and dissipative soliton in lyotropic and active LC systems, the relation between the solitons in LCs and the solitons in other physical systems, etc. All these questions remain elusive and require further experimental and theoretical investigations to answer.

After over five decades of research, various solitons have been created and described in different liquid crystalline systems. This not only broadens the research and understanding of LCs, but also enhances our understanding of solitons in other physical systems. Furthermore, the solitons in LCs may even lead to novel phenomena, such as emergent collective motion of solitons [92,93], and applications, such as micro-cargo transport [102,103,120], optic processing [84,85], or fast LC displays [14]. We hope this brief review can arouse more researchers' interest in the field of solitons in LC systems.

Author Contributions: Y.S. conceived and wrote the manuscript. I.D. contributed through discussing and writing the manuscript. All authors have read and agreed to the published version of the manuscript.

Funding: This research received no external funding.

Institutional Review Board Statement: Not applicable.

Informed Consent Statement: Not applicable.

Data Availability Statement: Not applicable.

Acknowledgments: Yuan Shen would like to thank the China Scholarship Council (CSC number: 201806310129).

Conflicts of Interest: The authors declare no conflict of interest.

References

1. Russell, J.S. *Report on Waves: Made to the Meetings of the British Association in 1842–1843*; Richard and John E Taylor: London, UK, 1845.
2. Korteweg, D.J.; De Vries, G. XLI. On the change of form of long waves advancing in a rectangular canal, and on a new type of long stationary waves. *Lond. Edinb. Dublin Philos. Mag. J. Sci.* **1895**, *39*, 422–443. [CrossRef]
3. Zabusky, N.J.; Kruskal, M.D. Interaction of "Solitons" in a Collisionless Plasma and the Recurrence of Initial States. *Phys. Rev. Lett.* **1965**, *15*, 240–243. [CrossRef]
4. Du, L.; Yang, A.; Zayats, A.V.; Yuan, X. Deep-subwavelength features of photonic skyrmions in a confined electromagnetic field with orbital angular momentum. *Nat. Phys.* **2019**, *15*, 650–654. [CrossRef]
5. Ray, M.W.; Ruokokoski, E.; Kandel, S.; Möttönen, M.; Hall, D. Observation of Dirac monopoles in a synthetic magnetic field. *Nature* **2014**, *505*, 657–660. [CrossRef]
6. Harada, K.; Matsuda, T.; Bonevich, J.; Igarashi, M.; Kondo, S.; Pozzi, G.; Kawabe, U.; Tonomura, A. Real-time observation of vortex lattices in a superconductor by electron microscopy. *Nature* **1992**, *360*, 51–53. [CrossRef]
7. Yu, X.; Onose, Y.; Kanazawa, N.; Park, J.; Han, J.; Matsui, Y.; Nagaosa, N.; Tokura, Y. Real-space observation of a two-dimensional skyrmion crystal. *Nature* **2010**, *465*, 901–904. [CrossRef] [PubMed]
8. Scott, A.C.; Chu, F.Y.F.; McLaughlin, D.W. The soliton: A new concept in applied science. *Proc. IEEE* **1973**, *61*, 1443–1483. [CrossRef]
9. Bullough, R. Solitons. *Phys. Bull.* **1978**, *29*, 78. [CrossRef]
10. Malomed, B.A.; Mihalache, D.; Wise, F.; Torner, L. Spatiotemporal optical solitons. *J. Opt. B Quantum Semiclassical Opt.* **2005**, *7*, R53. [CrossRef]
11. Dauxois, T.; Peyrard, M. *Physics of Solitons*; Cambridge University Press: Cambridge, UK, 2006.
12. De Gennes, P.-G.; Prost, J. *The Physics of Liquid Crystals*, 2nd ed.; Oxford University Press: Oxford, UK, 1993; Volume 83.
13. Shen, Y.; Dierking, I. Perspectives in Liquid-Crystal-Aided Nanotechnology and Nanoscience. *Appl. Sci.* **2019**, *9*, 2512. [CrossRef]
14. Lam, L.; Prost, J. *Solitons in Liquid Crystals*; Springer Science & Business Media: New York, NY, USA, 2012.
15. Helfrich, W. Alignment-Inversion Walls in Nematic Liquid Crystals in the Presence of a Magnetic Field. *Phys. Rev. Lett.* **1968**, *21*, 1518–1521. [CrossRef]
16. De Gennes, P. Mouvements de parois dans un nématique sous champ tournant. *J. De Phys.* **1971**, *32*, 789–792. [CrossRef]
17. Leger, L. Observation of wall motions in nematics. *Solid State Commun.* **1972**, *10*, 697–700. [CrossRef]
18. Brochard, F. Mouvements de parois dans une lame mince nématique. *J. De Phys.* **1972**, *33*, 607–611. [CrossRef]
19. Leger, L. Static and dynamic behaviour of walls in nematics above a Freedericks transition. *Solid State Commun.* **1972**, *11*, 1499–1501. [CrossRef]
20. Léger, L. Walls in Nematics. *Mol. Cryst. Liq. Cryst.* **1973**, *24*, 33–44. [CrossRef]
21. Shen, Y.; Dierking, I. Annihilation dynamics of reverse tilt domains in nematic liquid crystals. *J. Mol. Liq.* **2020**, *313*, 113547. [CrossRef]
22. Cladis, P.; Torza, S. Flow instabilities in Couette flow in nematic liquid crystals. In *Hydrosols and Rheology*; Elsevier: London, UK, 1976; pp. 487–499.
23. Guozhen, Z. Experiments on Director Waves in Nematic Liquid Crystals. *Phys. Rev. Lett.* **1982**, *49*, 1332–1335. [CrossRef]
24. Lei, L.; Changqing, S.; Juelian, S.; Lam, P.M.; Yun, H. Soliton Propagation in Liquid Crystals. *Phys. Rev. Lett.* **1982**, *49*, 1335–1338. [CrossRef]
25. Lei, L.; Changqing, S.; Gang, X. Generation and detection of propagating solitons in shearing liquid crystals. *J. Stat. Phys.* **1985**, *39*, 633–652. [CrossRef]
26. Lin, L.; Shu, C.; Xu, G. Comment on "on solitary waves in liquid crystals". *Phys. Lett. A* **1985**, *109*, 277–278. [CrossRef]
27. Shu, C.Q.; Shao, R.F.; Zheng, S.; Liang, Z.C.; He, G.; Xu, G.; Lam, L. Two-dimensional axisymmetric solitons in nematic crystals. *Liq. Cryst.* **1987**, *2*, 717–722. [CrossRef]
28. Ribotta, R. Critical Behavior of the Penetration Length of a Vortex into a Subcritical Region. *Phys. Rev. Lett.* **1979**, *42*, 1212–1215. [CrossRef]
29. Lowe, M.; Gollub, J.P. Solitons and the commensurate-incommensurate transition in a convecting nematic fluid. *Phys. Rev. A* **1985**, *31*, 3893–3897. [CrossRef] [PubMed]
30. Joets, A.; Ribotta, R. Localized bifurcations and defect instabilities in the convection of a nematic liquid crystal. *J. Stat. Phys.* **1991**, *64*, 981–1005. [CrossRef]
31. Joets, A.; Ribotta, R. Localized, Time-Dependent State in the Convection of a Nematic Liquid Crystal. *Phys. Rev. Lett.* **1988**, *60*, 2164–2167. [CrossRef]
32. Braun, E.; Faucheux, L.; Libchaber, A.; McLaughlin, D.; Muraki, D.; Shelley, M. Filamentation and undulation of self-focused laser beams in liquid crystals. *EPL (Europhys. Lett.)* **1993**, *23*, 239. [CrossRef]

33. Braun, E.; Faucheux, L.P.; Libchaber, A. Strong self-focusing in nematic liquid crystals. *Phys. Rev. A* **1993**, *48*, 611. [CrossRef]
34. Laudyn, U.A.; Kwaśny, M.; Karpierz, M.A.; Assanto, G. Electro-optic quenching of nematicon fluctuations. *Opt. Lett.* **2019**, *44*, 167–170. [CrossRef]
35. Assanto, G. *Nematicons: Spatial Optical Solitons in Nematic Liquid Crystals*; John Wiley & Sons: Hoboken, NJ, USA, 2012; Volume 74.
36. Izdebskaya, Y.V.; Shvedov, V.G.; Jung, P.S.; Krolikowski, W. Stable vortex soliton in nonlocal media with orientational nonlinearity. *Opt. Lett.* **2018**, *43*, 66–69. [CrossRef]
37. Laudyn, U.A.; Kwaśny, M.; Karpierz, M.A.; Assanto, G. Vortex nematicons in planar cells. *Opt. Express* **2020**, *28*, 8282–8290. [CrossRef]
38. Izdebskaya, Y.; Assanto, G.; Krolikowski, W. Observation of stable-vector vortex solitons. *Opt. Lett.* **2015**, *40*, 4182–4185. [CrossRef]
39. Assanto, G.; Karpierz, M.A. Nematicons: Self-localised beams in nematic liquid crystals. *Liq. Cryst.* **2009**, *36*, 1161–1172. [CrossRef]
40. Peccianti, M.; Assanto, G. Nematicons. *Phys. Rep.* **2012**, *516*, 147–208. [CrossRef]
41. Assanto, G. Nematicons: Reorientational solitons from optics to photonics. *Liq. Cryst. Rev.* **2018**, *6*, 170–194. [CrossRef]
42. Manton, N.; Sutcliffe, P. *Topological Solitons*; Cambridge University Press: Cambridge, UK, 2004.
43. Kauffman, L.H. *Knots and Physics*; World Scientific: Farrer Road, Singapore, 2001; Volume 1.
44. Hopf, H. Über die Abbildungen der dreidimensionalen Sphäre auf die Kugelfläche. *Math. Ann.* **1931**, *104*, 637–665. [CrossRef]
45. Finkelstein, D. Kinks. *J. Math. Phys.* **1966**, *7*, 1218–1225. [CrossRef]
46. Shuryak, E.V. The role of instantons in quantum chromodynamics:(I). Physical vacuum. *Nucl. Phys. B* **1982**, *203*, 93–115. [CrossRef]
47. Shuryak, E.V. The role of instantons in quantum chromodynamics:(II). Hadronic structure. *Nucl. Phys. B* **1982**, *203*, 116–139.
48. Abrikosov, A.A. Nobel Lecture: Type-II superconductors and the vortex lattice. *Rev. Mod. Phys.* **2004**, *76*, 975–979. [CrossRef]
49. O'dell, D.; Giovanazzi, S.; Kurizki, G. Rotons in gaseous Bose-Einstein condensates irradiated by a laser. *Phys. Rev. Lett.* **2003**, *90*, 110402. [CrossRef]
50. Skyrme, T.H.R. A unified field theory of mesons and baryons. *Nucl. Phys.* **1962**, *31*, 556–569. [CrossRef]
51. Oswald, P.; Baudry, J.; Pirkl, S. Static and dynamic properties of cholesteric fingers in electric field. *Phys. Rep.* **2000**, *337*, 67–96. [CrossRef]
52. Haas, W.E.L.; Adams, J.E. New optical storage mode in liquid crystals. *Appl. Phys. Lett.* **1974**, *25*, 535–537. [CrossRef]
53. Kawachi, M.; Kogure, O.; Kato, Y. Bubble domain texture of a liquid crystal. *Jpn. J. Appl. Phys.* **1974**, *13*, 1457. [CrossRef]
54. Haas, W.E.L.; Adams, J.E. Electrically variable diffraction in spherulitic liquid crystals. *Appl. Phys. Lett.* **1974**, *25*, 263–264. [CrossRef]
55. Nawa, N.; Nakamura, K. Observation of Forming Process of Bubble Domain Texture in Liquid Crystals. *Jpn. J. Appl. Phys.* **1978**, *17*, 219. [CrossRef]
56. Stieb, A. Structure of elongated and spherulitic domains in long pitch cholesterics with homeotropic boundary alignment. *J. De Phys.* **1980**, *41*, 961–969. [CrossRef]
57. Hirata, S.; Akahane, T.; Tako, T. New Molecular Alignment Models of Bubble Domains and Striped Domains in Cholesteric-Nematic Mixtures. *Mol. Cryst. Liq. Cryst.* **1981**, *75*, 47–67. [CrossRef]
58. Kerllenevich, B.; Coche, A. Bubble domain in cholesteric liquid crystals. *Mol. Cryst. Liq. Cryst.* **1981**, *68*, 47–55. [CrossRef]
59. Bouligand, Y.; Livolant, F. The organization of cholesteric spherulites. *J. De Phys.* **1984**, *45*, 1899–1923. [CrossRef]
60. Pirkl, S.; Ribière, P.; Oswald, P. Forming process and stability of bubble domains in dielectrically positive cholesteric liquid crystals. *Liq. Cryst.* **1993**, *13*, 413–425. [CrossRef]
61. Pirkl, S.; Oswald, P. From bubble domains to spirals in cholesteric liquid crystals. *J. De Phys. II* **1996**, *6*, 355–373. [CrossRef]
62. Baudry, J.; Pirkl, S.; Oswald, P. Looped finger transformation in frustrated cholesteric liquid crystals. *Phys. Rev. E* **1999**, *59*, 5562–5571. [CrossRef] [PubMed]
63. Smalyukh, I.I.; Lansac, Y.; Clark, N.A.; Trivedi, R.P. Three-dimensional structure and multistable optical switching of triple-twisted particle-like excitations in anisotropic fluids. *Nat. Mater.* **2009**, *9*, 139. [CrossRef] [PubMed]
64. Trushkevych, O.; Ackerman, P.; Crossland, W.A.; Smalyukh, I.I. Optically generated adaptive localized structures in confined chiral liquid crystals doped with fullerene. *Appl. Phys. Lett.* **2010**, *97*, 201906. [CrossRef]
65. Ackerman, P.J.; Qi, Z.; Smalyukh, I.I. Optical generation of crystalline, quasicrystalline, and arbitrary arrays of torons in confined cholesteric liquid crystals for patterning of optical vortices in laser beams. *Phys. Rev. E* **2012**, *86*, 021703. [CrossRef]
66. Smalyukh, I.I.; Kaputa, D.; Kachynski, A.V.; Kuzmin, A.N.; Ackerman, P.J.; Twombly, C.W.; Lee, T.; Trivedi, R.P.; Prasad, P.N. Optically generated reconfigurable photonic structures of elastic quasiparticles in frustrated cholesteric liquid crystals. *Opt. Express* **2012**, *20*, 6870–6880. [CrossRef] [PubMed]
67. Chen, B.G.-G.; Ackerman, P.J.; Alexander, G.P.; Kamien, R.D.; Smalyukh, I.I. Generating the Hopf fibration experimentally in nematic liquid crystals. *Phys. Rev. Lett.* **2013**, *110*, 237801. [CrossRef]
68. Ackerman, P.J.; Trivedi, R.P.; Senyuk, B.; van de Lagemaat, J.; Smalyukh, I.I. Two-dimensional skyrmions and other solitonic structures in confinement-frustrated chiral nematics. *Phys. Rev. E* **2014**, *90*, 012505. [CrossRef]
69. Ackerman, P.J.; Smalyukh, I.I. Diversity of Knot Solitons in Liquid Crystals Manifested by Linking of Preimages in Torons and Hopfions. *Phys. Rev. X* **2017**, *7*, 011006. [CrossRef]

70. Ackerman, P.J.; Smalyukh, I.I. Reversal of helicoidal twist handedness near point defects of confined chiral liquid crystals. *Phys. Rev. E* **2016**, *93*, 052702. [CrossRef]
71. Nych, A.; Fukuda, J.-i.; Ognysta, U.; Žumer, S.; Muševič, I. Spontaneous formation and dynamics of half-skyrmions in a chiral liquid-crystal film. *Nat. Phys.* **2017**, *13*, 1215. [CrossRef]
72. Fukuda, J.-I.; Nych, A.; Ognysta, U.; Žumer, S.; Muševič, I. Liquid-crystalline half-Skyrmion lattice spotted by Kossel diagrams. *Sci. Rep.* **2018**, *8*, 1–8. [CrossRef]
73. Duzgun, A.; Nisoli, C. Artificial spin ice of liquid crystal skyrmions. *arXiv* **2019**, arXiv:1908.03246.
74. Foster, D.; Kind, C.; Ackerman, P.J.; Tai, J.-S.B.; Dennis, M.R.; Smalyukh, I.I. Two-dimensional skyrmion bags in liquid crystals and ferromagnets. *Nat. Phys.* **2019**. [CrossRef]
75. Pandey, M.B.; Porenta, T.; Brewer, J.; Burkart, A.; Čopar, S.; Žumer, S.; Smalyukh, I.I. Self-assembly of skyrmion-dressed chiral nematic colloids with tangential anchoring. *Phys. Rev. E* **2014**, *89*, 060502. [CrossRef] [PubMed]
76. Porenta, T.; Čopar, S.; Ackerman, P.J.; Pandey, M.B.; Varney, M.C.M.; Smalyukh, I.I.; Žumer, S. Topological Switching and Orbiting Dynamics of Colloidal Spheres Dressed with Chiral Nematic Solitons. *Sci. Rep.* **2014**, *4*, 7337. [CrossRef]
77. Ackerman, P.J.; van de Lagemaat, J.; Smalyukh, I.I. Self-assembly and electrostriction of arrays and chains of hopfion particles in chiral liquid crystals. *Nat. Commun.* **2015**, *6*, 6012. [CrossRef]
78. Kim, Y.H.; Gim, M.-J.; Jung, H.-T.; Yoon, D.K. Periodic arrays of liquid crystalline torons in microchannels. *RSC Adv.* **2015**, *5*, 19279–19283. [CrossRef]
79. Sohn, H.R.O.; Liu, C.D.; Wang, Y.; Smalyukh, I.I. Light-controlled skyrmions and torons as reconfigurable particles. *Opt. Express* **2019**, *27*, 29055–29068. [CrossRef]
80. Ackerman, P.J.; Smalyukh, I.I. Static three-dimensional topological solitons in fluid chiral ferromagnets and colloids. *Nat. Mater.* **2016**, *16*, 426. [CrossRef]
81. Tai, J.-S.B.; Ackerman, P.J.; Smalyukh, I.I. Topological transformations of Hopf solitons in chiral ferromagnets and liquid crystals. *Proc. Natl. Acad. Sci. USA* **2018**, *115*, 921. [CrossRef] [PubMed]
82. Tai, J.-S.B.; Smalyukh, I.I. Three-dimensional crystals of adaptive knots. *Science* **2019**, *365*, 1449. [CrossRef]
83. Varanytsia, A.; Chien, L.-C. Photoswitchable and dye-doped bubble domain texture of cholesteric liquid crystals. *Opt. Lett.* **2015**, *40*, 4392–4395. [CrossRef] [PubMed]
84. Hess, A.J.; Poy, G.; Tai, J.-S.B.; Žumer, S.; Smalyukh, I.I. Control of light by topological solitons in soft chiral birefringent media. *Phys. Rev. X* **2020**, *10*, 031042. [CrossRef]
85. Poy, G.; Hess, A.J.; Smalyukh, I.I.; Žumer, S. Chirality-enhanced periodic self-focusing of light in soft birefringent media. *Phys. Rev. Lett.* **2020**, *125*, 077801. [CrossRef] [PubMed]
86. Loussert, C.; Iamsaard, S.; Katsonis, N.; Brasselet, E. Subnanowatt Opto-Molecular Generation of Localized Defects in Chiral Liquid Crystals. *Adv. Mater.* **2014**, *26*, 4242–4246. [CrossRef]
87. Varanytsia, A.; Chien, L.-C. P-134: A Spatial Light Modulator with a Two-Dimensional Array of Liquid Crystal Bubbles. *SID Symp. Dig. Tech. Pap.* **2014**, *45*, 1492–1495. [CrossRef]
88. Papič, M.; Mur, U.; Zuhail, K.P.; Ravnik, M.; Muševič, I.; Humar, M. Topological liquid crystal superstructures as structured light lasers. *Proc. Natl. Acad. Sci. USA* **2021**, *118*, 1–7. [CrossRef]
89. Mai, Z.; Yuan, Y.; Tai, J.-S.B.; Senyuk, B.; Liu, B.; Li, H.; Wang, Y.; Zhou, G.; Smalyukh, I.I. Nematic Order, Plasmonic Switching and Self-Patterning of Colloidal Gold Bipyramids. *Adv. Sci.* **2021**, *8*, 2102854. [CrossRef]
90. Ackerman, P.J.; Boyle, T.; Smalyukh, I.I. Squirming motion of baby skyrmions in nematic fluids. *Nat. Commun.* **2017**, *8*, 673. [CrossRef]
91. Sohn, H.R.O.; Ackerman, P.J.; Boyle, T.J.; Sheetah, G.H.; Fornberg, B.; Smalyukh, I.I. Dynamics of topological solitons, knotted streamlines, and transport of cargo in liquid crystals. *Phys. Rev. E* **2018**, *97*, 052701. [CrossRef]
92. Sohn, H.R.O.; Liu, C.D.; Smalyukh, I.I. Schools of skyrmions with electrically tunable elastic interactions. *Nat. Commun.* **2019**, *10*, 4744. [CrossRef] [PubMed]
93. Sohn, H.R.; Liu, C.D.; Voinescu, R.; Chen, Z.; Smalyukh, I.I. Optically enriched and guided dynamics of active skyrmions. *Opt. Express* **2020**, *28*, 6306–6319. [CrossRef]
94. Sohn, H.R.; Smalyukh, I.I. Electrically powered motions of toron crystallites in chiral liquid crystals. *Proc. Natl. Acad. Sci. USA* **2020**, *117*, 6437-6445. [CrossRef] [PubMed]
95. Shen, Y.; Dierking, I. Electrically Driven Formation and Dynamics of Skyrmionic Solitons in Chiral Nematics. *Phys. Rev. Appl.* **2021**, *15*, 054023. [CrossRef]
96. Smalyukh, I.I. knots and other new topological effects in liquid crystals and colloids. *Rep. Prog. Phys.* **2020**, *83*, 106601. [CrossRef]
97. Purwins, H.-G.; Bödeker, H.; Amiranashvili, S. Dissipative solitons. *Adv. Phys.* **2010**, *59*, 485–701. [CrossRef]
98. Bödeker, H.; Röttger, M.; Liehr, A.; Frank, T.; Friedrich, R.; Purwins, H.-G. Noise-covered drift bifurcation of dissipative solitons in a planar gas-discharge system. *Phys. Rev. E* **2003**, *67*, 056220. [CrossRef]
99. Li, B.-X.; Borshch, V.; Xiao, R.-L.; Paladugu, S.; Turiv, T.; Shiyanovskii, S.V.; Lavrentovich, O.D. Electrically driven three-dimensional solitary waves as director bullets in nematic liquid crystals. *Nat. Commun.* **2018**, *9*, 2912. [CrossRef]
100. Li, B.-X.; Xiao, R.-L.; Paladugu, S.; Shiyanovskii, S.V.; Lavrentovich, O.D. Three-dimensional solitary waves with electrically tunable direction of propagation in nematics. *Nat. Commun.* **2019**, *10*, 3749. [CrossRef]
101. Aya, S.; Araoka, F. Kinetics of motile solitons in nematic liquid crystals. *Nat. Commun.* **2020**, *11*, 1–10. [CrossRef]

102. Shen, Y.; Dierking, I. Dynamics of electrically driven solitons in nematic and cholesteric liquid crystals. *Commun. Phys.* **2020**, *3*, 1. [CrossRef]
103. Shen, Y.; Dierking, I. Dynamic dissipative solitons in nematics with positive anisotropies. *Soft Matter* **2020**, *16*, 5325. [CrossRef] [PubMed]
104. Shen, Y.; Dierking, I. Electrically driven formation and dynamics of swallow-tail solitons in smectic A liquid crystals. *Mater. Adv.* **2021**. [CrossRef]
105. Lavrentovich, O.D. Design of nematic liquid crystals to control microscale dynamics. *Liq. Cryst. Rev.* **2020**, *8*, 59–129. [CrossRef]
106. Carr, E.F. Influence of Electric Fields on the Molecular Alignment in the Liquid Crystal p-(Anisalamino)-phenyl Acetate. *Mol. Cryst.* **1969**, *7*, 253–268. [CrossRef]
107. Helfrich, W. Conduction-Induced Alignment of Nematic Liquid Crystals: Basic Model and Stability Considerations. *J. Chem. Phys.* **1969**, *51*, 4092–4105. [CrossRef]
108. Ibragimov, T.D. Influence of fullerenes C60 and single-walled carbon nanotubes on the Carr–Helfrich effect in nematic liquid crystal. *Optik* **2021**, *237*, 166768. [CrossRef]
109. Bodenschatz, E.; Zimmermann, W.; Kramer, L. On electrically driven pattern-forming instabilities in planar nematics. *J. De Phys.* **1988**, *49*, 1875–1899. [CrossRef]
110. Smith, I.; Galerne, Y.; Lagerwall, S.; Dubois-Violette, E.; Durand, G. Dynamics of electrohydrodynamic instabilities in nematic liquid crystals. *Le J. De Phys. Colloq.* **1975**, *36*, C1-237–C231-259. [CrossRef]
111. Barnik, M.I.; Blinov, L.M.; Grebenkin, M.F.; Trufanov, A.N. Dielectric Regime of Electrohydrodynamic Instability in Nematic Liquid Crystals. *Mol. Cryst. Liq. Cryst.* **1976**, *37*, 47–56. [CrossRef]
112. Kumar, P.; Heuer, J.; Tóth-Katona, T.; Éber, N.; Buka, Á. Convection-roll instability in spite of a large stabilizing torque. *Phys. Rev. E* **2010**, *81*, 020702. [CrossRef] [PubMed]
113. Barnik, M.; Blinov, L.; Pikin, S.; Trufanov, A. Instability mechanism in the nematic and isotropic phases of liquid crystals with positive dielectric anisotropy. *Sov Phys JETP* **1977**, *45*, 396–398.
114. Trufanov, A.; Barnik, M.; Blinov, L.; Chigrinov, V. Electrohydrodynamic instability in homeotropically oriented layers of nematic liquid crystals. *Sov Phys JETP* **1981**, *53*, 355–361.
115. Nakagawa, M.; Akahane, T. A new type of electrohydrodynamic instability in nematic liquid crystals with positive dielectric anisotropy. I. The existence of the charge injection and the diffusion current. *J. Phys. Soc. Jpn.* **1983**, *52*, 3773–3781.
116. Nakagawa, M.; Akahane, T. A new type of electrohydrodynamic instability in nematic liquid crystals with positive dielectric anisotropy. II. Theoretical treatment. *J. Phys. Soc. Jpn.* **1983**, *52*, 3782–3789. [CrossRef]
117. Monkade, M.; Martinot-Lagarde, P.; Durand, G. Electric polar surface instability in nematic liquid crystals. *EPL (Europhys. Lett.)* **1986**, *2*, 299. [CrossRef]
118. Lavrentovich, O.; Nazarenko, V.; Pergamenshchik, V.; Sergan, V.; Sorokin, V. Surface-polarization electrooptic effect in a nematic liquid crystal. *Sov. Phys. JETP* **1991**, *72*, 431–444.
119. Buka, A.; Éber, N.; Pesch, W.; Kramer, L. *Advances in Sensing with Security Applications*; Golovin, A.A., Nepomnyashchy, A.A., Eds.; Springer: Dordrecht, The Netherlands, 2006; pp. 55–82.
120. Li, B.-X.; Xiao, R.-L.; Shiyanovskii, S.V.; Lavrentovich, O.D. Soliton-induced liquid crystal enabled electrophoresis. *Phys. Rev. Res.* **2020**, *2*, 013178. [CrossRef]
121. Blinov, L.M. *Structure and Properties of Liquid Crystals*; Springer Science & Business Media: Dordrecht, The Netherlands, 2010; Volume 123.

Article

Ferroelectric Smectic Liquid Crystals as Electrocaloric Materials

Peter John Tipping and Helen Frances Gleeson *

School of Physics and Astronomy, University of Leeds, Leeds LS2 9JT, UK; py13pjt@leeds.ac.uk
* Correspondence: h.f.gleeson@leeds.ac.uk

Abstract: The 1980s saw the development of ferroelectric chiral smectic C (SmC*) liquid crystals (FLCs) with a clear focus on their application in fast electro-optic devices. However, as the only known fluid ferroelectric materials, they also have potential in other applications, one of which is in heat-exchange devices based on the electrocaloric effect. In particular, ferroelectric liquid crystals can be both the electrocaloric material and the heat exchanging fluid in an electrocaloric device, significantly simplifying some of the design constraints associated with solid dielectrics. In this paper, we consider the electrocaloric potential of three SmC* ferroelectric liquid crystal systems, two of which are pure materials that exhibit ferroelectric, antiferroelectric, and intermediate phases and one that was developed as a room-temperature SmC* material for electro-optic applications. We report the field-induced temperature changes of these selected materials, measured indirectly using the Maxwell method. The maximum induced temperature change determined, 0.37 K, is currently record-breaking for an FLC and is sufficiently large to make these materials interesting candidates for the development for electrocaloric applications. Using the electrocaloric temperature change normalised as a function of electric field strength, as a function of merit, the performances of FLCs are compared with ferroelectric ceramics and polymers.

Keywords: ferroelectric materials; smectic liquid crystals; electrocaloric effect

Citation: Tipping, P.J.; Gleeson, H.F. Ferroelectric Smectic Liquid Crystals as Electrocaloric Materials. *Crystals* **2022**, *12*, 809. https://doi.org/10.3390/cryst12060809

Academic Editor: Ingo Dierking

Received: 10 May 2022
Accepted: 4 June 2022
Published: 8 June 2022

Publisher's Note: MDPI stays neutral with regard to jurisdictional claims in published maps and institutional affiliations.

Copyright: © 2022 by the authors. Licensee MDPI, Basel, Switzerland. This article is an open access article distributed under the terms and conditions of the Creative Commons Attribution (CC BY) license (https://creativecommons.org/licenses/by/4.0/).

1. Introduction

The electrocaloric (EC) effect, discovered in 1930, is the induction of a reversible temperature change in a material via the adiabatic application of an electric field [1]. The EC effect has long been regarded as having potential as a cooling technology for cryogenic [2,3] and, more recently, for room temperature applications [4]. Interest has also grown as it is considered to be an environmentally friendly alternative to the ubiquitous vapour compression devices. This is because the typically high efficiency of vapour compression materials is offset by their large global warming impact. The unavoidable leakage of refrigerant results in a large environmental impact over the lifetime of the device [5]. In comparison, electrocaloric materials have a negligible direct impact on global warming as they are not volatile gases. Additionally, with the continuing rise in computing power, there is a growing demand for efficient and compact refrigeration technology in microelectronics. Vapour compression devices that have been proposed [6,7] are cumbersome with significantly reduced efficiency. Thus, there is a need for alternative refrigeration technologies that are efficient on a range of length scales and do not rely on the use of greenhouse gases.

For many years, EC materials have offered only small induced temperature (T) changes ($\Delta T \sim 2$ K) in response to very large applied voltages (several hundred volts), so the phenomenon was considered to be far from practical applications. However, in 2006 [8], Mischenko et al. reported $\Delta T \sim 12\ K$ in the ferroelectric ceramic lead zirconate titanate (PZT) for fields of 48 V μm^{-1} at a temperature of 220 °C, closely followed by a similarly large EC temperature change in ferroelectric polymers near room temperature [9]. Significant research into solid inorganic ceramics and fluorinated polymers as potential EC materials followed, but there has been only one report of a ferroelectric liquid crystal considered [10], despite some exciting potential advantages, discussed further below [4,11–13]. This paper examines three carefully selected FLCs for their electrocaloric potential, demonstrating both the current state-of-the-art research and offering insight into how to develop this exciting application area.

An EC material works as follows: the entropy of a dielectric material can be considered as being the sum of two contributions, one due to thermal vibrations and phonons and the second from the ordering of dipoles in the material. Upon adiabatic application of an electric field, the dipoles in the dielectric material align with the field, causing a decrease in the dipole entropy. The total entropy of the system is constant in an adiabatic process; therefore, the entropy due to molecular vibrations, and subsequently the temperature, increases. The converse occurs when the field is removed adiabatically, resulting in a temperature decrease. Clearly, for this to work in a device, heat exchange must also occur, and a common method of heat transfer is to pump a heat-exchanging liquid over the EC material and into a heat-exchange unit [14–18]. A major challenge with all electrocaloric device designs is the transfer of heat away from the refrigerated area. There are significant efficiency losses due to imperfect heat transfer between the heat exchange liquid and EC material. Alternative methods for heat exchange that do not use a liquid have been proposed, but there are still engineering challenges [19–22]. Therefore, the potential of using a dielectric liquid, rather than a solid, as the EC material is exciting as it could be pumped away from the refrigerated area. Liquid crystals (LCs) are obvious candidates worthy of serious consideration as novel EC materials; this paper specifically considers the potential of ferroelectric smectic liquid crystals as electrocaloric materials. In particular, we aim to: (i) deduce the electrocaloric performance of known ferroelectric, smectic, liquid crystals; (ii) elucidate the design rules for optimising the electrocaloric performance of new liquid crystal materials; and (iii) consider how their performance compares to solid-state ferroelectric electrocaloric materials (ceramics and polymers).

Most of the (rather few) measurements of the electrocaloric effect in LCs have explored commercially available materials and considered the phenomenon near the isotropic–nematic transition, with some promising results. Direct measurements of the EC effect in 5CB [23] showed an induced temperature change of $\Delta T = 0.36$ K for an applied field of 19 V μm^{-1}, while indirect measurements [24] suggest a peak change of $\Delta T = 5.26$ K for a field of 90 V μm^{-1}. A temperature change of $\Delta T \sim 1.4$ K has also been directly measured in 8CB [25] using a 6 V μm^{-1} field. The temperature change induced at the isotropic to SmA transition has been studied using 12CB, with a significant temperature change, $\Delta T \sim 6.5$ K, measured near the transition using an 8 V μm^{-1} field [25,26]. The results for 12CB suggest that the greatest EC effect is around a phase transition where the applied field can induce a comparatively large change in the order parameter. However, a disadvantage of measurements at an isotropic–LC phase transition is the very narrow temperature range over which the phenomenon can be exploited—typically a few tenths of a degree at most.

It seems an obvious step to consider the potential of ferroelectric liquid crystals as EC materials, given that the largest EC effect measured to date has been in solid ferroelectrics [12,13]. However, we are aware of only one report of the EC effect in ferroelectric liquid crystals, in which two commercial ferroelectric liquid crystal mixtures designed for electro-optic displays were investigated [10], with a peak temperature change $\Delta T = 0.16$ K. We selected three materials as follows: one is a well-known commercial material, SCE13, which is rather similar to the materials already studied, allowing us to perform a direct comparison with published data. As is explained in Section 2, a larger electrocaloric effect occurs for materials that exhibit a larger spontaneous polarisation, so we have also selected two pure, smectic, ferroelectric LCs (LC 1 and LC 2) with large values of P_S, each with different phase sequences. In all cases, we measured the EC effect indirectly using the Maxwell approach [27]. We also normalised the maximum EC absolute temperature change with respect to the applied field as $\frac{\Delta T}{\Delta E}$, offering a "figure of merit" for the electrocaloric effect, which allows a meaningful comparison of the effect across LCs, polymers, and ferroelectric ceramics. We demonstrate that simple material selection criteria that include the consideration of the P_S and the phase behaviour allowed us to record the largest EC temperature change to date for a ferroelectric LC, both in terms of absolute temperature change and normalised with respect to the applied field.

As already stated, in this paper, the electrocaloric temperature change is measured indirectly using the Maxwell approach [27]. An expression for the isothermal entropy change per unit volume, $\frac{\Delta S}{V}$, as a function of electric field can be derived using the Maxwell relation between the electric field and temperature,

$$\frac{\Delta S}{m^3} = \int_{E_1}^{E_2} \left(\frac{\partial P_S}{\partial T}\right)_E dE \qquad (1)$$

where E_1 and E_2 are the initial and final field strengths and $\left(\frac{\partial P_S}{\partial T}\right)_E$ is the rate of change of spontaneous polarisation with respect to temperature at a constant field strength. Assuming that the initial temperature and the volumetric heat capacity do not vary with the applied field, an estimate of the induced temperature change, ΔT, is

$$\Delta T \cong -\frac{T_1}{C_E(0, T_1)} \int_{E_1}^{E_2} \left(\frac{\partial P_S}{\partial T}\right)_E dE, \qquad (2)$$

where T_1 (K) is the temperature at which the field is applied and $C_E(0, T_1)$ (J K^{-1} m^{-3}) is the volumetric heat capacity at zero field, measured at T_1. Equation (2) offers a basis for indirectly measuring the temperature change of an FLC in response to an applied field, provided that both $\left(\frac{\partial P_S}{\partial T}\right)_E$ and $C_E(0, T_1)$ are known.

2. Materials and Methodology

The FLC materials chosen for this study were selected to (i) allow us to evaluate the influence of the magnitude of the P_S and (ii) examine the influence of the phase behaviour at the transition to the FLC phase. SCE13 is a ferroelectric mixture supplied by Merck, with the phase sequence shown in Figure 1. As this mixture was designed for use in electro-optic devices, the phase sequence includes a chiral nematic phase with very large pitch at the chiral nematic (N*)–smectic A (SmA) phase transition, and the material has a modest spontaneous polarization ($P_S \approx 25$ nC cm^{-2}) at room temperature. Liquid crystals 1 (LC 1) and 2 (LC 2) are shown in Figure 1, together with their phase sequences; they were originally designed as novel antiferroelectric materials and have a larger spontaneous polarization ($P_S \approx 70$ nC cm^{-2}). Full details of their properties are reported elsewhere [28,29]. Both LC 1 and LC 2 exhibit a narrow (~1 °C) smectic C alpha (SmC*$_\alpha$) phase directly above the SmC* phase. LC 1 also exhibits a twist-grain boundary A (TGB$_A$) phase extending for ~5 °C above the SmC* phase, before the material becomes isotropic. LC 2 has a relatively wide (~17 °C) SmA phase directly above the SmC*$_\alpha$ phase. The nature of the SmC*$_\alpha$ to SmC* phase transition has been discussed in detail elsewhere [30] and can be first or second order, a factor that will influence the field and temperature dependence of the spontaneous polarization. The subphases exhibited by LC 1 and LC 2 are all several degrees below ferroelectric to paraelectric (SmC*$_\alpha$ to TGB$_A$ or SmA phase) and do not contribute to the measurements reported in this paper.

The heat capacity, $C_E(0,T)$, is measured via differential scanning calorimetry using a TA Instruments Q2000 Different Scanning Calorimeter. All heat capacity measurements were made at zero field (the option of applying a field during the calorimetry was not available) and taken on cooling at 10 Kmin^{-1}. The data are quoted with respect to a critical temperature, T_O, defined as the temperature where the ferroelectric phase was first observed. To determine the volumetric heat capacity of a sample, the specific heat capacity is multiplied by the density of the sample. For this work, the density of the materials was estimated from literature values for calamitic LCs in the SmC [31–33] or SmA phase [34] at the phase transition. The range of values reported for density span 0.96 to 1.02 g cm^{-3}, with an average density of (0.97 $\pm$ 0.02) g cm^{-3}. We employed the average value to indirectly determine the electrocaloric temperature change and estimate that this contributes to ~2% uncertainty in our final measurement of ΔT.

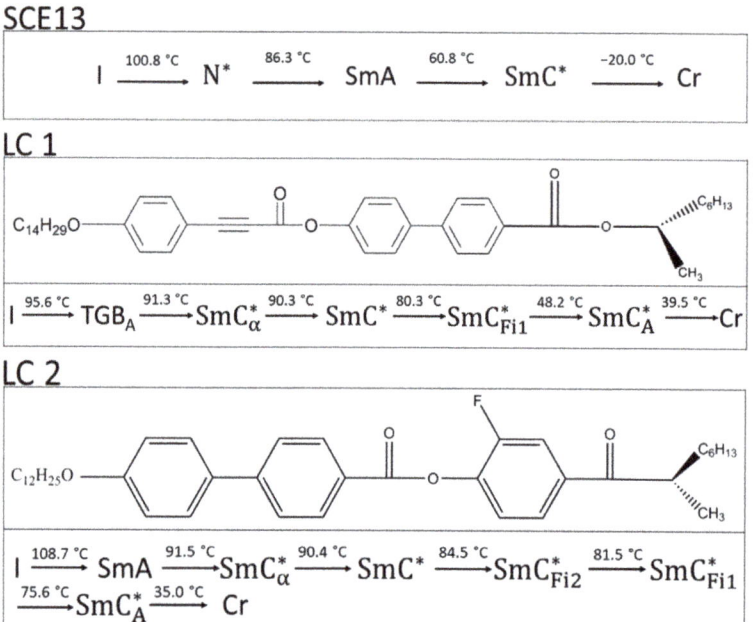

Figure 1. The phase transitions (measured on cooling) for all three materials studied, together with the chemical structure of LC 1 and LC 2. The notation of the phases is as defined in the text, with the following: Cr is crystal; SmC*$_{Fi1}$ is the 3-layer intermediate phase; SmC*$_{Fi2}$ is the 4-layer intermediate phase; SmC*$_A$ is the antiferroelectric phase.

The spontaneous polarisation is measured using the current reversal technique with an accuracy of ± 1 nC cm^{-2} [35]. All measurements were taken in cells approximately 1.8 µm thick treated for planar alignment, purchased from AWAT (Poland). An electric field with a triangular wave was applied to the cell, and the current associated with the change in sign of the P_S of the ferroelectric LC, I_P, was passed in series to a current-to-voltage amplifier. The resulting signal was then recorded on a Tektronix 2024C oscilloscope. P_S is determined by analysing the current peak using Equation (3),

$$P_S = \frac{1}{2A \cdot R} \int I_P dt \qquad (3)$$

where A is the surface area of the electrodes, R is the resistance of the current to voltage amplifier, and t is time.

Temperature control is achieved using a Linkam TMS 94 with an LTS 350 hot plate. The spontaneous polarisation is determined as a function of applied field at temperature intervals of 0.2 K spanning temperatures from 2 K above to 10 K below the SmA–SmC* transition for SCE13, TGB$_A$–SmC*$_\alpha$ for LC 1 and SmA–SmC*$_\alpha$ transition for LC 2. Results are plotted on a reduced temperature scale with the critical temperature, T_O, defined as the temperature where the ferroelectric phase was first observed using polarised microscopy.

The temperature change, ΔT, that occurs in the FLC as a consequence of applying a field is determined using the volumetric heat capacity and spontaneous polarisation data as outlined in the Introduction. For each material, P_S measurements were taken at 0.2 K temperature intervals across the phase transition while varying the electric field strength from 3–19 V µm^{-1} for SCE13 and LC 1 and from 5–19 V µm^{-1} for LC 2 in steps of ~3 V µm^{-1}.

The maximum electric field strengths applied to the samples are sufficient to fully saturate the P_S measurements whilst not inducing a chevron to bookshelf transition [36,37].

In order to determine $\left(\frac{\partial P_S}{\partial T}\right)_E$ and substitute into Equations (1) and (2), a numerical fit was applied to the experimental data to allow extrapolation between the data points as follows. The P_S measurements determined for SCE13 and Material 2 for each (constant) field strength were fit to a Curie–Weiss law ($P_S = P_0(T - T_C)^\gamma$) at temperatures up to the sample's respective critical temperature, T_C. The P_S results for LC 1 were fit to a third-order polynomial because LC 1 shows a more discontinuous transition into the SmC*$_\alpha$ phase. P_S values determined above the critical temperature for all materials were fit to an exponential decay curve.

The gradient of each of the numerical fits with respect to temperature $\left(\frac{\partial P_S}{\partial T}\right)$ in Equations (1) and (2) was thus found for each field strength, and this was then plotted as a function of temperature. Taking a vertical slice through the graph of $\frac{\partial P_S}{\partial T}$ as a function of temperature represents $\frac{\partial P_S}{\partial T}$ as a function of electric field strength at a constant temperature, $\left(\frac{\partial P_S}{\partial T}\right)_E$. Finally, a numerical fit can be made of $\frac{\partial P_S}{\partial T}$ as a function of electric field strength; integrating under the fitting curve results in the indirect measurement for isothermal entropy change per unit volume, as described by Equation (2). The electrocaloric temperature change was indirectly determined by multiplying the isothermal entropy change per unit volume at a given temperature, T_1, by the temperature in Kelvin, and dividing by the volumetric heat capacity at T_1.

3. Results

3.1. Heat Capacity

Figure 2 presents the volumetric heat capacity of SCE13, LC 1, and LC 2 around the transition into the ferroelectric phase of each material. The magnitude of the heat capacity is different for each of these materials, a factor that is important in their use as electrocaloric materials (Equation (2)). A peak is seen in the volumetric heat capacity around the ferroelectric to non-ferroelectric phase transitions for each of the materials. The relative magnitude of the peak is quite different for the three samples: LC 1 is the largest, 19% higher than the value 1 K above the transition; SCE13 shows less than a 1% increase, while LC 2 shows a 6% increase over the baseline value. The relative magnitude of the peaks is representative of the discontinuity at the phase transition. As will be explained further in Section 4, this subsequently affects the magnitude and applicable temperature range of the electrocaloric effect. The SmC*$_\alpha$–SmC transition is seen in the DSC traces for LC 1 and LC 2; however, the transition is convoluted with the TGB$_A$–SmC*$_\alpha$ transition in LC 1 and the SmA–SmC*$_\alpha$ transition in LC 2. The convolution broadens the peak; therefore, the heat capacity, C_E, around the transition stays larger over a wider temperature range. Subsequently, from Equation (2), the electrocaloric temperature change at all temperatures is deduced. For these measurements, the critical temperature for each sample is defined as the peak in the heat capacity.

3.2. Spontaneous Polarisation

The spontaneous polarization determined using Equation (3) as a function of reduced temperature is shown in Figure 3 over the temperature range around the transition. SCE13 has the smallest absolute spontaneous polarization, taking a value of ~14 nC cm^{-2} at 10 K below the transition. The field-induced P_S above T_C is very small, below the sensitivity of our experiment, as would be expected from the very small electroclinic effect [38] known for SCE13. Both LC 1 and LC 2 show evidence of a significant field-induced P_S above the phase transition, an extremely desirable phenomenon for electrocaloric applications, because this can extend the applicable temperature range. Both materials have a P_S of approximately 71 nC cm^{-2} 10 K below the transition, but as the rate of change is greater near the phase change for LC 1, a larger EC effect is expected.

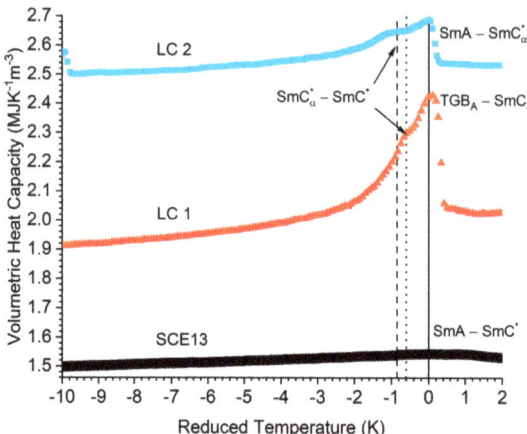

Figure 2. Volumetric heat capacity ($C_E(0, T_1)$) (MJ K^{-1} m^{-3}) measured from 2 K above to 10 K below the transition into the ferroelectric phase for each material. The critical temperature was defined as the peak in the heat capacity for each material. For SCE13 (black circles, the peak occurs across the SmA–SmC* transition. For LC 1 (red triangles), the peak occurs across the TGB$_A$–SmC$_\alpha$* transition, and for LC 2 (blue squares), the peak is across the SmA–SmC$_\alpha$* transition. In LC 1 and LC 2, the SmC$_\alpha$*–SmC* transition is marked using dotted and dashed lines (LC 1 and LC 2, respectively).

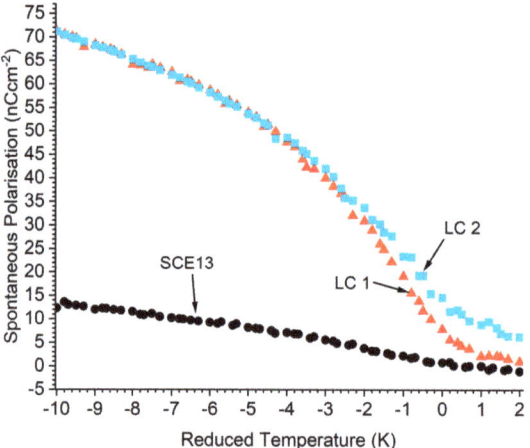

Figure 3. Maximum spontaneous polarisation (P_S) measured as a function of reduced temperature relative to the SmA–SmC* transition for SCE13 (black circles), TGB$_A$–SmC$_\alpha$* for LC 1 (red triangles), and SmA–SmC$_\alpha$* for LC 2 (blue squares). The uncertainty in P_S = ±1 nC cm^{-2}. LC 1 shows a discontinuous transition, while LC 2 and SCE13 show a continuous transition.

3.3. Field-Induced Isothermal Entropy Change per Unit Volume

Figure 4 shows the peak isothermal entropy change per unit volume, V, $\frac{\Delta S}{V}$, of the materials, determined using Equation (2), demonstrating the significantly better performance of LC 1 over that of the other systems considered. The maximum isothermal entropy change per unit volume of SCE13 is 0.4 kJ K^{-1} m^{-3}, less than 20% that of LC 1, which has a maximum of 2.3 kJ K^{-1} m^{-3}.

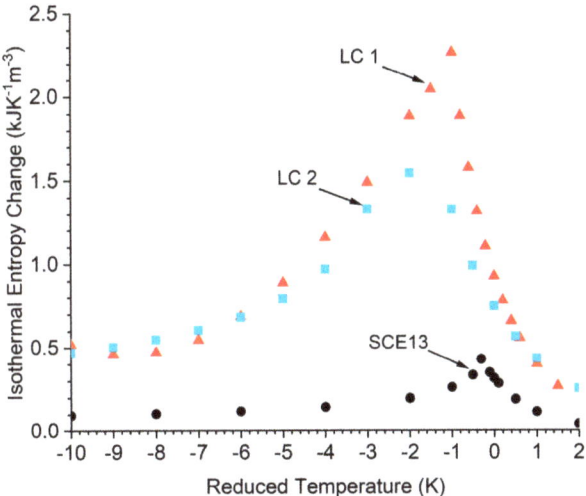

Figure 4. Isothermal entropy change per unit volume as a function of reduced temperature determined using Equation (1). The results for SCE13 and LC 1 are for an effective field of 16 V μm^{-1}, and an effective field of 14 V μm^{-1} for LC 2. The peak position of SCE13 (black circles) is ∼−0.3 K; for LC 1 (red triangles), the peak position is ∼−1 K; for LC 2 (blue squares), the peak position is ∼−2 K.

Although LC 1 and LC 2 have approximately the same value of P_S, LC 1 performs ∼25% better than LC 2. This is attributed to the more discontinuous TGB$_A$–SmC*$_\alpha$ transition in LC 1, which results in a larger $\frac{\partial P_S}{\partial T}$ at each field strength. The peak isothermal entropy changes of LC 1 and LC 2 occur at ∼1 K and ∼2 K, respectively, below the transition temperature. The temperature at which the maximum value occurs corresponds to the region where the gradient of the spontaneous polarisation, with respect to temperature, reaches a maximum for every electric field strength measured. These observations confirm the importance of the value of the parameter $\left(\frac{\partial P_S}{\partial T}\right)_E$ when considering which ferroelectric LCs will show the largest EC effect; although a large magnitude of P_S is important, having a large gradient is vital. It is also important to consider the range over which a useful EC effect is available; real systems would typically need to operate over ∼10 K. LC 1 maintains at least 90% of the value of its maximum entropy change over a 0.9 K range, while the useful range of LC 2 extends over 1.4 K, a noticeably wider temperature range.

3.4. Electrocaloric Temperature Change

An indirect measurement of the EC temperature change, ΔT, was obtained by multiplying the isothermal entropy change per unit volume by the scaling factor $\frac{T_1}{C_E(0,T_1)}$ (Equation (2), Figure 5). The importance of a low volumetric heat capacity can be seen in the relative differences in the maximum values for isothermal entropy, ΔS (Figure 4), and the EC temperature change ΔT (Figure 5). The lower heat capacity of SCE13 means that the electrocaloric temperature change is comparatively larger than the isothermal entropy change per unit volume alone would suggest. It is nonetheless still much smaller $\Delta T_{max} \sim 0.1$ K than the temperature change in LC 1 or LC 2 ($\Delta T_{max} \sim 0.37$ K and $\Delta T_{max} \sim 0.22$ K, respectively). The lower heat capacity of LC 1 serves to enhance the induced temperature change compared to LC 2.

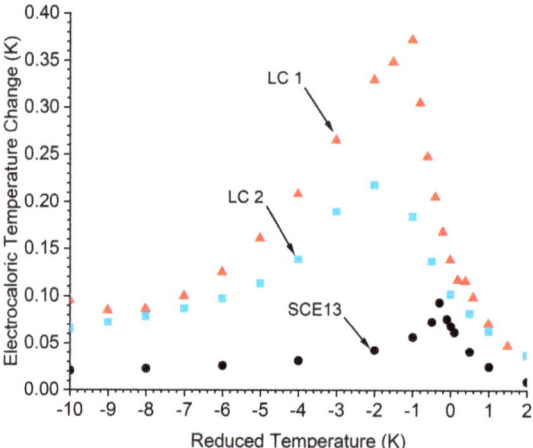

Figure 5. Indirect measurement of the electrocaloric temperature change as a function of reduced temperature for a field strength of 16 V μm^{-1} SCE13 (black circles) and LC 1 (red triangles) and 14 V μm^{-1} for LC 2 (blue squares). The peak position occurs at reduced temperatures of ~−0.3 K (SCE13), ~−1 K (LC 1), and ~−2 K (LC 2).

4. Discussion

Table 1 summarises the maximum values of the physical parameters relevant to the EC effect in the ferroelectric liquid crystals that are the subject of this paper and the few others reported elsewhere [10]. For comparison, the table also includes EC values for 12CB at the SmA–I transition [25] and for a solid-state ferroelectric ceramic device, which is currently considered state-of-the-art [39]. A figure of merit can be defined [40], $\frac{\Delta T_{max}}{\Delta E}$, where ΔE is the field required to induce the maximum temperature change, ΔT_{max}, which offers a measure of the efficiency of the EC effect in different materials.

Table 1. The maximum spontaneous polarisation (P_S), volumetric heat capacity (C_E), maximum EC temperature change (T_{max}), figure of merit, and temperature range over which the electrocaloric temperature change remains greater than 90% of the peak temperature change, for the materials studied in this paper and other systems chosen for comparison. The ferroelectric materials FELIX-017/000 and OB4HOB [10] were studied by Bsaibess et al., while Klemenčič et al. induced a temperature change at the isotropic to SmA phase transition in 12CB [25]. The ferroelectric ceramic, lead scandium tantalate, PST, [39] arranged in a multilayer capacitor MLC is also included. The temperature changes for 12CB and PST are direct measurements, and all other measurements are indirect.

Material	P_S (nC cm^{-2})	C_E (MJ K^{-1} m^{-3})	ΔT_{max} (K)	$\Delta T_{max}/\Delta E$ (K m MV^{-1})	Range Where $\Delta T_{EC}/\Delta T_{ECmax} > 0.9$ (K)
SCE13	26	1.6	0.09	0.006	0.2
LC 1	71	2.3	0.37	0.022	0.9
LC 2	71	2.6	0.22	0.016	1.4
FELIX-017/000	25	3.5	0.023	0.003	1.0
OB4HOB	60	7.1	0.17	0.021	0.1
12CB: I-SmA	-	-	6.5	0.8	<0.1
PST MLC (Ferroelectric ceramic)	30,000	2.7	3.3	0.19	73

LC 1 shows the largest EC temperature change reported to date for a ferroelectric liquid crystal, $\Delta T_{max} \sim 0.37$ K, compared to $\Delta T_{max} \sim 0.17$ K, which was the maximum reported by Bsaibess et al. for the ferroelectric liquid crystal OB4HOB [10]. It can be seen that the main reason for the significant improvement seen for LC 1 over OB4HOB is the much higher heat capacity (7.1 MJ K^{-1}m^{-3}) and slightly smaller spontaneous polarization (~60 nC cm^{-2}) of the latter material. Interestingly, the peak isothermal entropy change of OB4HOB can be estimated from data reported by Bsaibess et al. to be 3.1 kJ K^{-1} m^{-3}, which is ~ 50% larger than that of LC 1. An additional important factor in the EC response is the gradient, $\left(\frac{\partial P_S}{\partial T}\right)_{E'}$ which is considerably larger for OB4HOB due to its first-order isotropic to SmC* phase transition (the P_s saturates 4 K below the transition temperature). LC 1 has a much smaller gradient, and subsequently, the induced entropy change over the same temperature range is smaller.

As mentioned, the heat capacity is an important factor. Indeed, the heat capacity of OB4HOB is over three-times larger than that of LC 1 at the temperature where the EC peak occurs, which results in a lower induced temperature change, ΔT, for a given isothermal entropy change Equation (2). Clearly, the heat capacity, saturated P_S value, and the gradient of P_S with respect to temperature must be compared in ferroelectric LCs to determine the overall suitability of materials with respect to their electrocaloric effect. It is also important that the indirect methodology employed here uses the heat capacity measured at zero field. As discussed in a recent review [12], not considering the field or temperature dependence results in the heat capacity being smeared and overestimated. Therefore, the indirectly measured temperature change reported in Figure 5 is an underestimate of the expected temperature change.

Both of the ferroelectric liquid crystal mixtures designed for display devices, SCE13 and FELIX-017/000, behave unsurprisingly modestly in terms of their EC potential. As materials that were designed for a completely different application, namely electro-optic devices, it was desirable to have a relatively low P_s, and the phase sequences were designed to allow good alignment to be obtained. The optical properties of such commercial mixtures were also important, with the ideal tilt angle of 22.5° being carefully engineered in them. Their figures of merit are very poor, a consequence of the low values of P_s.

Although having a large gradient, $\left(\frac{\partial P_S}{\partial T}\right)_{E'}$ is clearly important in maximising ΔT, there are drawbacks to materials that reach maximum P_s over a shorter temperature range. Specifically, this will mean that the temperature span over which the electrocaloric effect decays is also smaller, giving such materials a poor useful range. Table 1 summarises this issue by considering the temperature range over which the electrocaloric temperature change remains greater than 90% of the peak temperature change, $\frac{\Delta T}{\Delta T_{max}} > 0.9$. As is shown in Figure 6, while the figure of merit, $\frac{\Delta T}{\Delta E}$ of OB4HOB is comparable to that of LC 1, $\frac{\Delta T}{\Delta T_{max}} > 0.9$ for OB4HOB is a factor of 9 smaller than LC 1. The rate of decay of the electrocaloric effect as a function of temperature is an important quantity to consider for engineering purposes, as any electrocaloric refrigeration device must be able to operate over a broad temperature span. For comparison, lead scandium tantalate, PST, a ferroelectric ceramic, arranged in a multilayer capacitor, is also shown in Figure 6. This ceramic demonstrates an electrocaloric temperature change of 3 K over one of the broadest temperature ranges reported, 73 K [39]. Although the figures of merit for ferroelectric liquid crystals are only around an order of magnitude lower than the very best ceramic EC devices, their useful temperature range is not yet comparable.

It is appropriate to discuss briefly the liquid crystal system that performs best in terms of ΔT_{max}, 12CB. This system was mentioned in the Introduction, and it can be seen that the field-induced isotropic to SmA phase change offers an enormous figure of merit, $\frac{\Delta T}{\Delta E} = 0.8$ K m MV^{-1}, and a giant maximum induced temperature change, $\Delta T_{max} = 6.5$ K. Unfortunately, this exceptional EC performance is unsuitable for applications because of the very narrow useful temperature range, less than 0.1 K. Both the relatively large ΔT_{max} and narrow temperature range are due the physical phenomena behind the EC effect in this material, i.e., factors that electrically drive the isotropic to smectic transition. The absorption of latent heat as the LC transitions dominates the electrocaloric effect when inducing a liquid crystal phase [25]. Consequently, the effect is only significant across the isotropic–SmA coexistence region, which is extremely narrow in a pure material, with the effect reducing significantly on cooling further into the phase. Extending the very limited usable temperature range in this system to ~ 2 K has been achieved by extending the coexistence region by mixing nanoparticles into the LC [26].

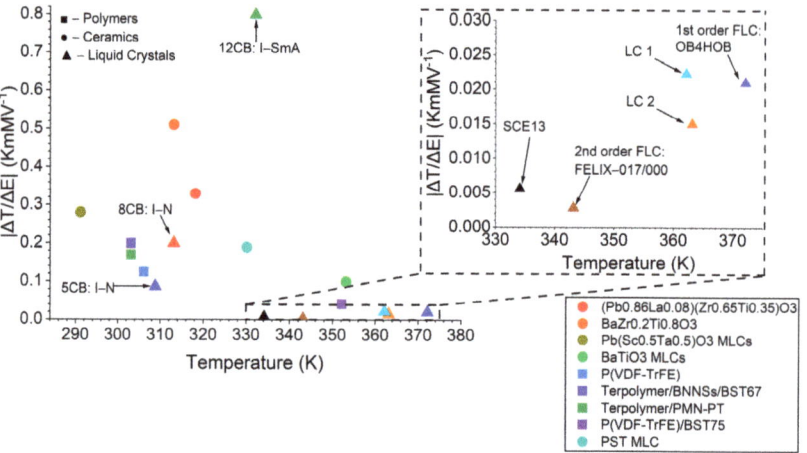

Figure 6. The EC figure of merit, $\Delta T / \Delta E$, plotted as a function of the temperature at which the maximum EC effect was recorded. Results for LCs (triangles and labelled) and a sample of solid EC materials, including polymers (circles) and ceramics [12,39] (squares), are shown. Data for 8CB and 12CB [25] and previously reported ferroelectric LCs [10] are also shown for comparison. The inset graph is an expansion of the dotted rectangle between 330 K and 375 K that expands the region containing ferroelectric LCs.

One final, but important point for consideration in applications is the actual temperature at which the maximum electrocaloric effect occurs. Figure 6 shows $\Delta T / \Delta E$ for the materials considered in this paper together with selected ferroelectric ceramics and polymers [12,39], the ferroelectric LCs previously reported [10,39], and the cyanobiphenyl nematic liquid crystals 8CB and 12CB (where the entropy changes at the isotropic to liquid crystal transitions were considered) [25]. The temperature on the ordinate axis is that where ΔT_{max} was recorded. The specific application will determine whether or not a particular ΔT_{max} and temperature range of the EC effect is suitable, but it is noteworthy that engineering phase transition temperatures and physical properties across wide temperature ranges is well known in liquid crystals. For example, SCE13 was designed to have a ferroelectric phase from ~−20–60 °C to make it suitable for display applications, a relatively low P_S to optimise switching speed, and a tilt angle of 22.5° over a wide temperature range to optimise the optical contrast of electro-optic devices. Thus, provided FLCs are considered promising for EC applications and the design rules are known, one might expect them to be serious contenders in the future. In this case, despite the fact that LC 1 shows the largest normalized EC temperature change of any ferroelectric LC to date, it is evident that these materials are currently an order of magnitude less efficient, in terms of $\frac{\Delta T}{\Delta E}$, than solid EC materials. Although a disappointing outcome, it is not a surprising one as the P_S of FLCs is two or three orders of magnitude lower than that of solid ferroelectrics.

5. Conclusions

This work showed that for the development of ferroelectric LCs for the EC effect, both the isothermal entropy change and volumetric heat capacity must be considered. A large P_S and small heat capacity are clearly important, but the maximum EC temperature change coincides with the maximum gradient in spontaneous polarisation with respect to temperature, $\left(\frac{\partial P_S}{\partial T}\right)_E$, bringing a new design rule to ferroelectric liquid crystals for this application. Furthermore, the variation in the gradient is the most significant factor affecting how the EC temperature change varies with temperature. Therefore, the development of new materials should focus on both maximising spontaneous polarisation and optimising the P_S gradient to occur over the broadest temperature range without significantly reducing the EC temperature change. These are very different design considerations than were relevant to the development of FLC electro-optic devices with microsecond response times. We suggest that materials with a large P_S designed for electroclinic devices would be interesting candidates for electrocaloric applications, but other systems have also been proposed, e.g., antiferroelectric bent-core liquid crystals [41]. However, it is clearly also important that the heat capacity of the LC material be considered; this changes by a factor of 3 even in the few liquid crystals we considered here. Finally, it is worth noting that although these current systems perform relatively poorly with respect to solid-state systems, the fact that we are considering fluids offers several significant advantages in electrocaloric applications. This is an exciting application area, especially in these times where sustainability and the efficiency of energy use are critical, and we demonstrated some important design considerations for developing liquid crystals for electrocaloric applications.

Author Contributions: Conceptualization, H.F.G. and P.J.T.; methodology, H.F.G. and P.J.T.; software, P.J.T.; validation, H.F.G. and P.J.T.; formal analysis, P.J.T.; investigation, P.J.T.; data curation, P.J.T.; writing—original draft preparation, P.J.T.; writing—review and editing, H.F.G. and P.J.T.; supervision, H.F.G.; project administration, H.F.G.; funding acquisition, H.F.G. All authors have read and agreed to the published version of the manuscript.

Funding: P.J.T. and H.F.G. acknowledge funding from the Engineering and Physical Sciences Research Council and from Merck Performance Materials Ltd. through a CASE award.

Institutional Review Board Statement: Not applicable.

Informed Consent Statement: Not applicable.

Data Availability Statement: The data associated with this paper are available from University of Leeds at https://doi.org/10.5518/1149.

Acknowledgments: P.J.T. and H.F.G. acknowledge the supply of materials from Merck Performance Materials Ltd.

Conflicts of Interest: The authors declare no conflict of interest.

References

1. Kobenko, P.; Kurtschatov, J.Z. Dielektrische Eigenschaften der Seignettesalzkristalle. *Z. Phys.* **1930**, *66*, 192–205. [CrossRef]
2. Lawless, W.N.; Morrow, A.J. Specific Heat and Electrocaloric Properties of a SrTiO$_3$ Ceramic at Low Temperatures. *Ferroelectrics* **1977**, *15*, 159–165. [CrossRef]
3. Lawless, W.N. Specific Heat and Electrocaloric Properties of KTaO$_3$ at Low Temperatures. *Phys. Rev. B* **1977**, *16*, 433–439. [CrossRef]
4. Greco, A.; Aprea, C.; Maiorino, A.; Masselli, C. A review of the state of the art of solid-state caloric cooling processes at room-temperature before 2019. *Int. J. Refrig* **2019**, *106*, 66–88. [CrossRef]
5. Shi, J.Y.; Han, D.L.; Li, Z.C.; Yang, L.; Lu, S.G.; Zhong, Z.F.; Chen, J.P.; Zhang, Q.M.; Qian, X.S. Electrocaloric Cooling Materials and Devices for Zero-Global-Warming-Potential, High-Efficiency Refrigeration. *Joule* **2019**, *3*, 1200–1225. [CrossRef]
6. Poachaiyapoom, A.; Leardkun, R.; Mounkong, J.; Wongwises, S. Miniature vapor compression refrigeration system for electronics cooling. *Case Stud. Therm. Eng.* **2019**, *13*, 100365. [CrossRef]
7. He, J.; Wu, Y.T.; Chen, X.; Lu, Y.W.; Ma, C.F.; Du, C.X.; Liu, G.; Ma, R. Experimental study of a miniature vapor compression refrigeration system with two heat sink evaporators connected in series or in parallel. *Int. J. Refrig* **2015**, *49*, 28–35. [CrossRef]
8. Mischenko, A.S.; Zhang, Q.; Scott, J.F.; Whatmore, R.W.; Mathur, N.D. Giant electrocaloric effect in thin-film PbZr$_{0.95}$Ti$_{0.05}$O$_3$. *Science* **2006**, *311*, 1270–1271. [CrossRef]
9. Neese, B.; Chu, B.J.; Lu, S.G.; Wang, Y.; Furman, E.; Zhang, Q.M. Large electrocaloric effect in ferroelectric polymers near room temperature. *Science* **2008**, *321*, 821–823. [CrossRef]
10. Bsaibess, E.; Sahraoui, A.H.; Boussoualem, Y.; Soueidan, M.; Duponchel, B.; Singh, D.P.; Nsouli, B.; Daoudi, A.; Longuemart, S. Study of the electrocaloric effect in ferroelectric liquid crystals. *Liq. Cryst.* **2019**, *46*, 1517–1526. [CrossRef]
11. Scott, J.F. Electrocaloric Materials. *Annu. Rev. Mater. Res.* **2011**, *41*, 229–240. [CrossRef]

12. Liu, Y.; Scott, J.F.; Dkhil, B. Direct and indirect measurements on electrocaloric effect: Recent developments and perspectives. *Appl. Phys. Rev.* **2016**, *3*, 031102. [CrossRef]
13. Moya, X.; Kar-Narayan, S.; Mathur, N.D. Caloric materials near ferroic phase transitions. *Nat. Mater.* **2014**, *13*, 439–450. [CrossRef] [PubMed]
14. Olsen, R.B. Ferroelectric Conversion of Heat to Electrical Energy—A Demonstration. *J. Energy* **1982**, *6*, 91–95. [CrossRef]
15. Plaznik, U.; Vrabelj, M.; Kutnjak, Z.; Malic, B.; Rozic, B.; Poredos, A.; Kitanovski, A. Numerical modelling and experimental validation of a regenerative electrocaloric cooler. *Int. J. Refrig* **2019**, *98*, 139–149. [CrossRef]
16. Plaznik, U.; Kitanovski, A.; Rozic, B.; Malic, B.; Ursic, H.; Drnovsek, S.; Cilensek, J.; Vrabelj, M.; Poredos, A.; Kutnjak, Z. Bulk relaxor ferroelectric ceramics as a working body for an electrocaloric cooling device. *Appl. Phys. Lett.* **2015**, *106*, 043903. [CrossRef]
17. Jia, Y.B.; Sungtaek, Y. A solid-state refrigerator based on the electrocaloric effect. *Appl. Phys. Lett.* **2012**, *100*, 242901. [CrossRef]
18. Sette, D.; Asseman, A.; Gerard, M.; Strozyk, H.; Faye, R.; Defay, E. Electrocaloric cooler combining ceramic multi-layer capacitors and fluid. *Apl. Mater.* **2016**, *4*, 091101. [CrossRef]
19. Gu, H.M.; Qian, X.S.; Li, X.Y.; Craven, B.; Zhu, W.Y.; Cheng, A.L.; Yao, S.C.; Zhang, Q.M. A chip scale electrocaloric effect based cooling device. *Appl. Phys. Lett.* **2013**, *102*, 122904. [CrossRef]
20. Zhang, T.; Qian, X.S.; Gu, H.M.; Hou, Y.; Zhang, Q.M. An electrocaloric refrigerator with direct solid to solid regeneration. *Appl. Phys. Lett.* **2017**, *110*, 243503. [CrossRef]
21. Wang, Y.D.; Smullin, S.J.; Sheridan, M.J.; Wang, Q.; Eldershaw, C.; Schwartz, D.E. A heat-switch-based electrocaloric cooler. *Appl. Phys. Lett.* **2015**, *107*, 134103. [CrossRef]
22. Meng, Y.; Zhang, Z.Y.; Wu, H.X.; Wu, R.Y.; Wu, J.H.; Wang, H.L.; Pei, Q.B. A cascade electrocaloric cooling device for large temperature lift. *Nat. Energy* **2020**, *5*, 996–1002. [CrossRef]
23. Lelidis, I.; Durand, G. Electrothermal effect in nematic liquid crystal. *Phys. Rev. Lett.* **1996**, *76*, 1868–1871. [CrossRef] [PubMed]
24. Qian, X.S.; Lu, S.G.; Li, X.Y.; Gu, H.M.; Chien, L.C.; Zhang, Q.M. Large Electrocaloric Effect in a Dielectric Liquid Possessing a Large Dielectric Anisotropy Near the Isotropic-Nematic Transition. *Adv. Funct. Mater.* **2013**, *23*, 2894–2898. [CrossRef]
25. Klemencic, E.; Trcek, M.; Kutnjak, Z.; Kralj, S. Giant electrocaloric response in smectic liquid crystals with direct smectic-isotropic transition. *Sci. Rep.* **2019**, *9*, 1721. [CrossRef] [PubMed]
26. Trček, M.; Lavrič, M.; Cordoyiannis, G.; Zalar, B.; Rožič, B.; Kralj, S.; Tzitzios, V.; Nounesis, G.; Kutnjak, Z. Electrocaloric and elastocaloric effects in soft materials. *Philos. Trans. R. Soc. A-Math. Phys. Eng. Sci.* **2016**, *374*, 1–11. [CrossRef]
27. Kutnjak, Z.; Rožič, B.; Pirc, R. Electrocaloric Effect: Theory, Measurements, and Applications. In *Wiley Encyclopedia of Electrical and Electronics Engineering*; Webster, J.G., Ed.; John Wiley & Sons: Hoboken, NJ, USA, Online; 2015; pp. 1–19. [CrossRef]
28. Robinson, W.K.; Miller, R.J.; Gleeson, H.F.; Hird, M.; Seed, A.J.; Styring, P. Antiferroelectricity in Novel Liquid Crystalline Materials. *Ferroelectrics* **1996**, *178*, 237–247. [CrossRef]
29. Mills, J.T.; Gleeson, H.F.; Goodby, J.W.; Hird, M.; Seed, A.; Styring, P. X-ray and optical studies of the tilted phases of materials exhibiting antiferroelectric, ferrielectric and ferroelectric mesophases. *J. Mater. Chem.* **1998**, *8*, 2385–2390. [CrossRef]
30. Hirst, L.S.; Watson, S.J.; Gleeson, H.F.; Cluzeau, P.; Barois, P.; Pindak, R.; Pitney, J.; Cady, A.; Johnson, P.M.; Huang, C.C.; et al. Interlayer structures of the chiral smectic liquid crystal phases revealed by resonant x-ray scattering. *Phys. Rev. E* **2002**, *65*, 041705. [CrossRef]
31. Rao, N.V.S.; Pisipati, V.G.K.M.; Alapati, P.R.; Potukuchi, D.M. Density Studies in TerephthalyIidene-bis-p-n-dodecylaniline. *Mol. Cryst. Liq. Cryst.* **1988**, *162*, 119–125.
32. Kiefer, R.; Baur, G. Density Studies on Various Smectic Liquid Crystals. *Liq. Cryst.* **1990**, *7*, 815–837. [CrossRef]
33. Lakshminarayana, S.; Prabhu, C.R.; Potukuchi, D.M.; Rao, N.V.S.; Pisipati, V.G.K.M. Pretransitional effects at the isotropic mesomorphic phase transitions in the TBAA series. *Liq. Cryst.* **1996**, *20*, 177–182. [CrossRef]
34. Leadbetter, A.J.; Durrant, J.L.A.; Rugman, M. Density of 4 n-Octyl-4-Cyano-Biphenyl (8CB). *Mol. Cryst. Liq. Cryst.* **1977**, *34*, 231–235. [CrossRef]
35. Lagarde, M.P. Direct Electrical Measurement of Permanent Polarization of a Ferroelectric Chiral Smectic C Liquid-Crystal. *J. Phys. Lett.* **1977**, *38*, L17–L19. [CrossRef]
36. Patel, J.S.; Lee, S.D.; Goodby, J.W. Electric Field Induced Layer Reorientation in Ferroelectric Liquid-Crystals. *Phys. Rev. A* **1989**, *40*, 2854–2856. [CrossRef]
37. Srajer, G.; Pindak, R.; Patel, J.S. Electric Field Induced Layer Reorientation in Ferroelectric Liquid-Crystals—An X-Ray Study. *Phys. Rev. A* **1991**, *43*, 5744–5747. [CrossRef]
38. Garoff, S.; Meyer, R.B. Electroclinic Effect at AC Phase-Change in a Chiral Smectic Liquid-Crystal. *Phys. Rev. Lett.* **1977**, *38*, 848–851. [CrossRef]
39. Nair, B.; Usui, T.; Crossley, S.; Kurdi, S.; Guzman-Verri, G.G.; Moya, X.; Hirose, S.; Mathur, N.D. Large electrocaloric effects in oxide multilayer capacitors over a wide temperature range. *Nature* **2019**, *575*, 468–472. [CrossRef]
40. Li, J.J.; Li, J.T.; Wu, H.H.; Qin, S.Q.; Su, X.P.; Wang, Y.; Lou, X.J.; Guo, D.; Su, Y.J.; Qiao, L.J.; et al. Giant Electrocaloric Effect and Ultrahigh Refrigeration Efficiency in Antiferroelectric Ceramics by Morphotropic Phase Boundary Design. *ACS Appl. Mater. Interfaces* **2020**, *12*, 45005–45014. [CrossRef]
41. Saha, R.; Feng, C.R.; Eremin, A.; Jakli, A. Antiferroelectric Bent-Core Liquid Crystal for Possible High-Power Capacitors and Electrocaloric Devices. *Crystals* **2020**, *10*, 652. [CrossRef]

Article

Molecular Simulation Approaches to the Study of Thermotropic and Lyotropic Liquid Crystals

Mark R. Wilson *, Gary Yu, Thomas D. Potter, Martin Walker, Sarah J. Gray, Jing Li and Nicola Jane Boyd

Chemistry Department, Durham University, Durham DH1 3LE, UK; gary.yu@durham.ac.uk (G.Y.); thomas.d.potter@durham.ac.uk (T.D.P.); martin.barugh@gmail.com (M.W.); sarah.j.gray@hotmail.co.uk (S.J.G.); jing.li@durham.ac.uk (J.L.); n.janeboyd@btinternet.com (N.J.B.)
* Correspondence: mark.wilson@durham.ac.uk

Abstract: Over the last decade, the availability of computer time, together with new algorithms capable of exploiting parallel computer architectures, has opened up many possibilities in molecularly modelling liquid crystalline systems. This perspective article points to recent progress in modelling both thermotropic and lyotropic systems. For thermotropic nematics, the advent of improved molecular force fields can provide predictions for nematic clearing temperatures within a 10 K range. Such studies also provide valuable insights into the structure of more complex phases, where molecular organisation may be challenging to probe experimentally. Developments in coarse-grained models for thermotropics are discussed in the context of understanding the complex interplay of molecular packing, microphase separation and local interactions, and in developing methods for the calculation of material properties for thermotropics. We discuss progress towards the calculation of elastic constants, rotational viscosity coefficients, flexoelectric coefficients and helical twisting powers. The article also covers developments in modelling micelles, conventional lyotropic phases, lyotropic phase diagrams, and chromonic liquid crystals. For the latter, atomistic simulations have been particularly productive in clarifying the nature of the self-assembled aggregates in dilute solution. The development of effective coarse-grained models for chromonics is discussed in detail, including models that have demonstrated the formation of the chromonic N and M phases.

Keywords: liquid crystals; molecular simulation; molecular dynamics; dissipative particle dynamics

Citation: Wilson, M.R.; Yu, G.; Potter, T.D.; Walker, M.; Gray, S.J.; Li, J.; Boyd, N.J. Molecular Simulation Approaches to the Study of Thermotropic and Lyotropic Liquid Crystals. *Crystals* **2022**, *12*, 685. https://doi.org/10.3390/cryst12050685

Academic Editor: Ingo Dierking

Received: 22 March 2022
Accepted: 30 April 2022
Published: 10 May 2022

Publisher's Note: MDPI stays neutral with regard to jurisdictional claims in published maps and institutional affiliations.

Copyright: © 2022 by the authors. Licensee MDPI, Basel, Switzerland. This article is an open access article distributed under the terms and conditions of the Creative Commons Attribution (CC BY) license (https://creativecommons.org/licenses/by/4.0/).

1. Introduction

In recent years, molecular simulation has become a powerful tool for studying a range of soft matter systems. Simulations have been extended to bulk polymers [1], polymer surfaces [2], proteins [3,4], membranes [5–7], self-assembly in solution [8], and molecules at water interfaces [9,10], in addition to many other systems. In these cases, simulations aim to provide quantitative predictions to compare with experiment and also to provide qualitative insights into local molecular structure and order, which are often difficult to obtain by experimental means. These comments are particularly true for liquid crystalline systems, where an important part of the contemporary picture of how molecules are ordered within liquid crystal phases comes from the insights that molecular simulation has been able to offer over the last few decades [11–14].

Molecular simulation models are part of a traditional hierarchy of simulation methodologies which cover different time and length scales from the microscopic to mesoscopic to continuum scales. For liquid crystals, this hierarchy is particularly significant (see Figure 1) because it covers the following:

- *The quantum mechanical regime*, where single-molecule calculations are valuable in determining molecular properties of single thermotropic mesogens to use, for example, in the development of materials for displays [15];

- *The atomistic regime* where atomistic simulation can, in principle, be used to study the detailed molecular structure of a liquid crystal and predict bulk properties for simpler thermotropic phases, such as nematics [16,17];
- *The coarse-grained regime* where simulations can be used to study the structure of smectics [18], twist grain boundary phases [19], and polymer liquid crystals [20–23] and more complex liquid crystal phases where molecular shape and packing (or often the shape of a larger object such as a colloid) are often significant in determining the structure of the phase;
- *The continuum regime* where simulations no longer consider a molecular description but instead considers a fluid description where the local orientation of the director can be followed in complex geometries [24].

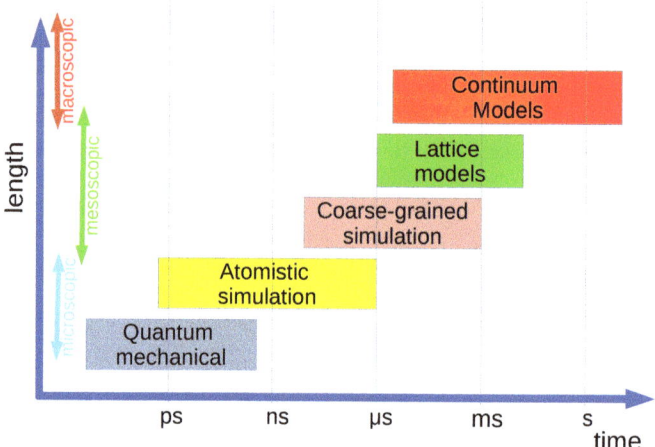

Figure 1. The time and length scale hierarchy for liquid crystal simulations covering microscopic, mesoscopic, and macroscopic regimes.

Over the last decade, the possibilities for modelling liquid crystals have been extended hugely with the realisation of readily accessible parallel computer architectures and simulation algorithms that can take advantage of them. This has led to a wide range of new studies [11]. This perspective article, which is part of a collection of articles showcasing liquid crystal research in the UK, highlights some of the recent progress made in studying thermotropic and lyotropic liquid crystal systems using molecular simulation methods. It covers both atomistic simulations and simulations carried out using different types of coarse-grained molecular simulation models. This article concentrates on new developments and also points to areas where the continued increase in computer power is opening up new possibilities.

2. Thermotropic Liquid Crystals

2.1. Atomistic Simulations of Thermotropics: Towards the Prediction of Accurate Transition Temperatures

A holy grail of atomistic simulation is to predict the phase sequence, transition temperatures, and local molecular structure of liquid crystal phases. However, this goal is extraordinarily difficult to realise and, in many ways, is one of the most challenging goals in soft matter simulation. Thermotropic liquid crystalline systems are often sensitive to small details in chemical structure. The addition of an extra functional group, or even just a small change in the length of an alkyl chain, can dramatically change transition temperatures or even change the sequence of thermodynamically stable liquid crystal phases that a molecule exhibits. Simulations have helped demonstrate the reasons for this but has not fully solved the prediction problem!

Firstly, the stability of thermotropic liquid crystals is a mix of two factors: shape and attractive interactions. For example, a small increase in molecular length will increase excluded volume effects and will drive molecular alignment of a thermotropic mesogen in a way that is better-known for colloidal systems and is well-understood by Onsager theory [25–27]. Moreover, an increase in density of packing (strictly a decrease in free volume) for a thermotropic system drives molecular alignment by a similar mechanism. However, changes in functional groups can disrupt molecular packing and also add preferred local interactions that can alter the phase behaviour. Add this to the further complication that many liquid crystal molecules can potentially exhibit a range of different unseen phases (i.e., unseen under standard thermodynamic conditions because they are marginally higher in free energy than the phases which are seen) and the phase diagram prediction problem becomes exceedingly challenging.

Nonetheless, considerable progress has been made in producing high-quality atomistic models that can represent structurally simpler liquid crystal phases and predict transition temperatures within 5–10 °C [28]. The key to the successes has been improvements in the accuracy of molecular force fields.

A typical molecular force field takes the following functional form:

$$E^{MM} = \sum_{bonds} K_r (r - r_{eq})^2 + \sum_{angles} K_\theta (\theta - \theta_{eq})^2$$
$$+ \sum_{torsions} \sum_{n=0}^{5} C_n (\cos(\psi))^n + \sum_{impropers} k_d (1 + \cos(n_d \omega - \omega_d)) \quad (1)$$
$$+ \sum_{i>j}^{N} \left[4\varepsilon_{ij} \left(\left(\frac{\sigma_{ij}}{r_{ij}} \right)^{12} - \left(\frac{\sigma_{ij}}{r_{ij}} \right)^{6} \right) + \frac{1}{4\pi\varepsilon_0} \frac{q_i q_j}{r_{ij}} \right],$$

where r_{eq} and θ_{eq} are, respectively, natural bond lengths and angles, K_r, K_θ, and C_n are, respectively, bond, angle, and torsional force constants, ψ is a dihedral (torsional) angle, k_d is a force constant associated with improper dihedral angles, ω, n_d, and ω_d, respectively, represent an improper dihedral angle, its periodicity, and the phase angle associated with it. σ_{ij} and ε_{ij} are the usual Lennard–Jones parameters modelling nonbonded interactions, and q_i, q_j are partial electronic charges. Here, we have written E^{MM} in the form used by the popular General AMBER Force Field (GAFF) [29].

In terms of liquid crystal systems, several things are crucial:

- A good charge distribution to represent the local electrostatic potential around a thermotropic mesogen;
- Accurate torsional potentials to represent internal rotations, as these can dramatically alter the average shape exhibited by a thermotropic mesogen within a liquid crystalline phase;
- Excellent nonbonded interactions to represent steric repulsion and attractive interactions.

In fact, the torsional potentials and nonbonded interactions are absolutely critical in determining transition temperatures. The majority of mesogens have alkyl chains that are "melted" in a mesophase (i.e., not in the lowest energy conformation but exhibit a range of conformations). If the torsional interactions are, for example, too stiff, then the average structure will tend to be elongated, phases will tend to be artificially stabilised; as a result, transition temperatures will be increased. Likewise, a very small error leading to an increase in density can dramatically promote mesophase stability by excluded volume effects and the Onsager mechanism [30], i.e., promoting translational entropy at the expense of rotational entropy.

With these factors in mind, Boyd and Wilson developed a version of GAFF, GAFF-LCFF (GAFF-Liquid Crystal Force Field), that was tuned for liquid crystalline systems [28,31]. Their work involved the careful optimisation of torsional potentials by fitting to high-quality density functional torsional scans, and (following the approach of previous workers [32–34]), careful optimisation of nonbonded parameters to reproduce densities and heats of vaporization of small-molecule fragments. Here, in particular, they aimed, where possible, to fit the densities of small molecules to better than 1%. Whilst such work has traditionally been extremely time consuming, in the future, it is likely to become much easier through developments such as the Open Force Fields project and automatic fitting tools such as the ForceBalance software [35].

Using GAFF-LCFF, liquid crystal state points and the isotropic to nematic transition are accessible for simulations extended over ∼240 ns per state point (although simulations times required are strongly system size-dependent and also depend on the viscosity of the system). These time scales are very short in comparison to director reorientation in bulk liquid crystals but on the nanoscale are sufficient to see the spontaneous alignment of a nematic liquid crystal from the isotropic phase. Figure 2 shows snapshots illustrating the molecular structure in the nematic phase of the bent-core mesogen C5-Ph-ODBP-Ph-OC12, including the beginnings of microphase phase separation between core and chains.

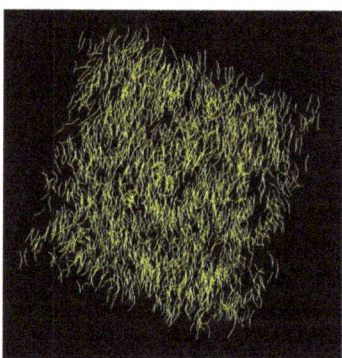 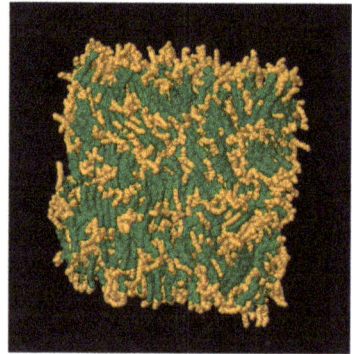

Figure 2. Simulation snapshots of 2048 molecules from the nematic phase of the bent-core mesogen C5-Ph-ODBP-Ph-OC12 at 480 K. (**Left**): Line drawing representation of the molecular bent core within the nematic phase. (**Right**): space-filling representation of C5-Ph-ODBP-Ph-OC12 molecules in the nematic phase showing molecular cores in green and alkyl tails in gold. The snapshot shows the beginnings of microphase separation between cores and tails that occurs in a pretransitional region before the phase transition to a DC phase at lower temperatures.

Orientational order parameters are typically calculated from suitable vectors within the molecule or, in the case of many calamitic mesogens, from the molecular long axis obtained from the moment of inertia tensor. The instantaneous average across molecular vectors in the simulations leads to both a liquid crystal director $\vec{n}$ and an orientational order parameter S_2. In practice, these are obtained by calculating the ordering tensor:

$$\mathbf{Q}_{\alpha\beta}(t) = \frac{1}{2N} \sum_{i=1}^{N} \left[3u_{i\alpha} u_{i\beta} - \delta_{\alpha\beta} \right], \quad \alpha, \beta = x, y, z, \tag{2}$$

where the sum runs over all N molecules. The largest eigenvalue of the **Q** tensor represents the following:

$$P_2(t) = \frac{1}{N} \sum_{i=1}^{N} P_2(\cos \theta_i), \tag{3}$$

where P_2 is the second Legendre polynomial, and the associated eigenvector is the director $\vec{n}(t)$. S_2 is defined as the time average of P_2 over a suitable time interval where $\langle P_2 \rangle$ is unchanging. However, in practice, to minimise system size effects in locating the phase transitions, S_2 is often obtained from $-2\times$ and the middle eigenvalue of **Q**, which fluctuates about a value of zero in the isotropic phase but equals $P_2(t)$ in the nematic phase. Here, we note in passing that in relatively small simulated systems, there is always a danger of some uncertainty connected with system size and the possibility of supercooling past phase transitions. Although this is less of a problem with nematic liquid crystals than many other soft matter systems where the enthalpy change associated with the phase transition is larger.

For the systems tested (molecular structures given in Figure 3), GAFF-LCFF provides excellent liquid crystal clearing point predictions, as shown in Table 1. Because of its success, GAFF (with GAFF-LCFF modifications) has been used for a range of liquid crystal systems, including recent studies into de Vries behaviour [36,37].

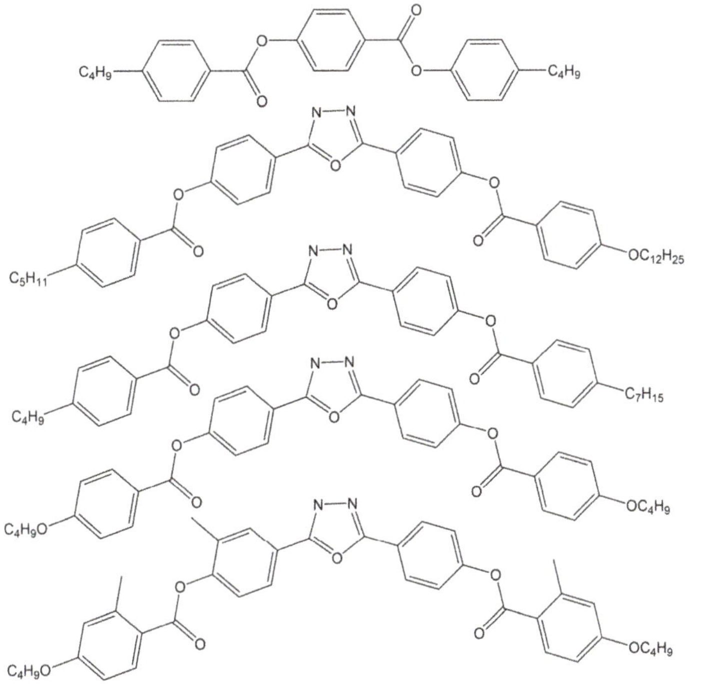

Figure 3. Structures of the five mesogens used in Table 1 (top to bottom: 1,3-benzenedicarboxylic acid,1,3-bis(4-butylphenyl)-ester, C5-Ph-ODBP-Ph-OC12, C4-Ph-ODBP-Ph-C7, C4O-Ph-ODBP, and C4O-Ph-ODBP (trimethylated)).

Table 1. Experimental and simulated clearing points for a series of liquid crystalline systems using GAFF-LCFF modifications [28,31] relative to the GAFF force field [29].

Molecule	T_{NI} (exp.) /K	T_{NI} (GAFF-LCFF) /K
1,3-benzenedicarboxylic acid,1,3-bis(4-butylphenyl)-ester	452	450–460
C5-Ph-ODBP-Ph-OC12	512.6	~510
C4-Ph-ODBP-Ph-C7	507	~500
C4O-Ph-ODBP	558	550–560
C4O-Ph-ODBP (trimethylated)	421	420–430

2.2. Simulation Insights into the Structure of New Phases

One of the most valuable features of atomistic simulations is that they can provide some insight into the formation of new liquid-crystalline phases and shed some light on liquid–crystalline structures where the arrangement of molecules is unclear from experiments. Figure 4 is from the preliminary work of Yu and Wilson and shows the structure of the twist-bend nematic (N_{TB}) phase for liquid crystal dimer CB7CB. Here, CB7CB is gradually cooled from a nematic phase into the underlying phase and equilibrated over periods of 100 ns per state point. There was originally considerable debate about the structure of the phases of CB7CB and similar dimers over a number of years, going back to the original synthetic work where an additional phase of unknown structure was identified below a conventional nematic. It was later suggested that this might correspond to the N_{TB} phase predicted by Meyer [38] and by Dozov [39]. Considerable experimental and modelling work (including atomistic simulations [40]) has now taken place, confirming, almost without doubt, that CB7CB shows a chiral phase structure with domains of opposite handedness [40–43] despite the fact that the molecules themselves are achiral. Up to a few years ago, atomistic simulation models were not possible for this type of phase, but good simulation models are now able to provide a molecular level picture of the order within such phases, subject to the usual constraints of system size and simulation time scale. Simulations confirm the transition from a nematic to a chiral twisted structure associated with an order parameter change, as shown in Figure 4, and confirm a helical pitch, order parameter, and conical tilt angle in agreement with experiment [44].

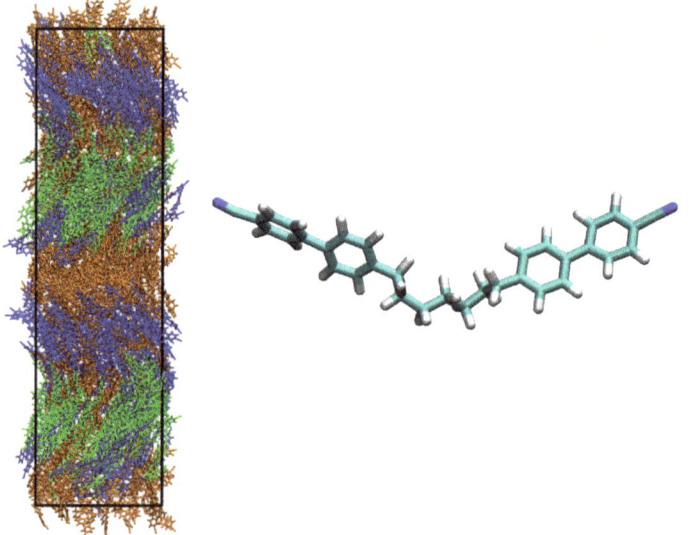

Figure 4. (**Left**): Snapshot, with orientational colour coding, showing the structure of the N_{TB} phase of CB7CB. (**Right**): The structure of a CB7CB molecule.

In recent work, Boyd and co-workers used atomistic simulations to study the local ordering of some very unusual liquid crystal phases where the structure of the phase ultimately arises from local molecular packing. These studies include the following:

- The dark conglomerate (DC) phase where atomistic simulations demonstrate the presence of molecular layers which undergo a saddle-splay layer deformation [31];
- Studies of B4 phase forming molecules [45,46] where subtle changes in molecular packing can be induced by the positioning of a (sterically significant) lateral methyl group at chiral centres in bent-core molecules with a long and short arm. These lead to changes in the splay, twist, and bend of molecular layers, which ultimately lead to the formation of twisted filaments and multi-level hierarchical self-assembled structures.

In the early days of liquid crystal simulation, simple coarse-grained models such as hard and soft spherocylinders, hard ellipsoids, the Gay–Berne model [11–13,47–49], and models composed of joined hard spheres [50,51] were instrumental in providing insights into simple nematic and smectic phases and polymer liquid crystals [20,22,52,53]. It is an interesting question whether a new generation of models can provide further insights into more complex phases, where phase structure appears to be dominated by local packing effects. Already considerable success has been achieved by exploring the influence of packing effects in non-linear rigid particles in a number of key areas:

- The formation of chiral superstructures from asymmetric bent-cores molecules composed of achiral tangential Lennard–Jones and WCA spheres [54];
- The formation of the N_{TB} phase from crescent-shaped particles composed of tangential hard spheres [55];
- The formation of biaxial, twist-bend, and splay-bend nematic phases from hard banana-shaped particles [56].

The latter work is particularly interesting because it shows that in this case biaxial, twist-bend, and splay-bend nematic phases are metastable with respect to smectic phases but stability can be induced by the presence of polydispersity in the particle's length or by curvature in the particle shape. This is immediately relevant to the design of new colloidal liquid crystals [57] but is also pertinent to understanding many thermotropic systems; i.e., many of the new phases that exhibit chiral arrangements of molecules form in systems where shape dominates the packing of molecules but where there is some "polydispersity" in molecular shape arising from conformational disorder that destabilises both crystal and smectic phase structures.

Figure 5 shows preliminary work from the work of the Wilson group exploring new coarse-grained models. Figure 5a shows a snapshot from a model composed of three spherocylinders interacting through the soft spherocylinder model of Lintuvuori and Wilson [23,58]. Here, some "polydispersity" in shape arises from flexibility built into the model through a torsional potential about the central spherocylinder. Figure 5b shows a snapshot from a rigid model of a chiral asymmetric-bent-core molecule. In this model, microphase separation is ensured through the use of Lennard–Jones sites to represent the central aromatic part of the molecule and WCA sites to represent aliphatic parts of the molecule, leading to smectic layer formation. Packing frustration is ensured by requiring a larger volume to pack aliphatic tails and through the presence of a chiral off-axis site that mimics the effects of lateral methyl substitution in the work of Hegmann and co-workers [45,46]. Here, twisted and splayed smectic layers are induced from these packing constraints, leading to a helical superstructure formed from twisted layers forming throughout the phase.

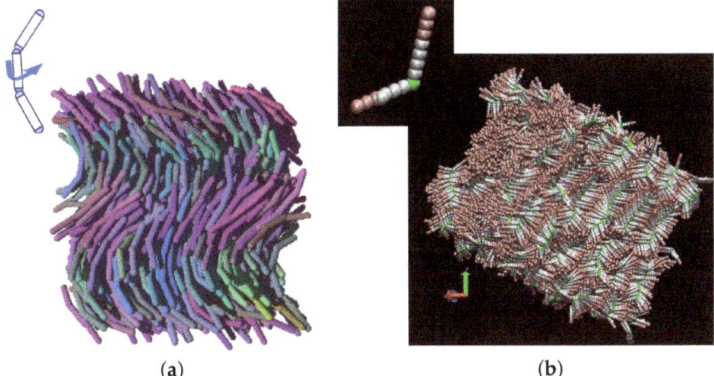

(a) (b)

Figure 5. (**a**): Snapshot from a simulation of a simple coarse-grained simulation model of jointed spherocylinders. 1728 molecules are used with each molecule composed of three bonded spherocylinders interacting through the soft spherocylinder model of Lintuvuori and Wilson [23,58]. The angle between adjacent spherocylinders is set at 150 degrees. A torsional potential about the central spherocylinder is used to control the orientation of the two arms to impose an average planar conformation corresponding to a typical bent-core mesogen. The snapshot shows a N_{TB} phase that arises spontaneously on cooling from a higher temperature nematic phase. (**b**): Twisted layer structures arising from simulations of 10,000 chiral-rigid-asymmetric-bent-core molecules. Green and white spheres represent a central aromatic core and interact through a Lennard–Jones potential with $\sigma/\sigma_0 = 0.75$, the pink spheres interact through a WCA potential with $\sigma/\sigma_0 = 0.75$. The two arms are positioned at an angle of 120 degrees.

2.3. Calculation of Material Properties Using Molecular Simulations

In addition to research aimed at predicting the structure of liquid crystal phases and their transition temperatures, considerable effort has been invested into developing methods that might be suitable for predicting material properties. The latter is potentially important for future screening applications, where simulation could act as a valuable tool in helping design appropriate liquid crystal molecules prior to synthesis based on predictions of their material properties. Much of the initial work in this area has been carried out on coarse-grained models, which can be simulated using large numbers of molecules and studied over long simulation times. Both of these criteria tend to be important for reliable calculations of a large number of key material properties.

Of particular note has been the work carried out in developing predictive methods for Frank elastic constants [59–61]. Here, the most successful approach [59] considers wavevector-dependent fluctuations in orientational order, which occur naturally during the course of a molecular dynamics simulation of a nematic liquid crystal at finite temperature. This approach has been successfully applied to nematic phases composed of hard prolate ellipsoids and hard spherocylinders [61,62] and to Gay–Berne mesogens [59,60]. However, the large system sizes required for these calculations make this methodology very challenging to implement for atomistic models at the current time. In principle, elastic constants can also be obtained from an approximate approach involving the direct simulation of the Freédericksz transition [63,64] (although this has proved to be inefficient compared to the fluctuation approach for simple systems and would be very challenging to carry out accurately for atomistic studies) or from calculations of the direct correlation function in the nematic phase [65].

Some progress has also been made in the calculation of other display-related material properties. An example is the rotational viscosity coefficient, γ_1, which is important (for example) in determining the on and off times of a twisted nematic display ($t_{off} \propto \gamma_1$, $t_{on} \propto \gamma_1$) and for other nematic devices. γ_1 can be determined by a number of techniques including equilibrium and non-equilibrium molecular dynamics methods [66–69]. While extensive work has been carried out on simple coarse-grained models, such as Gay–Berne potentials, these simulated values do not closely match the values of γ_1 measured for real thermotropic systems [69]. To some extent, this is because most Gay-Berne parametrisations more closely match colloidal liquid crystals than small thermotropic organic molecules. However, equilibrium molecular dynamics can be applied successfully, and relatively easily, to atomistic studies [16], and have the advantage of avoiding the need to couple molecules to an external field. Here, for example, γ_1 can be successfully obtained from the director angular velocity correlation or the director mean-squared displacement. Both methods rely on thermally excited fluctuations in orientational order, avoiding the need to perturb the system by application of an external field.

Atomistic studies have also been successful in calculating flexoelectric coefficients for a nematic liquid crystal [17]. Here, equilibrium molecular dynamics calculations have been performed for a series of temperatures in the nematic phase. The simulation data are used to calculate the flexoelectric coefficients e_s and e_b using the linear response formalism of Osipov and Nemtsov [70].

A further interesting area where atomistic simulations have been applied successfully is in the study of helical twisting powers. The helical twisting power (HTP) measures the ability of a chiral molecule to impart a chiral twist to a bulk nematic phase. Two successful approaches have been demonstrated involving the chirality order parameter approach of Ferrarini and co-workers [71–77] and a scaled chirality index arising from the work of Osipov and co-workers [78] that has been employed for molecular systems [79,80]. Part of the success of the two methods arises from them being single-molecule techniques; i.e., they do not require the simulation of bulk phases. Both approaches have been extensively studied and the results verified against experiments for a range of chiral molecules [81,82].

Interestingly, the chirality order parameter approach has also been applied to achiral bent-core molecules [83]. It is noted that some achiral bent-core molecules, acting as dopants, can increase the helical twisting power of a chiral host phase, i.e., the exact opposite of what would be expected. Calculations of the chirality order parameter, χ, show that although these molecules are on average achiral, they can exhibit "chiral conformations, which have extremely large values of χ in comparison to standard chiral dopants [83–85]. Hence, the suggestion is that these conformations have such large twists that they lead to a spontaneous asymmetry between right and left-handed forms in a chiral environment, through chiral templating with solvent molecules or other bent-core molecules (i.e., the handedness of the phase leads to there being a slight preference for either left-handed or right-handed conformations for the bent-core molecule). This then increases the overall chirality of the bulk phase [83]. This hypothesis was tested in bulk simulations of a simple coarse-grained model that exhibited high helical twisting powers for chiral conformations [86].

In principle, the HTP can also be deduced from the sampling of torques exerted on neighbouring molecules in a bulk phase [87,88]. This is a far harder calculation than a single-molecule based approach to calculating HTP, as it involves the simulation of bulk phases and the calculation of torques. Its application to atomistic models has been limited to rigid molecules within a coarse-grained Gay-Berne nematic solvent but the method does yield the correct sign of the helical twist induced by a molecule and provides a good approximation to the magnitude of HTP [87].

It is also possible to look at individual chiral dopant molecules within a uniformly twisted nematic phase of wavevector $k = 2\pi/P$ (where P is the pitch of the twisted phase). Here, there should be a chemical potential difference between enantiomers that want to induce a twist in the same direction as the host phase and those that want to induce a twist in the opposing direction. In the limit of infinite dilution this difference in chemical potential, $\Delta\mu$, is directly proportional to the helical twisting power, β, and the twist elastic constant K_2 [89].

$$\Delta\mu = \mu_- - \mu_+ = 8\pi\beta K_2 k \qquad (4)$$

This was exploited by Wilson and Earl using twisted periodic boundary conditions to calculate the helical twist power of five atomistic dopant molecules in simple coarse-grained solvents [90]. Relatively large errors arise with this method, as the dopants must be "grown" into the solvent in a series of separate simulation steps. In the original paper, this was performed for spherocylinder and Gay–Berne solvents. However, for a uniformly twisted nematic solvent, the key quantity of interest is the twist elastic constant, and the exact details of the solvent potential are not important, providing (of course) a uniform nematic is simulated. Therefore, in principle, soft-core model potentials [58] could be used to model the liquid crystal solvent, greatly reducing the difficulty of introducing a chiral dopant into a bulk phase in which there could initially be strong overlaps between the dopant and the solvent molecules.

Finally, it is worth noting that only very recently have advances in computer power and parallel algorithms opened up the possibility of studying large atomistic simulations of several thousand molecules. As computer time continues to increase in the future (and atomistic simulations increase further in size), it should be possible to look at other material properties that are currently difficult to study because of system size limitations. Dielectric anisotropy and ion conductivity in liquid crystals are good examples of such properties.

3. Lyotropic Liquid Crystals

3.1. Surfactant Models, Micelles, and the Formation of Lyotropic Phases

Conventional lyotropic liquid crystals have traditionally been very difficult to study using molecular simulation methods. At an atomistic level, the self-assembly of surfactants to give micelles and, subsequently, the self-assembly of micellar aggregates to give liquid crystal phases, occurs on time scales that are extremely challenging for atomistic simulation. However, advances in computer time mean that it is now possible to "see" the self-assembly process occurring for conventional amphiphiles for small aggregates if calculations are carried out well above the critical micelle concentration (CMC) and simulations are extended for tens, or (in some cases) hundreds, of nanoseconds. Moreover, coarse-grained simulation methods have become very useful in studying traditional lyotropic phase diagrams. Figure 6 shows a micelle that has spontaneously self-assembled from a dispersed group of monomers for the industrially important surfactant LAS (linear alkylbenzene sulfonate). Here, simulations were carried out at the fully atomistic level using the GAFF force field in combination with TIP3P water using the completely linear isomer of LAS shown in Figure 6a (noting that in most commercial applications LAS is used as a mixture of branched isomers a shown in Figure 6b). Here, micelles form within 100 ns of simulation for a system with a LAS:water ratio of 1:221, which is well above the CMC.

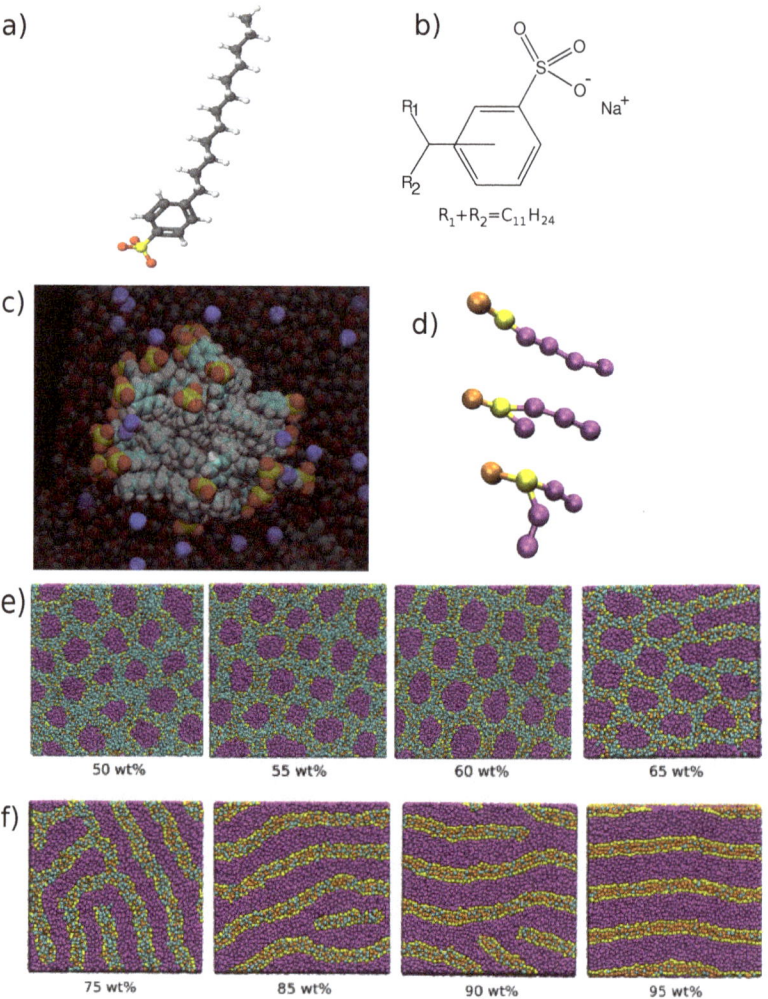

Figure 6. Structures and simulation models for versions of LAS (linear alkylbenzene sulfonates). (**a**) Molecular structure of a fully linear single chained version of the anionic surfactant LAS; (**b**) chemical structure of typical branched LAS molecules used in industry, (**c**) single micelle of LAS in water (sulfonate head groups are shown in yellow and red, and blue sites represent sodium counter ions, water molecules are shown in a partially transparent representation); (**d**) three DPD models of LAS with the orange bead representing the sulfonate head group, the yellow bead representing the phenyl group, and the purple beads representing parts of the alkyl chain; (**e,f**) simulations snapshots from the phase diagram of the linear form of LAS adapted with permission from Ref. [91], 2018, Sarah J. Gray; showing (**e**) the hexagonal phase composed of cylindrical micelles and (**f**) the lamellar phase.

For many surfactant systems, the CMC occurs at sufficiently low concentrations that atomistic studies are not feasible near the CMC because of the number of water molecules required and also the time scales (beyond the microsecond regime) required for diffusion of water molecules. However, good progress has been made using coarse-grained models. Figure 6 additionally shows models and snapshots from a typical dissipative particle dynamics (DPD) coarse-grained surfactant study. In DPD, soft spheres are used to represent coarse-grained sites corresponding to groups of atoms within the molecule. Here, soft spheres interact through a conservative force:

$$F_{C,ij} = a_{ij}(1 - \mathbf{r}_{ij}/r_{cut}) \quad \text{for} |\mathbf{r}_{ij}| < r_{cut}$$
$$F_{C,ij} = 0 \quad \text{for} |\mathbf{r}_{ij}| \geq r_{cut} \quad (5)$$

which leads to a harmonic repulsive potential. The chemical interactions are mainly controlled by the parameters a_{ij}, which govern the strength of the repulsive interactions between two species i and j. Part of the beauty of DPD is that simulations can be carried out using large dynamic time steps (due to the use of soft spheres and soft springs linking them) with a greatly reduced number of sites in comparison to atomistic simulation. This comes with some loss in terms of chemical specificity. However, in recent years DPD has become a powerful methodology for modelling surfactant systems and considerable work has been carried out in mapping the interactions in DPD models to "real" molecular systems [92–96].

Figure 6d shows three models for different LAS molecules that provide good phase diagram predictions for linear and branched LAS species [91]. Figure 6e shows snapshots taken from DPD simulations of a linear LAS model in the high-concentration regime, which corresponds to hexagonal phases composed of rod-shaped micelles and (at slightly higher concentrations) lamellar phases. The lamellar layers exhibit a spacing of between 3.1 and 3.3 nm, in excellent agreement with experiments (3.2–3.3 nm).

3.2. Models for Chromonic Liquid Crystals

Of significant interest over the last decade has been the study of chromonic lyotropic liquid crystals [97]. These systems arise from non-conventional amphiphiles in which solubilizing groups are attached to the periphery of "molecular discs" (see Figure 7). Chromonic mesogens have been seen in a variety of dye and drug molecules, which commonly exhibit these two structural motifs (discs and hydrophilic peripheral groups). Chromonics have recently found new potentials applications [98] in areas such as biosensors [99,100], in the fabrication of thin-film structures [101] and in controllable self-assembly of gold nanorods [102], and interest in chromonic has been extended to drug delivery systems and to new areas such as "living liquid crystals" [103]. Understanding chromonic self-assembly is important in all these areas.

For chromonics systems, simple coarse-grained models have demonstrated the formation of the most commonly seen mesophases: the chromonic N and M phases. Simulations have confirmed the structure of these mesophases and have allowed for the mapping out of chromonic phase diagrams as a function of concentration [104,105].

Alongside coarse-grained studies, atomistic simulations have been productive in the following:

- Clarifying the nature of self-assembled aggregates in dilute solution [106];
- Explaining the origin of different types of stacking motifs [107];
- Explaining the formation of "layer aggregates", which can lead to smectic or layered chromonic phases [107,108].

These are phenomena that are difficult to probe experimentally at a molecular level in dilute solutions and for which simulation thereby provides unique information.

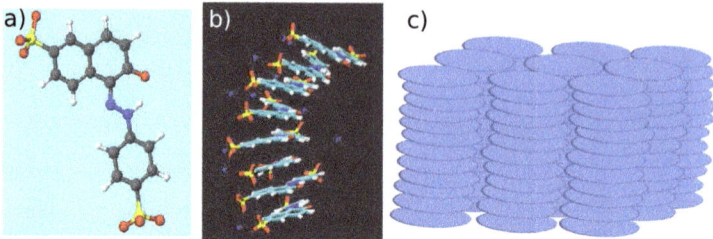

Figure 7. (a) The disc-shaped structure of the chromonic dianionic monoazo food dye sunset yellow in the NH hydrazone tautomer (the stable tautomeric form in aqueous solution); (b) stacking of sunset yellow molecules in solution into a chromonic aggregate (blue spheres represent sodium counterions) as seen in molecular dynamics simulations of sunset yellow [106]; (c) schematic diagram showing typical alignment of chromonic aggregates to form a liquid crystalline phase.

3.3. Studies of Nonionic Chromonics

The structure shown in Figure 8a is of 2,3,6,7,10,11-hexa-(1,4,7-trioxa-octyl)-triphenylene (TP6EO2M), which consists of a central polyaromatic core (a triphenylene ring) functionalised by six hydrophilic ethyleneoxy (EO) chains. TP6EO2M is the archetypal nonionic chromonic where the hydrophobic interactions arising from the aromatic rings induce stacking into a chromonic column but the hydrophilic interactions of the short EO chains are sufficient to solubilise the resulting aggregates.

TP6EO2M has become the "fruit fly" of chromonic liquid crystals simulations and has been simulated by a range of models, as shown in Figure 8. This is partly because the structure is relatively simple and symmetrical but also because the formation of chromonic stacks and chromonic phases for this molecule provides a major challenge for modern methods of multi-scale modelling.

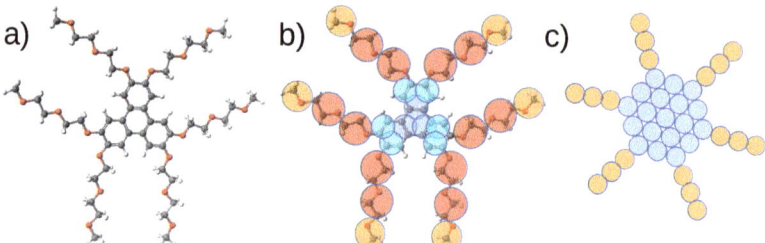

Figure 8. The nonionic chromonic mesogen TP6EO2M represented by three different levels of models. (a) All-atom model (b) a MARTINI-style coarse-grained model, and (c) a simple dissipative particle dynamics model. For coarse-grained models, bonds (not shown) link adjacent sites and angle interactions help to define molecular shape, blue beads represent different hydrophobic sites, and red and orange beads represent hydrophilic sites.

All-atom simulations of TP6EO2M (Figure 8a) using the OPLS-AA force field [109] demonstrate the formation of chromonic stacks. The free energy for association, $\Delta G^\circ_{\text{agg}}$, can be approximated from the depth of the "attractive well" obtained from potential of mean force (PMF) calculations, $U_{\text{PMF}}(r)$. Here, a molecule is pulled away from a dimer, or more generally from a n-mer, to a point where they are no longer interacting. In practice these can be performed by numerically integrating the average constraint force, f_c, obtained at a series of separation distances, s:

$$U_{\text{PMF}}(r) = \int_r^{r_{\max}} \left[\langle f_c \rangle_s + \frac{2k_\text{B}T}{s} \right] ds \qquad (6)$$

where r is the distance for the PMF and $2k_BT/s$ is a kinetic entropy term, which accounts for the increase in rotational volume at larger separation distances [110–112]. Many chromonic mesogens are assumed to undergo isodesmic association, with the same free energy change for each subsequent addition of a molecule to an n-mer. This leads to an exponential distribution of aggregate sizes. For TP6EO2M, the binding energy for a n-mer ($\Delta G^\circ_{agg} = -12RT$) is in good agreement with experiments. However, for TP6EO2M, and for many other chromonics, we find "quasi-isodesmic" association where the binding free energy, ΔG°_{agg}, for two molecules to form a dimer is slightly larger in magnitude (~ -2.5 RT lower for TP6EO2M) than the binding energy for an n-mer. Typically, for molecules in the interior of a stack, the entropy loss from chain confinement and orientational ordering upon aggregation is larger than on the formation of a dimer when the molecules are only confined by the presence of one neighbour.

Values of the enthalpy and the entropy of association can both be obtained from the temperature dependence of ΔG°_{agg}.

$$\Delta H^\circ = R \left[\frac{\partial \left(\Delta G^\circ_{agg}/RT \right)}{\partial (1/T)} \right]. \tag{7}$$

Interestingly, for TP6EO2M, $-T\Delta S^\circ$ and ΔH° both contribute favourably to ΔG°_{agg} and at 300 K it is the entropic contribution that is larger, indicating that the confinement of chains and orientations on molecular association is more than compensated for by the gain in entropy of the solvent that is released from interacting with the phenyl rings of triphenylene. This finding is very much in-line with traditional interpretations of the hydrophobic effect for small molecules at room temperature.

As might be expected, the relative balance of hydrophilic and hydrophobic interactions in these atomistic simulations is critical, and "out-of-the-box", the GAFF force field which normally performs well for a range of ionic chromonic systems behaves badly in comparison to OPLS, with the formation of disordered globular aggregates. With GAFF, the EO units are insufficiently hydrated to stabilise chromonic stacks.

At the other end of the scale of models in Figure 8c is a dissipative particle dynamics (DPD) representation of TP6EO2M: A model was designed by Walker et al. [104]. For TP6EO2M, two things are critical: the balance of interactions between hydrophobic and hydrophilic beads and also ΔG°_{agg} relative to the DPD energy scale, i.e., ϵ that controls the magnitude of the reduced temperature k_BT/ϵ. The former can be tuned for many DPD systems based on experimental infinite dilution activities or chemical potentials obtained from dilute atomistic simulations. For TP6EO2M, the latter is available from atomistic modelling (as discussed above).

DPD is sufficiently tractable to allow the full chromonic phase diagram to be simulated for TP6EO2M, producing an approximately exponential distribution of stacks in the isotropic phase (in-line with quasi-isodesmic association); a nematic N phase at intermediate concentrations, where columns are sufficiently long to align due to excluded volume interactions favouring translation entropy over orientational entropy; and a hexagonal M phase at higher concentrations, where the stacks pack on a hexagonally ordered lattice.

If EO chains are replaced in this model with hydrophobic–lipophobic chains [105] (e.g., as might occur, for example, with siloxanes or fluorinated systems) complex supramolecular aggregates form in which hydrophobic–lipophobic chains are excluded from water by the joining together of stacks with a single-molecule cross-section into dimers or trimers. This provides a mechanism for the formation of a novel chiral aggregate and also in the case of "Janus mesogens" with three adjacent hydrophobic–lipophobic chains, a novel smectic chromonic phase.

It is also interesting to ask whether an "in-between" coarse-grained model can provide further insights into the behaviour of TP6EO2M. This was the aim of recent work by Potter, Wilson, and co-workers who explored a MARTINI-style coarse-grained model (Figure 8b) where sites are grouped into beads representing two or three heavy atoms plus attached hydrogen atoms [113,114]. Such models are quite interesting as they aim to capture the individual interactions responsible for local order but also aim to capture the thermodynamics and correct self-assembly and self-organisation over larger length scales. They aim to perform this while also hitting a "sweet spot" where they employ only a fraction of the simulation time of an atomistic model, saving CPU cycles through a reduced number of sites, long dynamic time steps, and a speed-up in phase space exploration. While the idea of such a "Goldilocks" model is very appealing and has also many potential uses in related fields such as polymer simulation, membrane studies [5,7,115–119] and protein simulation, the challenges of making such a model are quite demanding. In fact, the very act of coarse-graining in an aqueous system alters the balance between entropy and enthalpy. The hydrophobic interaction in coarse-grained models of chromonics must incorporate additional enthalpic contributions to the free energy to make up for the loss of entropic ones.

There are a range of methods available to generate an intermediate coarse-grained model for a molecule such as TP6EO2M using both "bottom-up" and "top-down" routes [120], i.e., starting from an atomistic reference simulation, or starting from experimental structural and thermodynamic data (see the diagram in Figure 9).

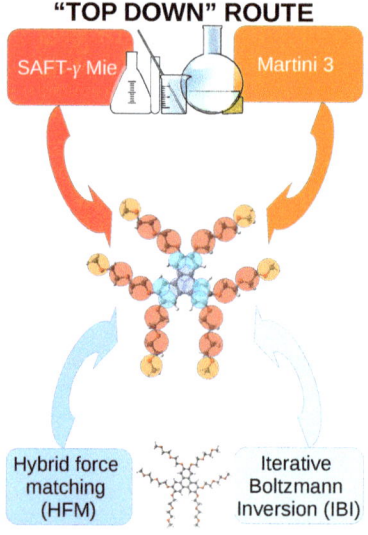

Figure 9. Routes to an ideal ("Goldilocks") coarse-grained model for chromonic mesogens using both bottom-up and top-down methodologies; ideally producing a model that is "just right" for molecular simulation. A Goldilocks model would capture the correct self-assembly behaviour in solution in terms of the structure of aggregates and the thermodynamics of self-assembly and would be sufficiently transferable to be used successfully over a range of concentrations and temperatures (providing of course these were not too "hot" or too "cold").

Traditionally, structure-based coarse-graining by methods such as multiscale coarse-graining (MS-CG) in the form of hybrid force matching (HFM) [120–123] or using iterative Boltzmann inversion [124], do a good job of capturing "molecular structure" and "order" but produce coarse-grained potentials that do not capture thermodynamics well (particularly mixing thermodynamics). They also tend to produce CG potentials that are extremely density and temperature-dependent and so have very limited transferability between state points. While HFM for TP6EO2M using an atomistic reference stack [113] yields quite good structures for the CG model, the very high binding energies make this rather unsuitable as a method to study changes in the aggregation of chromonics molecules with concentration.

It is worth noting that successful chemical engineering theories for mixtures, such as SAFT (Statistical Associating Fluid Theory), are often useful in predicting phase diagrams for simple molecular mixtures. These are starting to become useful in providing "top-down" pathways to thermodynamically consistent coarse-grained models. Of particular interest here is the SAFT-γ Mie model [2,125–130], which represents coarse-grained sites by generalised Mie potentials:

$$U_{\text{Mie},ij} = C\epsilon_{ij}\left[\left(\frac{\sigma_{ij}}{r}\right)^{\lambda_\text{r}} - \left(\frac{\sigma_{ij}}{r}\right)^{\lambda_\text{a}}\right] \quad (8)$$

where exponents λ_r and λ_a govern the ranges of the repulsion and attraction of the potential, and the following is the case.

$$\epsilon_{ij} = (1 - k_{ij})\frac{\sqrt{\sigma_{ii}^3\sigma_{jj}^3}}{\sigma_{ij}^3}\sqrt{\epsilon_{ii}\epsilon_{jj}}. \quad (9)$$

For species that are chemically incompatible for entropic or enthalpic reasons, $k_{ij} > 0$, and for species where one group is well solvated by another, $k_{ij} < 0$ (recalling that within a coarse-grained model an entropically unfavourable interaction is often partially translated into an unfavourable enthalpic interaction due to the reduction in the degrees of freedom). SAFT was exploited by Potter et al. to develop a SAFT-γ Mie model for TP6EO2M [114]. The balance of k_{ij} values was found to be crucial. If ethylene oxide units are poorly solvated, chromonic stacks do not form and instead phase separation, or a conglomerate of aggregates, occurs. Moreover, k_{ij} must be greater than zero for both aromatic-water interactions and for aromatic-ethylene oxide interactions in order to see stable chromonic stacks. In other cases, no aggregation occurs, or the system phase separates with the formation of large non-structured aggregates depending on specific k_{ij} combinations. Chromonic aggregation, therefore, appears only in a small region of parameter space where the hydrophilic–hydrophobic balance between aromatic, ethylene oxide, and water is just right.

It is perhaps not surprising that the most successful intermediate coarse-grained model for TP6EO2M comes from a Martini 3 model of the mesogen [113]. Martini models have been parametrised to reproduce good solvation free energies for different coarse-grained sites and, hence, are able to capture the correct hydrophilic–hydrophobic balance required to see chromonic behaviour. Figure 10 shows two snapshots from simulations of TP6EO2M using the tuned Martini 3 model of Potter [113] and co-workers, showing the structure of chromonic N and M phases. The latter occurs at high mesogen concentrations when, as shown in Figure 10, the columns are sufficiently close to undergo hexagonal packing.

3.4. Ionic Chromonics: A Rich Variety of Aggregation Motifs

Yu and Wilson [107] have used atomistic models to explore the rich variety of chromonic aggregation that occurs in cyanine dye systems, studying four dyes: pseudoisocyanine chloride (PIC), pinacyanol chloride (PCYN), 5,5′,6,6′-tetrachloro-1,1′,3,3′-tetraethylbenzimidazolylcarbocyanine chloride (TTBC), and 1,1′-disulfopropyl-3,3′-diethyl-5,5′,6,6′-tetrachloro-benzimidazolylcarbocyanine sodium salt (BIC). Simulations with the GAFF/TIP3P force field combination allowed a range of aggregate behaviour to be observed and rationalised based on potentials of mean force obtained from pulling molecules from the end of stacks of two, three, or four molecules. Both H-aggregates (where molecules exhibit face-to-face stacking and show hypsochromic (blue) spectra shifts in comparison to monomers) and J-aggregates (where adjacent molecules exhibit staggered stacking and show bathochromic (red) spectra shifts) are seen depending on the system, with molecular stacking arrangements leading to Y-junction and shift defects and sheet-like assemblies of molecules.

Examples of the aggregate structures seen are shown in Figure 11. What is particularly interesting is the layered structures seen by Yu and Wilson and also by Thind and co-workers [108], which correspond to a different type of chromonic aggregate to that normally reported. This layered structure can form the basis for a chromonic smectic phase. Moreover, it is likely that the layered structures reported represent an intermediate state that precedes the organisation of molecules into tubular or other higher-order architectures. In experimental studies, large scale tubular structures have been reported for some cyanine dyes using electron microscopy techniques [131–133].

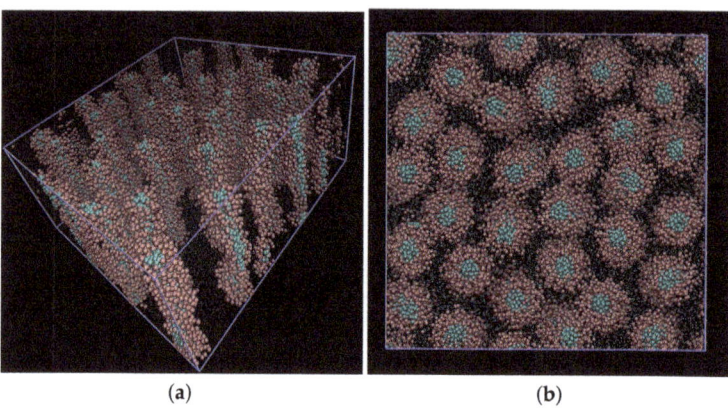

Figure 10. (a) Snapshot from the N phase composed of 900 TP6EO2M molecules simulated at a coarse-grained level using the Martini 3 model from reference [113] (54,909 Martini water sites representing 219,636 water molecules). The snapshot shows short chromonic stacks that are joined across the periodic boundary conditions. Small blue dots between columns represent Martini water sites. (b) Snapshot from the M phase of TP6EO2M showing the hexagonal packing of columns in this phase (9266 Martini water sites).

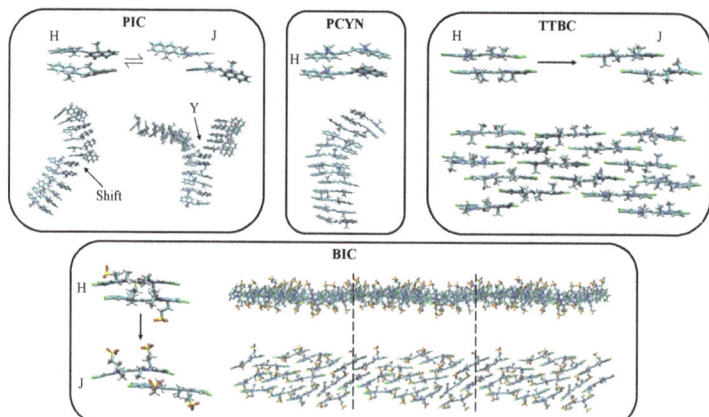

Figure 11. (**Top left**): H and J aggregate structures for PIC, plus shift defects and Y-defects seen in large aggregates. (**Top middle**): H aggregate structures for PCYN. (**Top right**): H and J aggregate structures for TTBC and a brick-like layer structure seen for TTBC at high concentrations, where J-aggregation dominates over H-aggregation. (**Bottom**): H and J aggregate structures for BIC and a layer structure observed for BIC molecules (side and top view). Figure adapted from Ref. [107], 2021 CC-BY licence, with permission from the Royal Society of Chemistry.

It is extremely challenging to develop coarse-grained molecular models that are able to capture the same types of aggregation seen with PIC, PCYN, TTBC, and BIC, including the thermodynamic equilibrium between monomers, H-aggregates and J-aggregates: M ⇌ H ⇌ J. However, such models are required to explore larger-scale aggregation beyond the time and length scales that are accessible to atomistic studies.

Yu and Wilson have explored this for one perylene dye molecule bis-(N,N-diethylaminoethyl)perylene-3,4,9,10-tetracarboxylic diimide dihydrochloride, PER (Figure 12) [8]. While bottom-up models fail to capture the correct behaviour of PER molecules in solution, it is possible to tune a top-down Martini 3 model to capture the same aggregation behaviour seen by atomistic studies. Moreover, a proof-of-concept hybrid model combining hybrid force matching (HFM) and Martini 3 was successfully developed. Here, a combined approach is able to capture the aggregate structure and provide a good representation of short-range electrostatic interactions that often control the local structure of chromonic aggregates. The use of Martini water and ions provides a large reduction in the number of sites within the model and improves the thermodynamics of hydration (which is usually poorly represented within a force matching approach). In Figure 12, we show the atomistic and coarse-grained structures seen by Yu and Wilson, together with the potential of mean force corresponding to the binding of dimer molecules. The phase structures shown in the lower part of Figure 12 are seeded to produce columns that connect over the periodic boundary conditions, i.e., effectively columns of infinite length. This makes it possible to investigate the concentration corresponding to the N to M phase transition, when the concentration is sufficiently high to see hexagonal packing of columns.

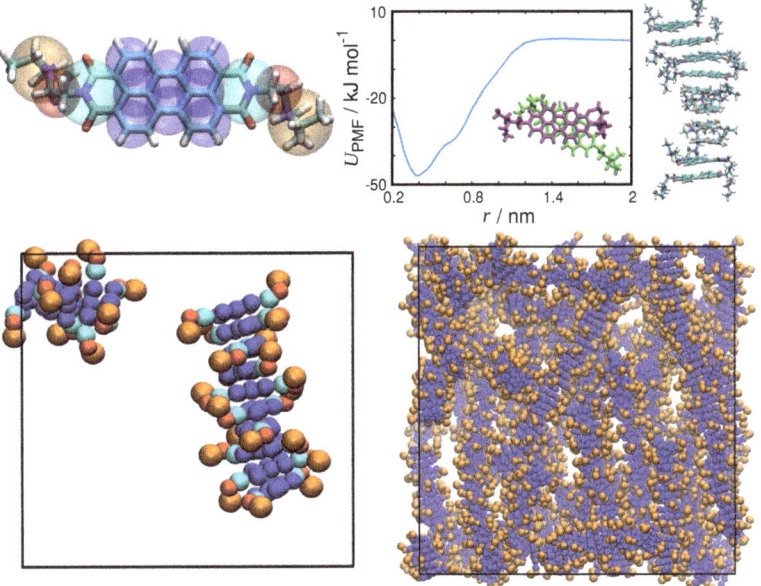

Figure 12. *Cont.*

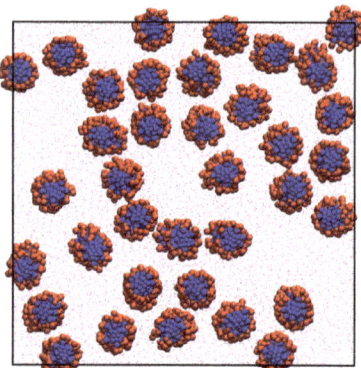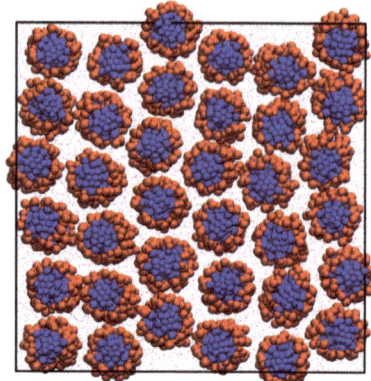

Figure 12. (**Top left**): the atomistic structure of PER and a coarse-grained representation suitable for use with bottom-up coarse-graining methods and the top-down Martini 3 approach. (**Top middle**): the potential of mean force for the binding of a PER dimer calculated from atomistic simulations, together with the lowest energy structure seen at the bottom of the potential well. (**Top right**): A snapshot from atomistic simulations showing the typical aggregate structure for PER. (**Middle left**): Aggregates of PER from a tuned Martini 3 model. (**Middle right**): A snapshot from the chromonic N phase of PER (water molecules not shown) obtained from long simulation runs using the tuned Martini 3 model developed in reference [8]. (**Bottom left**): Plan view of the N phase at 30 wt%. (**Bottom right**): Plan view of the M phase at 50 wt%. The top row sub-figures are redrawn from Figures 2 and 3 of reference [8]. Parts of this figure are adapted from reference [8], 2022 CC-BY licence.

4. Conclusions

The huge increase in computer time that has become available over the last decade, together with new algorithms capable of exploiting parallel computer architectures, has opened up many possibilities for the modelling of both thermotropic and lyotropic liquid crystal systems. This perspective article, which is part of a collection of articles showcasing developments within the UK, has pointed to several exciting areas of research emerging.

For thermotropic nematics, it is now possible to simulate a few thousand molecules, and, with the advent of improved molecular force fields, provide predictions for nematic clearing temperatures, T_{NI} within a 10 °C (or better) range (see Section 2.1). Current studies are limited to a few hundred nanoseconds of simulation time for this size of system. Nonetheless, as shown in Section 2.2, this type of simulation is also able to provide valuable insights into the structure of more complex liquid crystalline phases, where the details of molecular arrangements may be very difficult to probe experimentally. Section 2.2 also points to the possibilities of using simpler coarse-grained models to study larger systems of molecules as a method of understanding the structure of complex layered phases and/or thermotropic phases where chirality arises from achiral systems. Here, packing arrangements between molecules are often very important but there are many challenges arising from the complex interplay of packing constraints, microphase separation and favourable local interactions arising from the presence of specific functional groups.

Considerable progress has now been made in the development of methods for the calculation of liquid crystalline material properties for thermotropics. Undoubtedly, coarse-grained models have proved extremely useful as a testbed for this work, and it has been possible to test the convergence of results over long simulations with large simulation sizes. It should be noted that some of these studies point to the need to carry out simulations that are larger than can reasonably be simulated at an atomistic resolution. However, as computer time increases, this restriction will gradually be removed. Already good progress has been made in the calculation of helical twisting powers, rotational viscosities, and flexoelectric coefficients at an atomistic level.

For conventional lyotropic systems, formed through surfactants, atomistic simulations are often limited to the study of single micelles, which can be seen to form spontaneously at concentrations well above the CMC. However, recent years have seen many developments in coarse-grained models for these systems, particularly in the area of dissipative particle dynamics. These models are able to provide a good link to underlying thermodynamic data measured experimentally or determined by theory or modelling. They can also provide good predictions for experimentally measured observables, such as the lamellar layer spacing (see Section 3.1).

Chromonic liquid crystals were also discussed in detail. For these, non-conventional amphiphiles atomistic simulations have been particularly productive in clarifying the nature of the self-assembled aggregates in dilute solution (Section 3.2), which have the potential to be more complex than those seen for conventional surfactants. In recent years, different forms of self-assembly have been noted particularly in chromonic dye molecules, from the most common form of association involving stacking into columns (H-aggregate formation), to defect stacking with shift-defects and Y-defects (leading to branched structures), and self-assembly into brickwork layers (Section 3.4).

Considerable progress has also been made in the development of different levels of coarse-grained models for chromonic liquid crystals (see Sections 3.3 and 3.4). Of significant interest here is the challenge of capturing the correct local stacking arrangements at the same time as capturing the correct thermodynamics of self-assembly. Noting that in coarse-graining chromonic systems, the balance of entropy and enthalpy shifts as degrees of freedom are removed from the system. Coarse-grained models at DPD and bead-spring level have both proved successful in visualising the chromonic N and M phases.

This last decade has seen the modelling of liquid crystalline systems mature considerably. The decade ahead promises much in terms of developments in modelling. For liquid crystals, machine learning methods are likely to further improve molecular force fields. Moreover, such methods are likely to lead to a new generation of coarse-grained models. Better connection of models across time and lengths scales may well lead to truly multi-scale modelling approaches for understanding liquid crystal systems.

Author Contributions: Conceptualization, M.R.W.; methodology, M.R.W., G.Y., T.D.P., M.W., S.J.G., J.L. and N.J.B.; software, M.R.W., G.Y., T.D.P. and M.W.; validation, M.R.W., G.Y., T.D.P., M.W., S.J.G., J.L. and N.J.B.; formal analysis, M.R.W., G.Y., T.D.P., M.W., S.J.G., J.L. and N.J.B.; investigation, M.R.W., G.Y., T.D.P., M.W., S.J.G., J.L. and N.J.B.; resources, M.R.W.; data curation, M.R.W., G.Y., T.D.P., M.W., S.J.G., J.L. and N.J.B.; writing—original draft preparation, M.R.W.; writing—review and editing, M.R.W., G.Y., T.D.P., M.W., S.J.G., J.L. and N.J.B.; visualization, M.R.W., G.Y., T.D.P., M.W., S.J.G., J.L. and N.J.B.; supervision, M.R.W. and M.W.; project administration, M.R.W.; funding acquisition, M.R.W. All authors have read and agreed to the published version of the manuscript.

Funding: This research was part-funded by EPSRC in the UK: EPSRC grant EP/R513039/1 funding a studentship for Gary Yu, grants EP/J004413/1 and EP/P007864/1 providing funding for Dr. Martin Walker, and a studentship award (grant EP/M507854/1, award reference 1653213) funding Thomas Potter. Sarah Gray was supported by an EPSRC Case studentship, reference 1482182, with funding from Procter and Gamble, and Jing Li was partially funded by a research grant from Procter and Gamble. Prof. Mark Wilson was part funded by EPSRC by grant EP/V056891/1. This work made use of the facilities of the N8 Centre of Excellence in Computationally Intensive Research (N8 CIR) provided and funded by the N8 research partnership and EPSRC (Grant No. EP/T022167/1). The Centre is coordinated by the Universities of Durham, Manchester, and York.

Data Availability Statement: Not applicable.

Acknowledgments: The authors acknowledge support from the Advanced Research Computing Division at Durham University for the provision of computer time through its Hamilton high performance computer system. The authors are grateful for computer time on the Bede supercomputer at the N8 Centre of Excellence in Computationally Intensive Research.

Conflicts of Interest: The authors declare no conflicts of interest. The funders had no role in the design of the study; in the collection, analyses, or interpretation of data; in the writing of the manuscript; or in the decision to publish the results.

References

1. Güryel, S.; Walker, M.; Geerlings, P.; De Proft, F.; Wilson, M.R. Molecular dynamics simulations of the structure and the morphology of graphene/polymer nanocomposites. *Phys. Chem. Chem. Phys.* **2017**, *19*, 12959–12969. [CrossRef] [PubMed]
2. Tasche, J.; Sabattié, E.F.D.; Thompson, R.L.; Campana, M.; Wilson, M.R. Oligomer/Polymer Blend Phase Diagram and Surface Concentration Profiles for Squalane/Polybutadiene: Experimental Measurements and Predictions from SAFT-γ Mie and Molecular Dynamics Simulations. *Macromolecules* **2020**, *53*, 2299–2309. [CrossRef] [PubMed]
3. Rodgers, T.L.; Townsend, P.D.; Burnell, D.; Jones, M.L.; Richards, S.A.; McLeish, T.C.B.; Pohl, E.; Wilson, M.R.; Cann, M.J. Modulation of Global Low-Frequency Motions Underlies Allosteric Regulation: Demonstration in CRP/FNR Family Transcription Factors. *PLoS Biol.* **2013**, *11*, e1001651. [CrossRef] [PubMed]
4. McLeish, T.C.B.; Rodgers, T.L.; Wilson, M.R. Allostery without conformation change: Modelling protein dynamics at multiple scales. *Phys. Biol.* **2013**, *10*, 056004. [CrossRef]
5. Marrink, S.J.; de Vries, A.H.; Tieleman, D.P. Lipids on the move: Simulations of membrane pores, domains, stalks and curves. *Biochim. Biophys. Acta (BBA)-Biomembr.* **2009**, *1788*, 149–168. [CrossRef]
6. Catte, A.; White, G.F.; Wilson, M.R.; Oganesyan, V.S. Direct Prediction of EPR Spectra from Lipid Bilayers: Understanding Structure and Dynamics in Biological Membranes. *ChemPhysChem* **2018**, *19*, 2183–2193. [CrossRef]
7. Catte, A.; Wilson, M.R.; Walker, M.; Oganesyan, V.S. Antimicrobial action of the cationic peptide, chrysophsin-3: A coarse-grained molecular dynamics study. *Soft Matter* **2018**, *14*, 2796–2807. [CrossRef]
8. Yu, G.; Wilson, M.R. Molecular simulation studies of self-assembly for a chromonic perylene dye: All-atom studies and new approaches to coarse-graining. *J. Mol. Liq.* **2022**, *345*, 118210. [CrossRef]
9. Prasitnok, K.; Wilson, M.R. A coarse-grained model for polyethylene glycol in bulk water and at a water/air interface. *Phys. Chem. Chem. Phys.* **2013**, *15*, 17093–17104. [CrossRef]
10. Anderson, P.M.; Wilson, M.R. Molecular dynamics simulations of amphiphilic graft copolymer molecules at a water/air interface. *J. Chem. Phys.* **2004**, *121*, 8503–8510. [CrossRef]
11. Allen, M.P. Molecular simulation of liquid crystals. *Mol. Phys.* **2019**, *117*, 2391–2417. [CrossRef]
12. Wilson, M.R. Progress in computer simulations of liquid crystals. *Int. Rev. Phys. Chem.* **2005**, *24*, 421–455. [CrossRef]
13. Wilson, M.R. Molecular simulation of liquid crystals: Progress towards a better understanding of bulk structure and the prediction of material properties. *Chem. Soc. Rev.* **2007**, *36*, 1881–1888. [CrossRef] [PubMed]
14. Care, C.M.; Cleaver, D.J. Computer simulation of liquid crystals. *Rep. Prog. Phys.* **2005**, *68*, 2665–2700. [CrossRef]
15. Bremer, M.; Kirsch, P.; Klasen-Memmer, M.; Tarumi, K. The TV in Your Pocket: Development of Liquid-Crystal Materials for the New Millennium. *Angew. Chem. Int. Ed.* **2013**, *52*, 8880–8896. [CrossRef]
16. Cheung, D.L.; Clark, S.J.; Wilson, M.R. Calculation of the rotational viscosity of a nematic liquid crystal. *Chem. Phys. Lett.* **2002**, *356*, 140–146. [CrossRef]
17. Cheung, D.L.; Clark, S.J.; Wilson, M.R. Calculation of flexoelectric coefficients for a nematic liquid crystal by atomistic simulation. *J. Chem. Phys.* **2004**, *121*, 9131–9139. [CrossRef]
18. Berardi, R.; Lintuvuori, J.S.; Wilson, M.R.; Zannoni, C. Phase diagram of the uniaxial and biaxial soft–core Gay–Berne model. *J. Chem. Phys.* **2011**, *135*, 134119. [CrossRef]
19. Allen, M.P.; Warren, M.A.; Wilson, M.R. Molecular-dynamics simulation of the smectic-A* twist grain-boundary phase. *Phys. Rev. E* **1998**, *57*, 5585–5596. [CrossRef]
20. Lyulin, A.; Al-Barwani, M.; Allen, M.; Wilson, M.; Neelov, I.; Allsopp, N. Molecular dynamics simulation of main chain liquid crystalline polymers. *Macromolecules* **1998**, *31*, 4626–4634. [CrossRef]
21. Al Sunaidi, A.; Den Otter, W.K.; Clarke, J.H.R. Liquid-crystalline ordering in rod-coil diblock copolymers studied by mesoscale simulations. *Philos. Trans. R. Soc. Lond. A* **2004**, *362*, 1773–1781. [CrossRef] [PubMed]
22. Stimson, L.M.; Wilson, M.R. Molecular dynamics simulations of side chain liquid crystal polymer molecules in isotropic and liquid-crystalline melts. *J. Chem. Phys.* **2005**, *123*, 034908. [CrossRef] [PubMed]
23. Lintuvuori, J.S.; Wilson, M.R. A coarse-grained simulation study of mesophase formation in a series of rod-coil multiblock copolymers. *Phys. Chem. Chem. Phys.* **2009**, *11*, 2116–2125. [CrossRef] [PubMed]
24. Walton, J.; Mottram, N.; McKay, G. Nematic liquid crystal director structures in rectangular regions. *Phys. Rev. E* **2018**, *97*, 022702. [CrossRef] [PubMed]
25. Al-Barwani, M.S.; Allen, M.P. Isotropic-nematic interface of soft spherocylinders. *Phys. Rev. E* **2000**, *62*, 6706–6710. [CrossRef] [PubMed]
26. Camp, P.J.; Mason, C.P.; Allen, M.P.; Khare, A.A.; Kofke, D.A. The isotropic-nematic phase transition in uniaxial hard ellipsoid fluids: Coexistence data and the approach to the Onsager limit. *J. Chem. Phys.* **1996**, *105*, 2837–2849. [CrossRef]
27. McGrother, S.C.; Williamson, D.C.; Jackson, G. A re-examination of the phase diagram of hard spherocylinders. *J. Chem. Phys.* **1996**, *104*, 6755–6771. [CrossRef]

28. Boyd, N.J.; Wilson, M.R. Optimization of the GAFF force field to describe liquid crystal molecules: The path to a dramatic improvement in transition temperature predictions. *Phys. Chem. Chem. Phys.* **2015**, *17*, 24851–24865. [CrossRef]
29. Wang, J.M.; Wolf, R.M.; Caldwell, J.W.; Kollman, P.A.; Case, D.A. Development and testing of a general amber force field. *J. Comp. Chem.* **2004**, *25*, 1157–1174. [CrossRef]
30. Onsager, L. The effects of shape on the interaction of colloidal particles. *Ann. N. Y. Acad. Sci.* **1949**, *51*, 627–659. [CrossRef]
31. Boyd, N.J.; Wilson, M.R. Validating an optimized GAFF force field for liquid crystals: TNI predictions for bent-core mesogens and the first atomistic predictions of a dark conglomerate phase. *Phys. Chem. Chem. Phys.* **2018**, *20*, 1485–1496. [CrossRef] [PubMed]
32. Cheung, D.L.; Clark, S.J.; Wilson, M.R. Parametrization and validation of a force field for liquid- crystal forming molecules. *Phys. Rev. E* **2002**, *65*, 051709. [CrossRef] [PubMed]
33. Wang, J.; Hou, T. Application of Molecular Dynamics Simulations in Molecular Property Prediction. 1. Density and Heat of Vaporization. *J. Chem. Theory Comp.* **2011**, *7*, 2151–2165. [CrossRef] [PubMed]
34. Caleman, C.; van Maaren, P.J.; Hong, M.; Hub, J.S.; Costa, L.T.; van der Spoel, D. Force Field Benchmark of Organic Liquids: Density, Enthalpy of Vaporization, Heat Capacities, Surface Tension, Isothermal Compressibility, Volumetric Expansion Coefficient, and Dielectric Constant. *J. Chem. Theory Comp.* **2012**, *8*, 61–74. [CrossRef] [PubMed]
35. Qiu, Y.; Smith, D.G.; Boothroyd, S.; Jang, H.; Hahn, D.F.; Wagner, J.; Bannan, C.C.; Gokey, T.; Lim, V.T.; Stern, C.D.; et al. Development and Benchmarking of Open Force Field v1. 0.0—The Parsley Small-Molecule Force Field. *J. Chem. Theory Comput.* **2021**, *17*, 6262–6280. [CrossRef]
36. Poll, K.; Sims, M.T. An insight into de Vries behaviour of smectic liquid crystals from atomistic molecular dynamics simulations. *J. Mater. Chem. C* **2020**, *8*, 13040–13052. [CrossRef]
37. Poll, K.; Sims, M.T. Sub-layer rationale of anomalous layer-shrinkage from atomistic simulations of a fluorinated mesogen. *Mater. Adv.* **2022**, *3*, 1212–1223. [CrossRef]
38. Meyer, R.B. *Les Houches Summer School in Theoretical Physics*; Balian, R.G., Weil, G., Eds.; Gordon and Breach: New York, NY, USA, 1976; pp. 273–373.
39. Dozov, I. On the spontaneous symmetry breaking in the mesophases of achiral banana-shaped molecules. *Europhys. Lett.* **2001**, *56*, 247. [CrossRef]
40. Chen, D.; Porada, J.H.; Hooper, J.B.; Klittnick, A.; Shen, Y.; Tuchband, M.R.; Korblova, E.; Bedrov, D.; Walba, D.M.; Glaser, M.A.; Maclennan, J.E.; Clark, N.A. Chiral heliconical ground state of nanoscale pitch in a nematic liquid crystal of achiral molecular dimers. *Proc. Natl. Acad. Sci. USA* **2013**, *110*, 15931–15936. [CrossRef]
41. Paterson, D.A.; Gao, M.; Kim, Y.K.; Jamali, A.; Finley, K.L.; Robles-Hernández, B.; Diez-Berart, S.; Salud, J.; de la Fuente, M.R.; Timimi, B.A.; et al. Understanding the twist-bend nematic phase: The characterisation of 1-(4-cyanobiphenyl-4′-yloxy)-6-(4-cyanobiphenyl-4′-yl)hexane (CB6OCB) and comparison with CB7CB. *Soft Matter* **2016**, *12*, 6827–6840. [CrossRef]
42. Cestari, M.; Diez-Berart, S.; Dunmur, D.A.; Ferrarini, A.; de La Fuente, M.R.; Jackson, D.J.B.; Lopez, D.O.; Luckhurst, G.R.; Perez-Jubindo, M.A.; Richardson, R.M.; et al. Phase behavior and properties of the liquid-crystal dimer 1″,7″-bis (4-cyanobiphenyl-4′-yl) heptane: A twist-bend nematic liquid crystal. *Phys. Rev. E* **2011**, *84*, 031704. [CrossRef] [PubMed]
43. Borshch, V.; Kim, Y.K.; Xiang, J.; Gao, M.; Jákli, A.; Panov, V.P.; Vij, J.K.; Imrie, C.T.; Tamba, M.G.; Mehl, G.H.; et al. Nematic twist-bend phase with nanoscale modulation of molecular orientation. *Nat. Commun.* **2013**, *4*, 2635. [CrossRef] [PubMed]
44. Yu, G.; Wilson, M.R. All-atom simulations of bent liquid crystal dimers: The twist-bend nematic phase and insights into conformational chirality. *Soft Matter* **2022**, *18*, 3087–3096. [CrossRef] [PubMed]
45. Shadpour, S.; Nemati, A.; Boyd, N.J.; Li, L.; Prévôt, M.E.; Wakerlin, S.L.; Vanegas, J.P.; Salamończyk, M.; Hegmann, E.; Zhu, C.; et al. Heliconical-layered nanocylinders (HLNCs)–hierarchical self-assembly in a unique B4 phase liquid crystal morphology. *Mater. Horizons* **2019**, *6*, 959–968. [CrossRef]
46. Shadpour, S.; Nemati, A.; Salamończyk, M.; Prévôt, M.E.; Liu, J.; Boyd, N.J.; Wilson, M.R.; Zhu, C.; Hegmann, E.; Jákli, A.I.; et al. Missing Link between Helical Nano- and Microfilaments in B4 Phase Bent-Core Liquid Crystals, and Deciphering which Chiral Center Controls the Filament Handedness. *Small* **2020**, *16*, 1905591. [CrossRef]
47. Berardi, R.; Emerson, A.P.J.; Zannoni, C. Monte Carlo investigations of a Gay—Berne liquid crystal. *J. Chem. Soc. Faraday Trans.* **1993**, *89*, 4069–4078. [CrossRef]
48. Zannoni, C. Molecular Design and Computer Simulations of Novel Mesophases. *J. Mater. Chem.* **2001**, *11*, 2637–2646. [CrossRef]
49. Berardi, R.; Muccioli, L.; Orlandi, S.; Ricci, M.; Zannoni, C. Computer simulations of biaxial nematics. *J. Phys.-Condens. Matter* **2008**, *20*, 463101. [CrossRef]
50. Wilson, M.; Allen, M. Computer simulation study of liquid crystal formation in a semi-flexible system of linked hard spheres. *Mol. Phys.* **1993**, *80*, 277–295. [CrossRef]
51. Wilson, M.R. Molecular dynamics simulation of semi-flexible mesogens. *Mol. Phys.* **1994**, *81*, 675–690. [CrossRef]
52. AlSunaidi, A.; den Otter, W.K.; Clarke, J.H.R. Inducement by directional fields of rotational and translational phase ordering in polymer liquid-crystals. *J. Chem. Phys.* **2013**, *138*, 154904. [CrossRef] [PubMed]
53. Skačej, G.; Zannoni, C. Molecular simulations elucidate electric field actuation in swollen liquid crystal elastomers. *Proc. Natl. Acad. Sci. USA* **2012**, *109*, 10193–10198. [CrossRef] [PubMed]
54. Yan, F.; Hixson, C.A.; Earl, D.J. Self-assembled chiral superstructures composed of rigid achiral molecules and molecular scale chiral induction by dopants. *Phys. Rev. Lett.* **2008**, *101*, 157801. [CrossRef] [PubMed]

55. Greco, C.; Ferrarini, A. Entropy-driven chiral order in a system of achiral bent particles. *Phys. Rev. Lett.* **2015**, *115*, 147801. [CrossRef]
56. Chiappini, M.; Drwenski, T.; Van Roij, R.; Dijkstra, M. Biaxial, twist-bend, and splay-bend nematic phases of banana-shaped particles revealed by lifting the "smectic blanket". *Phys. Rev. Lett.* **2019**, *123*, 068001. [CrossRef]
57. Fernández-Rico, C.; Chiappini, M.; Yanagishima, T.; de Sousa, H.; Aarts, D.G.; Dijkstra, M.; Dullens, R.P. Shaping colloidal bananas to reveal biaxial, splay-bend nematic, and smectic phases. *Science* **2020**, *369*, 950–955. [CrossRef]
58. Lintuvuori, J.S.; Wilson, M.R. A new anisotropic soft-core model for the simulation of liquid crystal mesophases. *J. Chem. Phys.* **2008**, *128*, 044906. [CrossRef]
59. Humpert, A.; Allen, M.P. Elastic constants and dynamics in nematic liquid crystals. *Mol. Phys.* **2015**, *113*, 2680–2692. [CrossRef]
60. Allen, M.P.; Warren, M.A.; Wilson, M.R.; Sauron, A.; Smith, W. Molecular dynamics calculation of elastic constants in Gay-Berne nematic liquid crystals. *J. Chem. Phys.* **1996**, *105*, 2850–2858. [CrossRef]
61. Allen, M.P.; Frenkel, D. Calculation of liquid-crystal Frank constants by computer simulation. *Phys. Rev. A* **1988**, *37*, 1813–1816. [CrossRef]
62. Fischermeier, E.; Bartuschat, D.; Preclik, T.; Marechal, M.; Mecke, K. Simulation of a hard-spherocylinder liquid crystal with the pe. *Comput. Phys. Commun.* **2014**, *185*, 3156–3161. [CrossRef]
63. Cleaver, D.; Allen, M. Computer Simulations of the elastic properties of liquid crystals. *Phys. Rev. A* **1991**, *43*, 1918–1931. [CrossRef]
64. Gruhn, T.; Hess, S. Monte Carlo simulation of the director field of a nematic liquid crystal with three elastic coefficients. *Z. Naturf.* **1996**, *51*, 1–9. [CrossRef]
65. Phuong, N.H.; Germano, G.; Schmid, F. Elastic constants from direct correlation functions in nematic liquid crystals: A computer simulation study. *J. Chem. Phys.* **2001**, *115*, 7227–7234. [CrossRef]
66. Sarman, S.; Evans, D.J. Statistical mechanics of viscous flow in nematic fluids. *J. Chem. Phys.* **1993**, *99*, 9021–9036. [CrossRef]
67. Kuwajima, S.; Manabe, A. Computing the rotational viscosity of nematic liquid crystals by an atomistic molecular dynamics simulation. *Chem. Phys. Lett.* **2000**, *332*, 105–109. [CrossRef]
68. Sarman, S. Molecular dynamics of liquid crystals. *Physica A* **1997**, *240*, 160–172. [CrossRef]
69. Cuetos, A.; Ilnytskyi, J.M.; Wilson, M.R. Rotational viscosities of Gay-Berne mesogens. *Mol. Phys.* **2002**, *100*, 3839–3845. [CrossRef]
70. Osipov, M.; Nemtsov, V. On the statistical theory of the flexoelectric effect in liquid crystals. *Sov. Phys. Crystallogr.* **1986**, *31*, 125–130.
71. Ferrarini, A.; Moro, G.; Nordio, P. A shape model for the twisting power of chiral solutes in nematics. *Liq. Cryst.* **1995**, *19*, 397–399. [CrossRef]
72. Feltre, L.; Ferrarini, A.; Pacchiele, F.; Nordio, P. Numerical prediction of twisting power for chiral dopants. *Mol. Cryst. Liq. Cryst. Sci. Technol. Sect. A* **1996**, *290*, 109–118. [CrossRef]
73. di Matteo, A.; Todd, S.M.; Gottarelli, G.; Solladié, G.; Williams, V.E.; Lemieux, R.P.; Ferrarini, A.; Spada, G.P. Correlation between molecular structure and helicity of induced chiral nematics in terms of short-range and electrostatic-induction interactions. The case of chiral biphenyls. *J. Am. Chem. Soc.* **2001**, *123*, 7842–7851. [CrossRef] [PubMed]
74. Ferrarini, A.; Gottarelli, G.; Nordio, P.L.; Spada, G.P. Determination of absolute configuration of helicenes and related biaryls from calculation of helical twisting powers by the surface chirality model. *J. Chem. Soc. Perkin Trans. 2* **1999**, 411–418. [CrossRef]
75. Ferrarini, A.; Nordio, P.; Shibaev, P.; Shibaev, V. Twisting power of bridged binaphthol derivatives: Comparison of theory and experiment. *Liq. Cryst.* **1998**, *24*, 219–227. [CrossRef]
76. Ferrarini, A.; Moro, G.; Nordio, P. Shape model for ordering properties of molecular dopants inducing chiral mesophases. *Mol. Phys.* **1996**, *87*, 485–499. [CrossRef]
77. Ferrarini, A.; Moro, G.; Nordio, P. Simple molecular model for induced cholesteric phases. *Phys. Rev. E* **1996**, *53*, 681–688. [CrossRef]
78. Osipov, M.; Pickup, B.; Dunmur, D. A new twist to molecular chirality: Intrinsic chirality indices. *Mol. Phys.* **1995**, *84*, 1193–1206. [CrossRef]
79. Solymosi, M.; Low, R.J.; Grayson, M.; Neal, M.P. A generalized scaling of a chiral index for molecules. *J. Chem. Phys.* **2002**, *116*, 9875–9881. [CrossRef]
80. Solymosi, M.; Low, R.J.; Grayson, M.; Neal, M.P.; Wilson, M.R.; Earl, D.J. Scaled chiral indices for ferroelectric liquid crystals. *Ferroelectrics* **2002**, *277*, 483–490. [CrossRef]
81. Earl, D.J.; Wilson, M.R. Predictions of molecular chirality and helical twisting powers: A theoretical study. *J. Chem. Phys.* **2003**, *119*, 10280–10288. [CrossRef]
82. Neal, M.P.; Solymosi, M.; Wilson, M.R.; Earl, D.J. Helical twisting power and scaled chiral indices. *J. Chem. Phys.* **2003**, *119*, 3567–3573. [CrossRef]
83. Earl, D.; Osipov, M.; Takezoe, H.; Takanishi, Y.; Wilson, M. Induced and spontaneous deracemization in bent-core liquid crystal phases and in other phases doped with bent-core molecules. *Phys. Rev. E* **2005**, *71*, 021706. [CrossRef] [PubMed]
84. Jo, S.Y.; Kim, B.C.; Jeon, S.W.; Bae, J.H.; Walker, M.; Wilson, M.; Choi, S.W.; Takezoe, H. Enhancement of the helical twisting power with increasing the terminal chain length of nonchiral bent-core molecules doped in a chiral nematic liquid crystal. *RSC Adv.* **2017**, *7*, 1932–1935. [CrossRef]

85. Kim, B.C.; Walker, M.; Jo, S.Y.; Wilson, M.R.; Takezoe, H.; Choi, S.W. Effect of terminal chain length on the helical twisting power in achiral bent-core molecules doped in a cholesteric liquid crystal. *RSC Adv.* **2018**, *8*, 1292–1295. [CrossRef]
86. Lintuvuori, J.S.; Yu, G.; Walker, M.; Wilson, M.R. Emergent chirality in achiral liquid crystals: Insights from molecular simulation models of the behaviour of bent-core mesogens. *Liq. Cryst.* **2018**, *45*, 1996–2009. [CrossRef]
87. Earl, D.J.; Wilson, M.R. Calculations of helical twisting powers from intermolecular torques. *J. Chem. Phys.* **2004**, *120*, 9679–9683. [CrossRef]
88. Germano, G.; Allen, M.P.; Masters, A.J. Simultaneous calculation of the helical pitch and the twist elastic constant in chiral liquid crystals from intermolecular torques. *J. Chem. Phys.* **2002**, *116*, 9422–9430. [CrossRef]
89. Allen, M.P. Calculating the helical twisting power of dopants in a liquid crystal by computer simulation. *Phys. Rev. E* **1993**, *47*, 4611–4614. [CrossRef]
90. Wilson, M.R.; Earl, D.J. Calculating the helical twisting power of chiral dopants. *J. Mater. Chem.* **2001**, *11*, 2672–2677. [CrossRef]
91. Gray, S.J. Dissipative Particle Dynamics Simulations of Surfactant Systems: Phase Diagrams, Phases and Self-Assembly. Ph.D. Thesis, University of Durham, Durham, UK, 2018.
92. Groot, R.D.; Warren, P.B. Dissipative particle dynamics: Bridging the gap between atomistic and mesoscopic simulation. *J. Chem. Phys.* **1997**, *107*, 4423–4435. [CrossRef]
93. Lavagnini, E.; Cook, J.L.; Warren, P.B.; Hunter, C.A. Translation of Chemical Structure into Dissipative Particle Dynamics Parameters for Simulation of Surfactant Self-Assembly. *J. Phys. Chem. B* **2021**, *125*, 3942–3952. [CrossRef] [PubMed]
94. Eslami, H.; Khani, M.; Müller-Plathe, F. Gaussian charge distributions for incorporation of electrostatic interactions in dissipative particle dynamics: Application to self-assembly of surfactants. *J. Chem. Theory Comput.* **2019**, *15*, 4197–4207. [CrossRef] [PubMed]
95. McDonagh, J.L.; Shkurti, A.; Bray, D.J.; Anderson, R.L.; Pyzer-Knapp, E.O. Utilizing machine learning for efficient parameterization of coarse grained molecular force fields. *J. Chem. Inf. Model.* **2019**, *59*, 4278–4288. [CrossRef] [PubMed]
96. Johnston, M.A.; Duff, A.I.; Anderson, R.L.; Swope, W.C. Model for the Simulation of the C_nE_m Nonionic Surfactant Family Derived from Recent Experimental Results. *J. Phys. Chem. B* **2020**, *124*, 9701–9721. [CrossRef] [PubMed]
97. Lydon, J. Chromonic review. *J. Mater. Chem.* **2010**, *20*, 10071–10099. [CrossRef]
98. Bosire, R.; Ndaya, D.; Kasi, R.M. Recent progress in functional materials from lyotropic chromonic liquid crystals. *Polym. Int.* **2021**, *70*, 938–943. [CrossRef]
99. Shiyanovskii, S.V.; Lavrentovich, O.D.; Schneider, T.; Ishikawa, T.; Smalyukh, I.I.; Woolverton, C.J.; Niehaus, G.D.; Doane, K.J. Lyotropic chromonic liquid crystals for biological sensing applications. *Mol. Cryst. Liq. Cryst.* **2005**, *434*, 587–598. [CrossRef]
100. Shiyanovskii, S.V.; Schneider, T.; Smalyukh, I.I.; Ishikawa, T.; Niehaus, G.D.; Doane, K.J.; Woolverton, C.J.; Lavrentovich, O.D. Real-time microbe detection based on director distortions around growing immune complexes in lyotropic chromonic liquid crystals. *Phys. Rev. E* **2005**, *71*, 020702. [CrossRef]
101. Kaznatcheev, K.V.; Dudin, P.; Lavrentovich, O.D.; Hitchcock, A.P. X-ray microscopy study of chromonic liquid crystal dry film texture. *Phys. Rev. E* **2007**, *76*, 61703. [CrossRef]
102. Park, H.S.; Agarwal, A.; Kotov, N.A.; Lavrentovich, O.D. Controllable Side-by-Side and End-to-End Assembly of Au Nanorods by Lyotropic Chromonic Materials. *Langmuir* **2008**, *24*, 13833–13837. [CrossRef]
103. Zhou, S.; Sokolov, A.; Lavrentovich, O.D.; Aranson, I.S. Living liquid crystals. *Proc. Natl. Acad. Sci. USA* **2014**, *111*, 1265–1270. [CrossRef] [PubMed]
104. Walker, M.; Masters, A.J.; Wilson, M.R. Self-assembly and mesophase formation in a non-ionic chromonic liquid crystal system: Insights from dissipative particle dynamics simulations. *Phys. Chem. Chem. Phys.* **2014**, *16*, 23074–23081. [CrossRef] [PubMed]
105. Walker, M.; Wilson, M.R. Formation of complex self-assembled aggregates in non-ionic chromonics: Dimer and trimer columns, layer structures and spontaneous chirality. *Soft Matter* **2016**, *12*, 8588–8594. [CrossRef] [PubMed]
106. Chami, F.; Wilson, M.R. Molecular Order in a Chromonic Liquid Crystal: A Molecular Simulation Study of the Anionic Azo Dye Sunset Yellow. *J. Am. Chem. Soc.* **2010**, *132*, 7794–7802. [CrossRef] [PubMed]
107. Yu, G.; Walker, M.; Wilson, M.R. Atomistic simulation studies of ionic cyanine dyes: Self-assembly and aggregate formation in aqueous solution. *Phys. Chem. Chem. Phys.* **2021**, *23*, 6408–6421. [CrossRef]
108. Thind, R.; Walker, M.; Wilson, M.R. Molecular Simulation Studies of Cyanine-Based Chromonic Mesogens: Spontaneous Symmetry Breaking to Form Chiral Aggregates and the Formation of a Novel Lamellar Structure. *Adv. Theory Simul.* **2018**, *1*, 1800088. [CrossRef]
109. Carbone, P.; Varzaneh, H.A.K.; Chen, X.; Müller-Plathe, F. Transferability of coarse-grained force fields: The polymer case. *J. Chem. Phys.* **2008**, *128*, 064904. [CrossRef]
110. Villa, A.; Peter, C.; van der Vegt, N.F.A. Self-assembling dipeptides: Conformational sampling in solvent-free coarse-grained simulation. *Phys. Chem. Chem. Phys.* **2009**, *11*, 2077–2086. [CrossRef]
111. Villa, A.; van der Vegt, N.F.A.; Peter, C. Self-assembling dipeptides: Including solvent degrees of freedom in a coarse-grained model. *Phys. Chem. Chem. Phys.* **2009**, *11*, 2068–2076. [CrossRef]
112. Li, C.; Shen, J.; Peter, C.; van der Vegt, N.F.A. A Chemically Accurate Implicit-Solvent Coarse-Grained Model for Polystyrenesulfonate Solutions. *Macromolecules* **2012**, *45*, 2551–2561. [CrossRef]
113. Potter, T.D.; Walker, M.; Wilson, M.R. Self-assembly and mesophase formation in a non-ionic chromonic liquid crystal: Insights from bottom-up and top-down coarse-grained simulation models. *Soft Matter* **2020**, *16*, 9488–9498. [CrossRef] [PubMed]

114. Potter, T.D.; Tasche, J.; Barrett, E.L.; Walker, M.; Wilson, M.R. Development of new coarse-grained models for chromonic liquid crystals: Insights from top-down approaches. *Liq. Cryst.* **2017**, *44*, 1979–1989. [CrossRef]
115. Saiz, L.; Klein, M.L. Computer simulation studies of model biological membranes. *Acc. Chem. Res.* **2002**, *35*, 482–489. [CrossRef] [PubMed]
116. Talandashti, R.; Mehrnejad, F.; Rostamipour, K.; Doustdar, F.; Lavasanifar, A. Molecular Insights into Pore Formation Mechanism, Membrane Perturbation, and Water Permeation by the Antimicrobial Peptide Pleurocidin: A Combined All-Atom and Coarse-Grained Molecular Dynamics Simulation Study. *J. Phys. Chem. B* **2021**, *125*, 7163–7176. [CrossRef] [PubMed]
117. Souza, L.M.; Souza, F.R.; Reynaud, F.; Pimentel, A.S. Tuning the hydrophobicity of a coarse grained model of 1, 2-dipalmitoyl-sn-glycero-3-phosphatidylcholine using the experimental octanol-water partition coefficient. *J. Mol. Liq.* **2020**, *319*, 114132. [CrossRef]
118. Bertrand, B.; Garduño-Juárez, R.; Munoz-Garay, C. Estimation of pore dimensions in lipid membranes induced by peptides and other biomolecules: A review. *Biochim. Biophys. Acta Biomembr.* **2021**, *1863*, 183551. [CrossRef]
119. Potter, T.D.; Barrett, E.L.; Miller, M.A. Automated Coarse-Grained Mapping Algorithm for the Martini Force Field and Benchmarks for Membrane–Water Partitioning. *J. Chem. Theory Comput.* **2021**, *17*, 5777–5791. [CrossRef]
120. Potter, T.D.; Tasche, J.; Wilson, M.R. Assessing the transferability of common top-down and bottom-up coarse-grained molecular models for molecular mixtures. *Phys. Chem. Chem. Phys.* **2019**, *21*, 1912–1927. [CrossRef]
121. Izvekov, S.; Voth, G.A. A multiscale coarse-graining method for biomolecular systems. *J. Phys. Chem. B* **2005**, *109*, 2469–2473. [CrossRef]
122. Noid, W.G.; Chu, J.W.; Ayton, G.S.; Krishna, V.; Izvekov, S.; Voth, G.A.; Das, A.; Andersen, H.C. The multiscale coarse-graining method. I. A rigorous bridge between atomistic and coarse-grained models. *J. Chem. Phys.* **2008**, *128*, 244114. [CrossRef]
123. Noid, W.G.; Liu, P.; Wang, Y.; Chu, J.W.; Ayton, G.S.; Izvekov, S.; Andersen, H.C.; Voth, G.A. The multiscale coarse-graining method. II. Numerical implementation for coarse-grained molecular models. *J. Chem. Phys.* **2008**, *128*, 244115. [CrossRef] [PubMed]
124. Reith, D.; Pütz, M.; Müller-Plathe, F. Deriving effective mesoscale potentials from atomistic simulations. *J. Comput. Chem.* **2003**, *24*, 1624–1636. [CrossRef] [PubMed]
125. Lafitte, T.; Apostolakou, A.; Avendaño, C.; Galindo, A.; Adjiman, C.S.; Müller, E.A.; Jackson, G. Accurate statistical associating fluid theory for chain molecules formed from Mie segments. *J. Chem. Phys.* **2013**, *139*, 154504. [CrossRef] [PubMed]
126. Avendaño, C.; Lafitte, T.; Galindo, A.; Adjiman, C.S.; Jackson, G.; Müller, E.A. SAFT-γ Force Field for the Simulation of Molecular Fluids. 1. A Single-Site Coarse Grained Model of Carbon Dioxide. *J. Phys. Chem. B* **2011**, *115*, 11154–11169. [CrossRef] [PubMed]
127. Avendaño, C.; Lafitte, T.; Adjiman, C.S.; Galindo, A.; Müller, E.A.; Jackson, G. SAFT-γ Force Field for the Simulation of Molecular Fluids: 2. Coarse-Grained Models of Greenhouse Gases, Refrigerants, and Long Alkanes. *J. Phys. Chem. B* **2013**, *117*, 2717–2733. [CrossRef]
128. Müller, E.A.; Jackson, G. Force Field Parameters from the SAFT-γ Equation of State for use in Coarse-Grained Molecular Simulations. *Annu. Rev. Chem. Biomol. Eng.* **2014**, *5*, 405–427. [CrossRef]
129. Fayaz-Torshizi, M.; Müller, E.A. Coarse-Grained Molecular Simulation of Polymers Supported by the Use of the SAFT-γ Mie Equation of State. *Macromol. Theory Simul.* **2021**, *31*, 2100031. [CrossRef]
130. Fayaz-Torshizi, M.; Müller, E.A. Coarse-grained molecular dynamics study of the self-assembly of polyphilic bolaamphiphiles using the SAFT-γ Mie force field. *Mol. Syst. Des. Eng.* **2021**, *6*, 594–608. [CrossRef]
131. von Berlepsch, H.; Böttcher, C.; Dähne, L. Structure of J-Aggregates of Pseudoisocyanine Dye in Aqueous Solution. *J. Phys. Chem. B* **2000**, *104*, 8792–8799. [CrossRef]
132. Bricker, W.P.; Banal, J.L.; Stone, M.B.; Bathe, M. Molecular model of J-aggregated pseudoisocyanine fibers. *J. Chem. Phys.* **2018**, *149*, 024905. [CrossRef]
133. Kirstein, S.; Daehne, S. J-aggregates of amphiphilic cyanine dyes: Self-organization of artificial light harvesting complexes. *Int. J. Photoenergy* **2006**, *5*, 020363. [CrossRef]

Article

Defect Dynamics in Anomalous Latching of a Grating Aligned Bistable Nematic Liquid Crystal Device

J. C. Jones [1,*], S. A. Jones [1], Z. R. Gradwell [1], F. A. Fernandez [2] and S. E. Day [2]

1 Soft Matter Physics Group, School of Physics and Astronomy, The University of Leeds, Leeds LS2 9JT, UK
2 Department of Electronic and Electrical Engineering, University College London, London WC1E 6BT, UK
* Correspondence: j.c.jones@leeds.ac.uk

Abstract: Deliberate manipulation of topological defects is of particular interest for liquid crystal applications. For example, surface bistability occurs in the grating aligned Zenithal Bistable Device due to the stabilisation of $\pm\frac{1}{2}$ defects at the points of high surface curvature. Conventional latching between continuous and defect states has previously been simulated satisfactorily using Q-tensor models that include the effect of weak-anchoring and flexoelectricity. However, experimental studies show that some arrangements lead to anomalous latching regimes. The Q-tensor model is used to show that such effects occur when the defects become detached from the surface and have more complex paths in the bulk of the sample.

Keywords: topological defects; nematic liquid crystals; gratings; defect dynamics; bistability; LCD; ZBD

1. Introduction

Topological defects are phenomenologically important for a range of physical sciences, [1] from condensed matter to cosmology [2]. As birefringent soft matter materials, nematic liquid crystals are an ideal medium for the study of topological defects, readily forming them in confined systems at standard temperatures and pressures, and easily visualised using polarised light microscopy. For example, they have been used as a model system for the study of cosmic strings [3]. However, topological defects in nematics are also used deliberately and advantageously in commercially available display devices [4–7]. The Zenithal Bistable Display, or ZBD, uses a sub-micron surface relief grating to support two mutually stable alignment states of the nematic director. In one state, termed the Defect or D state, defects of strength $\pm\frac{1}{2}$ are formed at the points of high surface curvature at the grating peaks and troughs, whereas the other Continuous C state is devoid of such defects [4,5]. Since this invention was made, and the realisation that the self-assembly of colloidal materials can be controlled through the interaction of nematic topological defects [8] in the mid-1990s, the interest in the effect of defects on liquid crystal systems has grown considerably, in nematic colloids [9–11], in display devices [12,13], and for fundamental understanding [14–17]. In this current work, the dynamics of flexo-electrically driven $+\frac{1}{2}$ and $-\frac{1}{2}$ defects moving between C and D states are considered using a Q-tensor simulation and the results are compared to experiment for both conventional behaviour, as reported previously, and anomalous switching of a ZBD device is reported for the first time. Such anomalous switching is avoided in commercial displays through an empirical optimization of the device design. However, the analysis reported here shows the route to controlling the switching behaviour and thereby has the potential to lead to novel device modes.

The application of liquid crystals in display, and other electro-optic and optical modulators, requires uniform alignment of the liquid crystal director in both the quiescent and switched states. Insufficient care in device design leads to unwanted defects between domains with different director profiles, causing scattering and unnecessary variations [18].

For example, a 90° twisted nematic display potentially has topologically distinct director profiles with opposite senses of twist and tilt: either state cannot be transformed to the other without the formation of a defect wall [18,19], as illustrated schematically in Figure 1a. Doping the nematic with a chiral material to give a preferred sense of twist, coupled with surface pretilts that favour that same sense of twist allows one sign of twist and splay/bend to form on removal of the applied field. Without these measures, the contrast ratio of a TN device would be too low and variable for display purposes due to scattering. Even so, unwanted defects can still form at pixel edges [20], a factor that decreases the maximum resolution possible for nematic displays and spatial light modulators.

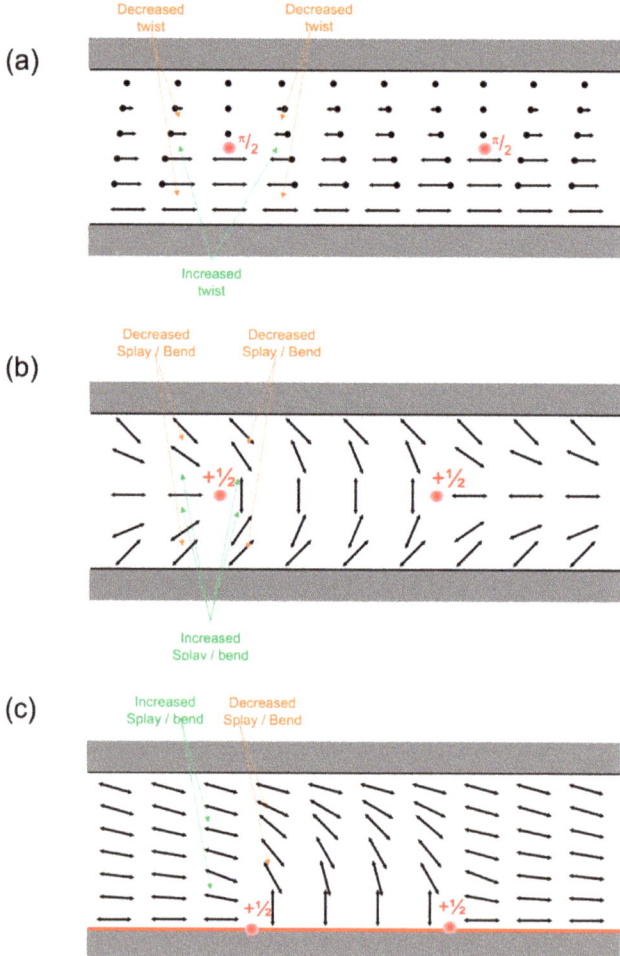

Figure 1. Defect loops of strength $+\frac{1}{2}$ in homotopically inequivalent states with equivalent boundary conditions: (**a**) Left- and right-handed twist; (**b**) Parallel-aligned surfaces with 45° pretilt giving horizontal and vertical states. Shown in (**c**) is a defect loop caused by a surface alignment transition, such as would occur with a zenithally bistable surface, indicated in red. Here, the upper surface has a significant pretilt so that one sign of tilt is maintained throughout the device. Without the pretilt on the upper surface, the central HAN state could also form a homotopically inequivalent state with opposing sign of tilt separated by a $+\frac{1}{2}$ defect loop in the bulk of the cell.

Rather than consider the defects as unwanted artefacts of poor device design, Bigelow and Kashnow [19] considered the deliberate use of energetically equivalent but homotopically distinct twist states to form a bistable device. At that time, bistability showed substantial promise for highly multiplexed displays [21] with more than a score of rows to be addressed because of the inherent non-linearity of the liquid crystal arrangement, and without the use of an additional semiconductor element such as a thin-film-transistor (TFT). In addition to showing the usual transient switching response to an applied electric field, bistable devices can also be latched between at least two optically distinct states. Latching occurs with some appropriate signal that provides sufficient impulse to overcome the activation energy that separates the bistable states. After the impulse, the new state is retained, and the device is said to be latched. Thus, an image is built up row by row; a scanning impulse is applied to the addressed row that is at, or near, the threshold value whilst a data signal of information is synchronously applied on the columns that either increases the impulse (latching) or decrease (non-latching, where the previous state is retained). In the work of [19], surface inhomogeneities were used to pin the $\pi/2$ twist wall between states of opposing twist sense, and the states were selected electrically with fields applied using orthogonal interdigitated electrodes on the opposing surfaces. Although these initial attempts proved impractical, the use of topological defects to separate bistable states with different director profiles was a subject of much interest for several decades: the defect-mediated transition between states of different homotopy created a threshold voltage, and hence promised unlimited addressing without TFT. This drive stimulated attempts to use defects to induce vertical and horizontal bistable states in a nematic liquid crystal [21,22], such as those that form in parallel-aligned high pretilt nematic devices shown in Figure 1b. Such devices were the first to show zenithal bistability, where the term zenithal refers to the out of plane angle that the director makes with respect to the device plane in the two bistable states. They did not receive any commercial development, due to the complexities of construction and latching mechanism. Following the successful commercialisation of low-cost TFT over large areas at high yield, the focus of bistable nematic display research moved to ultra-low power devices with image storage [21].

ZBD is a commercially successful bistable nematic display invented in the UK [4] and used in electronic shelf edge labels for the retail sector [6,7]. The technology has recently been acquired by New Vision Displays Ltd. [23], and it continues to operate from its Malvern laboratories in the UK. Unlike the earlier zenithal bistable devices that relied on containment from the two opposing surfaces, the ZBD uses a single zenithally bistable surface, where the alignment surface itself has two different but stable pretilt states. Figure 1c shows an example arrangement where a bistable surface that is either planar or homeotropic creates surface defects at the boundary between the states. In ZBD, the zenithally bistable surface is provided by a (sub)optical homeotropic grating that can stabilise $-\frac{1}{2}$ strength defects at the grating ridges and $+\frac{1}{2}$ defects at the grating troughs. Arranging the grating pitch P to be approximately equal to its amplitude a ($a/P \approx 1.0 \pm 0.3$) leads to both a defect-free (continuous, C) state and a defect (or D) state being stable and separated by an activation energy typically of the order 10^{-5} J/m^2. An example of the grating used in a commercial device [23] is shown in Figure 2, wherein $a/P \approx 1.3$.

Close to a homeotropic grating surface with sufficiently strong anchoring, an elastic distortion of the director field is induced if that field is to be continuous. In this continuous C state, the distortion dies down after a few hundred nanometres from the grating peaks, and the director is uniformly normal to the plane of the grating. Alternatively, the elastic distortion can be reduced by inducing the half-strength defects stabilised by the surface curvature, leading to a uniform and low tilt at a small distance from the grating plane. Examples of the C and D states are shown for a typical grating in Figure 2a. This grating surface may be placed opposite either a homeotropic or planar homogeneous surface aligned perpendicular to the grating vector to give vertical-aligned to hybrid aligned nematic (VAN/HAN) or hybrid aligned to 90° twisted nematic (HAN/TN) display modes, respectively [24], or opposing a second zenithally bistable grating surface to give VAN/HAN/TN [25].

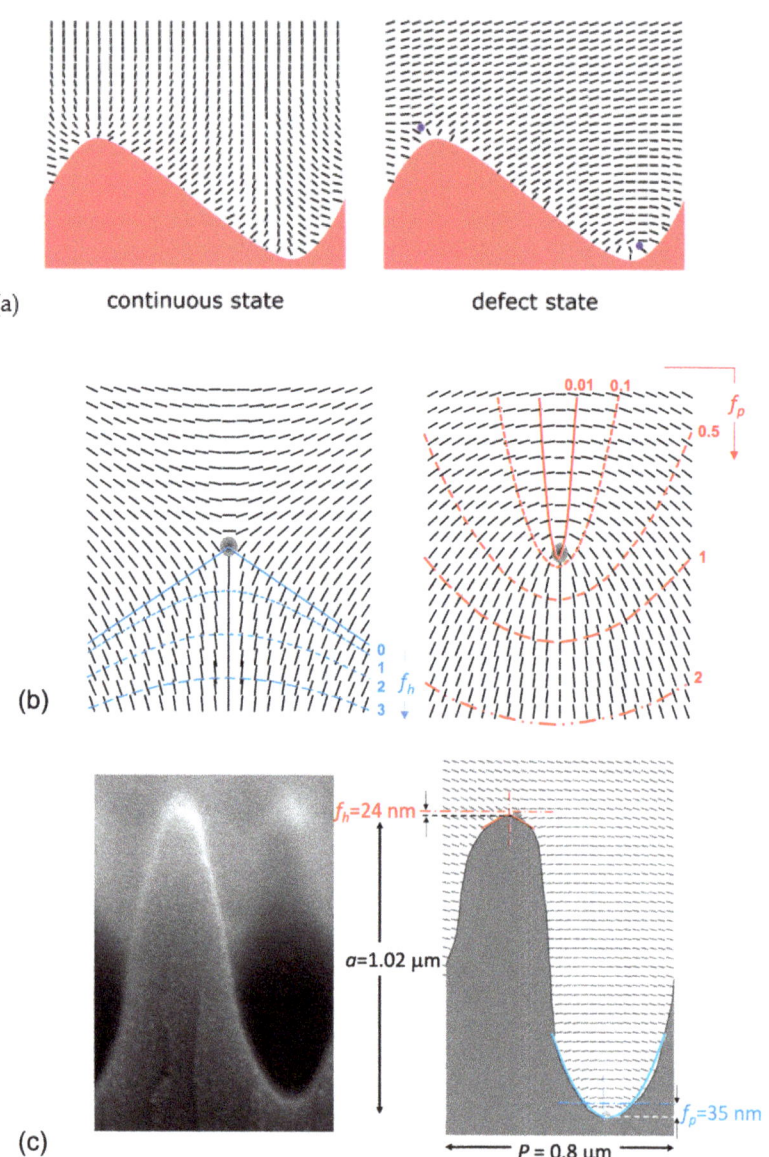

Figure 2. Schematic view of nematic director alignments around a homeotropic grating. (**a**) The continuous and defect state. Two defects are indicated as dots in the latter image: $+\frac{1}{2}$ defect at the base of the grating and $=\frac{1}{2}$ at the peak. Line segments indicate average director alignment at specific points. (**b**) Fits to the threefold hyperbolic and one-fold parabolic functions for the director normal surrounding the $-\frac{1}{2}$ and $+\frac{1}{2}$ defects for an elastically isotropic nematic. (**c**) SEM and Q-tensor model from [26], illustrating the positive hyperbolic and negative parabolic surface curvatures at the top and bottom of the grating grooves, respectively.

2. Surface Stabilised Topological Defects

The director profiles surrounding half-strength defects are illustrated in Figure 2a, where an elastically isotropic nematic ($K_{11} = K_{33}$) confined to the plane of the paper is assumed. In actuality, the defects are not points but lines that run perpendicular to the plane shown. The core is approximately a 50 nm diameter cylinder with uniaxial symmetry for the local director around the $-\frac{1}{2}$ defect and with biaxial symmetry around the $+\frac{1}{2}$ defect. These defects form loops that terminate at the boundary between D and C regions. Away from the core, the **n**-director of an elastically isotropic nematic is a simple parabola for the $+\frac{1}{2}$ defect and hyperbolic with three-fold symmetry for the $-\frac{1}{2}$ defect. For an infinitely bound homeotropic surface with negative parabolic local surface curvature, the $+\frac{1}{2}$ defect will be at centered at the focal point f_p, as illustrated in Figure 2b and given by:

$$y = \frac{1}{f_p}\left(\frac{x}{2}\right)^2 - f_p \tag{1}$$

where x and y are the orthogonal spatial coordinates with y parallel to the defect axis. Similarly, the three-fold hyperbolic $-\frac{1}{2}$ defect has a focus at f_h for a positive hyperbolic surface with locus

$$y = \frac{1}{2\sqrt{3}}\sqrt{\left(3f_h^2 + 4x^2\right)}, \tag{2}$$

and the defect is centered at f_h. Thus, for infinitely bound homeotropic surfaces, the topological defects sit near to the surface at a distance that is related to the degree of surface curvature. Figure 2c shows a typical grating that can give zenithal bistability, together with the Q-tensor modelling from reference [26]. If the grating were to have infinite anchoring, the $+\frac{1}{2}$ and $-\frac{1}{2}$ defects would sit at 35 nm and 24 nm, respectively. In the model, zenithal anchoring of $W_\theta \approx 2 \times 10^{-4}$ Jm^{-2}, which is realistic of actual devices [27], and the defects then sit at the grating surface (as indicated by the grey spots in Figure 2c. The relationship between surface curvature and anchoring energy plays a critical role in these devices, and as shown in the current work, can lead to unusual switching and latching mechanisms.

Conventional voltage-dependent switching of the dielectrically anisotropic nematic will occur. For example, for a positive $\Delta\varepsilon$ material, the bulk director will increase its tilt angle to lie (anti) parallel to the applied field. Although this can lead to unwanted latching to the continuous C state, it will only act in one direction and cannot be used to give discrimination between the high tilt C and low tilt D states. Instead, polar switching from the flexoelectric polarisation is utilised to latch between these bistable states, where the term latching is used to denote a switching impulse that is sufficiently high to overcome some threshold value and the new state be retained after removal of the impulse [28]. In the ZBD, the elastic distortion is most strongly localised close to the topological defects, as illustrated in Figure 3a [29]. The sum of the flexoelectric components is small, being typically ($e_{11} + e_{33}$) = 0.035 nC/m for the conventional twisted nematic mixture E7 [30], and 0.2 nC/m for novel bent dimers specifically designed for flexoelectric applications [31]. For comparison, a typical ferroelectric liquid crystal would have a spontaneous polarisation of 100 nC/cm^2. However, the elastic distortions in both the C and D states are concentrated to within $\approx$100 nm of the grating surface, from which we can estimate that the local polarisation is around 35–200 nC/cm^2 and the latching electro-optics of the ZBD is similar to that of an FLC (albeit with the slower optical response time associated with the much larger cell gaps used). This polarity may lead to forces and torques on the defects with an applied electric field. In the bulk liquid crystal, a $-\frac{1}{2}$ defect has three-fold symmetry and only a rotational electric torque, whereas the unipolar $+\frac{1}{2}$ defect has directionality and also undergoes a translational electric force [32]. Thus, in the bulk of a nematic sample, electric field-induced movement of the defects is largely due to the $+\frac{1}{2}$ defects. However, these symmetries are broken when the defects are in proximity to a surface, as shown in Figure 3b. Here, both defects have a net translational force that tends to move the defects along the surfaces to the points of maximum curvature, and then into the bulk of the device in the direction roughly parallel to the surface normal.

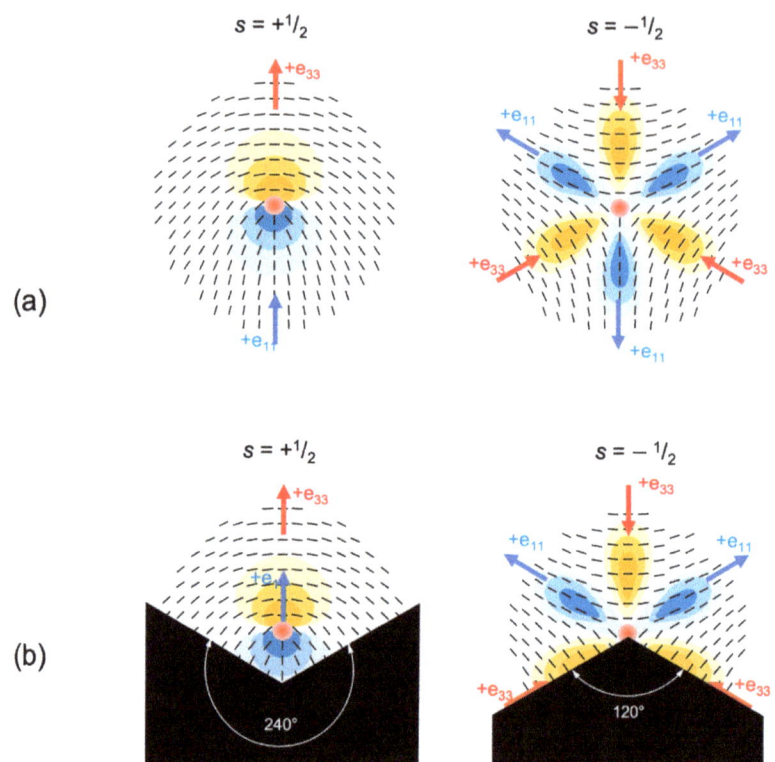

Figure 3. Schematic representations of the flexoelectric polarity associated with $+\frac{1}{2}$ (LHS) and $-\frac{1}{2}$ (RHS) defects (adapted from [17]). In the bulk (**a**), the $+\frac{1}{2}$ defect has polar symmetry and can be moved with an applied electric field, but the $-\frac{1}{2}$ defect has tripolar symmetry and will not translate under the influence of an applied electric field. When stabilised by surface curvature (**b**), the defects both can be propelled either parallel or antiparallel to the surface normal by the applied field. Bend distortions characterised by e_{33} are shown in orange, whereas the splay distortions and e_{11} shown in blue.

3. Conventional and Anomalous Latching between Bistable States

In practice, selective discrimination of either of the states is done using a bipolar electric pulse, wherein the polarity of the trailing pulse determines the final state. This helps ensure that there is no DC offset that may damage the liquid crystal compounds or lead to unwanted latching from the data applied to adjacent rows in a passive matrix addressing scheme [28]. Moreover, the leading pulse of the opposite polarity to that of the latched state will elastically stress the director closer to the grating surface due to the switching effect of $\varepsilon_0 \Delta \varepsilon E^2$. This in turn increases the flexoelectric polarisation and reduces the electrical impulse needed for latching [28,33]. When viewed on a microscopic level, the latching process occurs between an onset impulse (τV, where minimal domains of the new state are retained) to the fully latched state. The partial latching region is typically 1–2 V between the initial and full latch; this dictates the minimum data voltage required in the addressing schemes to give complete discrimination [28].

Latching is a complex, three-dimensional process. The C-D transition involves nucleation of a defect pair at a discontinuity on the grating surface, usually deliberately defined using "slips" in the grating structure [6,7]. The defects form a loop that initially moves vertically across the grating surface to the apices of the structure, and then spreads in the direction of the grating peaks and troughs until joining with other loops on the same trajectory or another discontinuity. The full dynamics of this transition have not been simulated. Instead, literature studies [26,34–37] use a two-dimensional Q-tensor model of the plane containing the normals to the device surfaces and the surface relief grating. Of these, reference [26] is the only one that compares the predictions with experimental results. However, that work also shows that a simple, one-dimensional (1D) model proposed by Davidson and Mottram [38] describes the behaviour adequately. That simple model uses a digital anchoring condition at the lower surface of Figure 1c, wherein application of a torque sufficient to break the surface anchoring energy W_θ causes latching. The threshold pulse duration τ and voltage V are related though [26]:

$$\tau = \frac{\gamma_1 l_s d'}{(e_{11}+e_{33})(|V-V_{th}|)} = \frac{A}{|V-V_{th}|}, \qquad (3)$$

where γ_1 is the rotational viscosity of the liquid crystal, l_s is the liquid slip length at the bistable surface, d' is the cell gap adjusted to include the dielectric contribution of the grating, $(e_{11}+e_{33})$ is the flexoelectric sum, V is the applied voltage, and V_{th} is the threshold voltage:

$$|V_{th}| = \frac{2W_\theta d'}{(e_{11}+e_{33})+\sqrt{\epsilon_0 \Delta\epsilon K_{33}}}, \qquad (4)$$

ϵ_0 is the vacuum permittivity, $\Delta\epsilon$ is the dielectric anisotropy of the liquid crystal, and K_{33} is the bend elastic constant. Visualisation of fits is improved by rearranging Equation (3) to the linear form:

$$|V| = \frac{A}{\tau} + |V_{th}|. \qquad (5)$$

Figure 4 shows typical fits for the experimental latching thresholds for a ZBD test device to Equations (3)–(5) where the fitting parameters are given in Table 1. All results in this work were taken using devices constructed by embossing an ITO-coated glass plate with a homeotropic photopolymer using proprietary grating film provided by Displaydata Ltd. The surface anchoring energy W_θ was controlled through mixing of weak and strong photopolymers, also provided by Displaydata Ltd., as described in [27]. The grating is spaced from a second ITO-coated glass plate, onto which the commercially available homeotropic polyimide SE1211 had been deposited, using 7 µm plastic spacer beads. This arrangement gives the VAN director profile for the C state and the HAN profile for the D state; when oriented with the grating at 45° to crossed polarisers, these states are dark and transmissive, respectively. Electrical waveforms were provided to the electrodes using an arbitrary waveform generator (WFG500 from FLC Electronics). Other than at short pulse widths, where unaccounted electrode losses begin to play a role, the simple $(V - V_{th})^{-1}$ behaviour of the latching pulse width is evident in Figure 4. It should be noted that V_{th} is similar for both the C-D and D-C transitions, as expected. However, the large difference in A between the two transitions suggests different routes are taken by the defects in either case.

Table 1. Fitting parameters for Figure 4.

Transition		$10^3 \times$ A (Vs)	V_{th} (V)
DC	Onset	2.8	−4.3
	Full	4.7	−4.2
CD	Onset	0.89	+4.5
	Full	0.88	+4.9

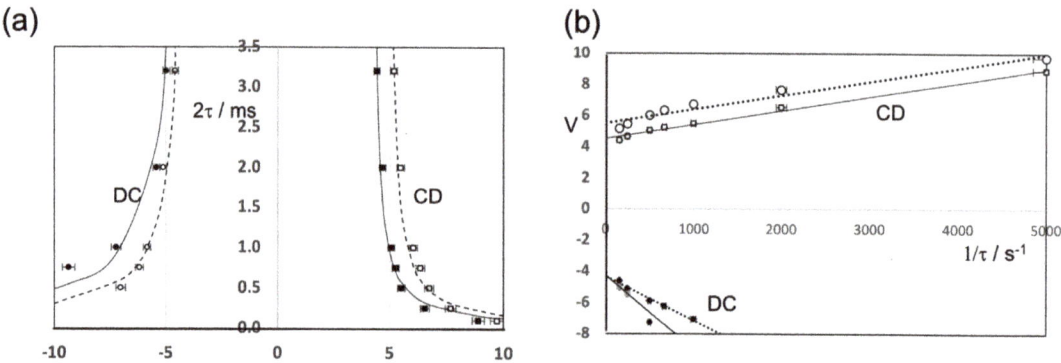

Figure 4. Experimental latching curves for the high flexoelectric liquid crystal nematic mixture MLC6204-000 in a 5.0 μm ZBD cell at 30 °C showing (**a**) applied pulse width (2τ) versus the amplitude of pulse required to latch and (**b**) the $(2t)^{-1}$ V plot for the same data. Both the onset of latch and the total latch are indicated, by empty for mainly D and filled markers for mainly C, respectively. Positive pulses indicate the latch from continuous to defect state and negative pulses indicate the latch from defect to continuous. Error bars in voltage are taken as ±3% and the temporal errors are not visible at this scale. The best fits to Equation (5) are shown as continuous lines, with the fitted parameters listed in Table 1.

Various grating designs have been studied using this model, and the results were successfully compared with the experimental findings [39,40]. However, in studying different liquid crystal materials and surface anchoring energies as part of this study, it was found that the form of the latching deviates occasionally from this expected behaviour. For example, Figure 5b is the C to D latch curve for single compound pentyl-cyanobiphenyl (5CB) at 30 °C in a 7 μm cell constructed with a high anchoring strength photopolymer grating. Instead of diverging at a threshold voltage V_{th}, latching at the longer times requires a higher voltage, not lower. This leads to degenerate behaviour at a particular voltage, wherein pulses of three different durations all lead to full latching.

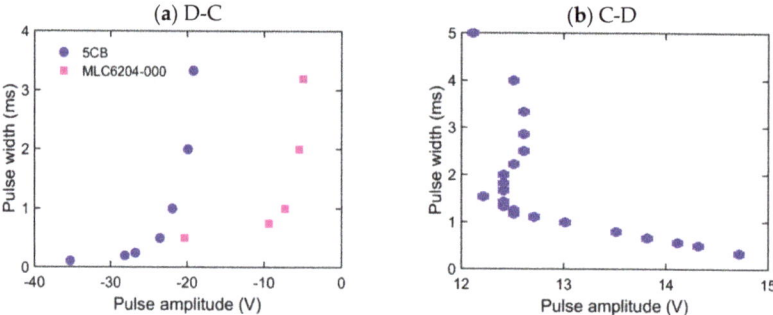

Figure 5. Experimental τV of the (**a**) defect to continuous state (D-C) latching and (**b**) continuous to defect state (C-D) latching for the compound 5CB in a 7 μm ZBD cell (HAN mode) at 30 °C. Strong anchoring of $2.4 \pm 0.5 \times 10^{-4}$ N/m was obtained using the method of reference [27]. Markers indicate the threshold for a full latching. Error bars indicate the uncertainty in the pulse amplitude for latch. Also shown in magenta are the results for D-C latching for MLC6204-000 for comparison.

Understanding the dynamics that leads to this behaviour required 2D finite element modelling, which is reviewed in the following sections.

4. Q-Tensor Modelling

A 2D finite element discretisation of the Landau-de Gennes free energy is performed to examine the dynamics of the device. A finite element solution to the Qian and Sheng formalism [41] was found using the program *Q-LCSolver*, as described in [35,36]. For a complete treatment of the software, the interested reader is referred to [37], but a simple review of the approach is included below. Although the program solved in 3D, the third (y-) dimension was disregarded for most of the simulations, to save computational time and a 2D model used. Time saving measures were particularly important to find the latching thresholds and defect dynamics both above and below the threshold. Sub-microsecond time steps were used to allow for the defect dynamics, but applied over sufficient time following the applied fields to allow the bulk director reorientation to complete (usually tens of milliseconds). Moreover, many simulations of both time and voltage are done for each point plotted to allow the onset and completion of latching to be calculated. It should be noted that the model has previously been used to investigate the behaviour of defect loops [42], and in particular the path of the defects at phase shifts deliberately added to the grating structure in commercial devices [6]. For the qualitative description of the current work, the 2D model was found to be sufficient.

The program uses the Q-tensor representation for liquid crystals, which accounts for the head–tail symmetry of the nematic director and changes to the order parameter, such as those that occur close to defects and surfaces. In the uniaxial case, this is written:

$$Q_{ij} = \tfrac{S}{2}(3n_i n_j - \delta_{ij}), \tag{6}$$

where i and j are integers from 1 to 3, S is the scalar order parameter, n is the director, and δ_{ij} is the Kronecker delta. The total free energy F_{tot} of the system in a device of volume Ω with surfaces Γ is taken as:

$$F_{tot} = \int_\Omega (f_b + f_d - f_E) d\Omega + \int_\Gamma f_s d\Gamma. \tag{7}$$

This equation includes the thermotropic bulk free energy density f_b, the elastic distortion energy density f_d, the electric energy density f_E, and the surface elastic energy contribution f_s. The thermotropic energy density is written as:

$$f_b = \tfrac{1}{2} a\, Tr(Q^2) + \tfrac{1}{3} b\, Tr(Q^3) + \tfrac{1}{4} c\, Tr(Q^4), \tag{8}$$

and the Landau de Gennes critical coefficients a, b, and c were taken from [26]. Elastic distortion energy density takes the form:

$$f_d = \tfrac{1}{2} L_1 Q_{ij,k} Q_{ij,k} + \tfrac{1}{2} L_2 Q_{ij,j} Q_{ik,k}, \tag{9}$$

where L_1 and L_2 are constants related to the Frank elastic constants K_{11}, K_{22}, and K_{33} and the splay-bend and chiral terms have been omitted.

The electrical free energy density includes the flexoelectric term:

$$f_E = \tfrac{\epsilon_0}{2}(E \cdot \varepsilon_{ij} \cdot E) + P \cdot E, \tag{10}$$

Here, the flexoelectric polarisation vector P can be written in terms of the director as:

$$P = e_{11}(n\nabla . n) - e_{33}((\nabla \times n) \times n), \tag{11}$$

with e_{11} and e_{33} representing the flexoelectric coefficients for splay and bend deformations of the director, respectively. In the limit of constant uniaxial order parameter, the polarisation can be written in terms of the Q tensor [43]:

$$P_i = \tfrac{2}{9S_0}(e_{11} + 2e_{33}) Q_{ij,j} + \tfrac{4}{9S_0^2}(e_{11} - e_{33}) Q_{ij} Q_{jk,k} \tag{12}$$

The relative dielectric tensor ε_{ij} in Equation (11) is:

$$\varepsilon_{ij} = \varepsilon_\perp \delta_{ij} + \Delta\varepsilon \left(\frac{2}{3S_0} Q_{ij} + \frac{1}{3}\delta_{ij}\right) \tag{13}$$

where $\Delta\varepsilon$ is the dielectric anisotropy of the liquid crystal material and S_0 is the equilibrium scalar order parameter (approximately $S_0 \approx 0.6$ in this instance). The electric field was found from the potential ϕ, which satisfies the Poisson equation:

$$\epsilon_0 \nabla \cdot (\varepsilon_{ij} \cdot \nabla \phi) = -\rho \tag{14}$$

where the charge density ρ is considered solely due to the flexoelectric polarisation, $\nabla \cdot \mathbf{P}$ from Equation (11). For the surface aspect of the free energy, the model includes the possibility of weak anchoring at the cell walls. In this case, the surface elastic energy contribution takes the form:

$$f_S(Q_{ij}) = A_S Tr(\mathbf{Q}^2) + W_\theta(\hat{\xi}_1 \cdot \mathbf{Q} \cdot \hat{\xi}_1) + W_\varphi(\hat{\xi}_2 \cdot \mathbf{Q} \cdot \hat{\xi}_2) \tag{15}$$

where $\hat{\xi}_1$ and $\hat{\xi}_2$ are orthogonal unit vectors perpendicular to the alignment direction. Anchoring is then controlled through three parameters: A_S and the anchoring strengths W_θ and W_φ, which are associated with the surface tilt and twist deformations.

The dynamic behaviour of $\mathbf{Q}$ is determined by the frictional losses in a Raleigh dissipation equation:

$$\delta \dot{F}_{tot} + \delta D = 0 \tag{16}$$

following the approach of [44], where δ represents infinitesimal changes in the power $\dot{F}_{tot}$ and the frictional losses D. Equation (16) represents a conservation condition where changes in total energy are balanced by the dissipation effects. The dissipation function D is a function of $\mathbf{Q}$, $\dot{\mathbf{Q}}$, and the flow velocity of the liquid crystal and represents the dissipation effects due to rotational and translational flow [37]. In the current work, the effects of flow were not included and only the rotational viscosity γ_1 used.

The modelling volume is that of a single pitch of the grating surface at the base of a 5 µm device, Figure 6. It is assumed to be infinitely repeating through periodic boundaries in x and y where a 1 µm depth is modelled in the y direction. Both the top surface (in z) and the grating surface are given homeotropic boundary conditions, thereby matching the configuration of the experimental cells. The top surface anchoring is assumed infinite, whereas the grating surface anchoring was varied between weak 5.0×10^{-5} J/m^2 and strong 2.8×10^{-4} J/m^2, as determined in [27]. Note, the term strong is often used to mean infinite anchoring in the prior literature, whereas here it denotes a relative magnitude. Unlike [26], which used the spline of the grating surface from an experimental SEM of the commercial device shown in Figure 2c, a simple but representative blazed sinusoidal surface was used with an 800 nm pitch and 1.0 µm amplitude. The blaze was high enough to induce pretilts of 25° and 89.5° in the D and C states, respectively. These were sufficient to prevent any degeneracy induced by the fixed top surface. To save on computational time, an adaptive mesh was used, wherein the initial density of mesh points was high close to the grating surface. Results of the simulations were analysed in two dimensions using slices of the x-z plane. The simulations were set to output data on every tenth iteration, giving a clear view of the defect behaviour while reducing data handling. Results were evaluated in the data analysis and visualisation program *ParaView*. The surface slip values were estimated by a linear extrapolation of the near surface director over several frames using *ImageJ*, with an error calculated according to the variation found at different parts of the surface.

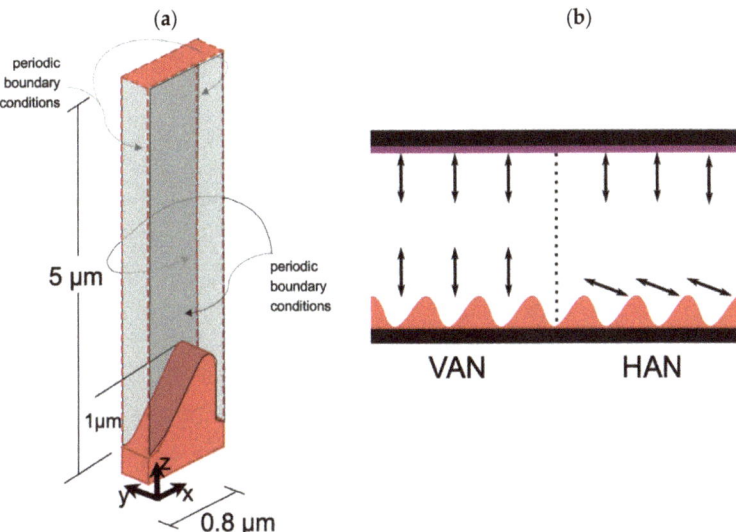

Figure 6. The simulation geometry. (**a**) Schematic of the modelling volume of a 5 µm ZBD device. Boundaries in x and y were periodic, giving an infinite grating surface. In z, both the top cell boundary and the grating surface were given homeotropic anchoring conditions. (**b**) The bulk geometry of the device with vertically aligned nematic (VAN) for the C state and hybrid aligned nematic (HAN) for the D state.

Liquid crystal parameters were chosen to coincide with that of previous modelling [26] and measurements [39] and are listed in Table 2. Although the parameters have not all been determined for the commercial nematic mixture MLC6204-000, but rather a bespoke formulation made to give low voltage and fast flexoelectric latching (mixture B from [45]), they are expected to be typical of the highly positive $\Delta \varepsilon$ nematics used in commercial devices. This enables a qualitative study at least of the latching behaviour and the corresponding defect dynamics.

Table 2. Liquid crystal parameters used in the simulations [26].

		Viscoelastic Coefficients			
K_{11}	12.5	pN	K_{33}	17.9	pN
γ_1	0.155	kg m^{-1} s^{-1}			
		Electrical Coefficients			
$\varepsilon_\parallel$	62.5		e_{11}	69	pC m^{-1}
$\varepsilon_\perp$	23.5		e_{33}	45	pC m^{-1}
$\Delta \varepsilon$	+39				
		Landau- de Gennes Critical Coefficients			
a	65,000	J m^{-3} K^{-1}			
b	530,000	J m^{-3}			
c	980,000	J m^{-3}			

For each anchoring energy, the two equilibrium profiles were first found for the continuous and defect states. For the C state, the director was initially set to vertical throughout the bulk of the cell and allowed to reorient over 5 ms. An analogous process was used for the D state, allocating a horizontal planar bulk orientation, and allowing 50 ms for the reorientation. This required a longer time due to the hybrid alignment of the D state and the elastic distortion throughout the bulk of the device. The resulting director profiles for a given anchoring condition were used as the starting conditions for the subsequent runs. For example, Figure 7 shows the resultant quiescent director profiles for the $W_\theta = 2.8 \times 10^{-4}$ J/m^2.

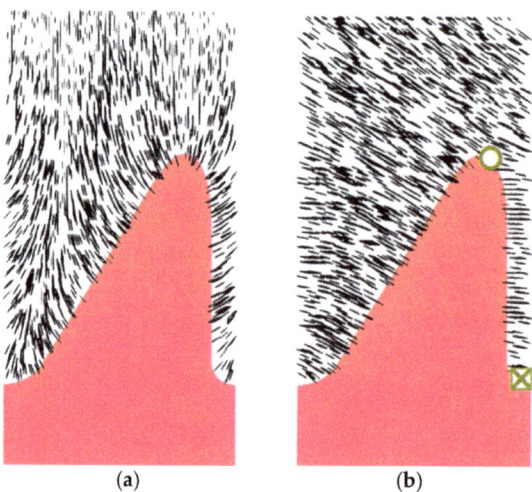

Figure 7. Modelled director alignments near the grating surface for continuous (**a**) and defect (**b**) states at equilibrium. The grating profile was added in the graphics software Inkscape to aid visibility. The director orientations are produced by the modelling software. Indicated on the defect state are the $+\frac{1}{2}$ defect (circle) and the $-\frac{1}{2}$ defect (cross).

Electrically induced latching between the states was studied for bipolar pulses of the form shown in Figure 8. The polarity of the trailing pulse determines the selected state, with the trailing pulse being positive on the grating surface tending to give the D state. For instance, to latch from continuous to defect state, first a negative pulse is applied, followed by a positive pulse of equal magnitude, followed by a period of between 10 ms and 20 ms with no field applied, to allow the bulk director profile to relax to its final state. For each voltage and time, the final state is determined, and the process repeated for a new impulse τV until the change of state is found. Latching may be either successful or unsuccessful, depending on the impulse of the trailing pulse. For a given time, the threshold pulse height is then recorded as the midpoint between the highest-voltage unsuccessful latch and the lowest-voltage successful latch. The error bars are then set as the width between those two values.

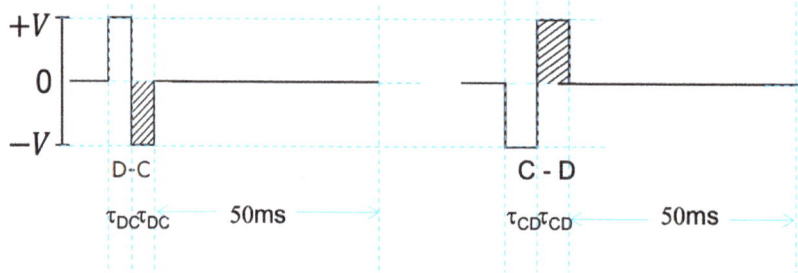

Figure 8. Schematic of the waveforms applied to latch between states, from D to C and from C to D. Shading indicates the latching pulse.

5. Simulations of Anomalous Bistable Latching

Latching from C-D and D-C was investigated for anchoring energies ranging from 1×10^{-5} N/m $\leq W_\theta \leq 2.8 \times 10^{-4}$ N/m. For anchoring at and below 5×10^{-5} N/m, the continuous state did not stabilise, and the device was monostable. The defects were submerged into the surface and the director profile was uniformly at the 25° pretilt defined by the grating blaze. Experimental observations agree that the bistable states are lost for weak anchoring of this magnitude. However, unlike the 2D simulation, the director orients into the xy plane to have a higher component parallel to the grooves. That is, when the defects are significantly below the surface, the director tends towards the continuous state for a planar-aligned grating. However, it is instructive to note that the failure to give the D state in the 2D simulation occurs at a similar anchoring strength to that found experimentally, which adds some confidence to the validity of the model and (static) parameters used.

The τV characteristics were modelled for a d = 5 μm ZBD using the liquid crystal properties of Table 2, for weak ($W_\theta = 1.0 \times 10^{-4}$ N/m) and strong ($W_\theta = 2.8 \times 10^{-4}$ N/m) anchoring conditions. The results are summarised in Table 3, and the full characteristics for the strong anchoring case shown in Figure 9. Table 3 includes the fits of the simulated data to Equation (5), using the surface slips found in the simulations. For the weak anchoring, the reciprocal behaviour was followed for both D-C and C-D transitions across the whole range, but the quality of fit to the form of Equation (5) was poor. For example, the threshold voltages from the fits were consistently higher than that indicated by the low voltage latching. Moreover, inputting the liquid crystal parameters used for the simulations into Equations (3) and (4) also gave substantial differences. The simulated behaviour of the CD latching characteristic for the strong anchoring deviated from the reciprocal form of Equation (5) quite dramatically, as evident in Figure 9b. Here, the reciprocal fit was satisfactory for short pulse widths and high voltages, but at lower voltages, the threshold swapped to higher voltages than expected. This type of behaviour is reminiscent of the anomalous latching seen in some experimental results, such as those shown in Figure 5. Hence, we investigate the latching mechanisms in more detail, in particular the defects trajectories during the switching process at five points shown in Figure 9: two of conventional latching behaviour (1 and 2) and three in the region of anomalous C-D latching (3, 4, and 5).

Table 3. Summary of the fits of the 2D Q-tensor simulation to Equation (5), compared with the anticipated values for weak and strong anchoring conditions.

W_θ (10^4 N/s)		Results from 2D Q-Tensor Simulation				Predicted Values from 1D Model (Equations (3) and (4))	
		Surface Slip (nm)	A (10^3 Nms/C)	V_{th} (V)	Correlation R^2	A (10^3 Nms/C)	$\|V_{th}\|$ (V)
1.0	DC	178 ± 10	0.8	−1.3	0.94	1.19 ± 0.07	5.1
	CD		1.6	+6.5	0.97		
2.8	DC	56 ± 15	1.1	−3.7	0.96	0.38 ± 0.1	14.5
	CD		0.7	+11.0	0.99		

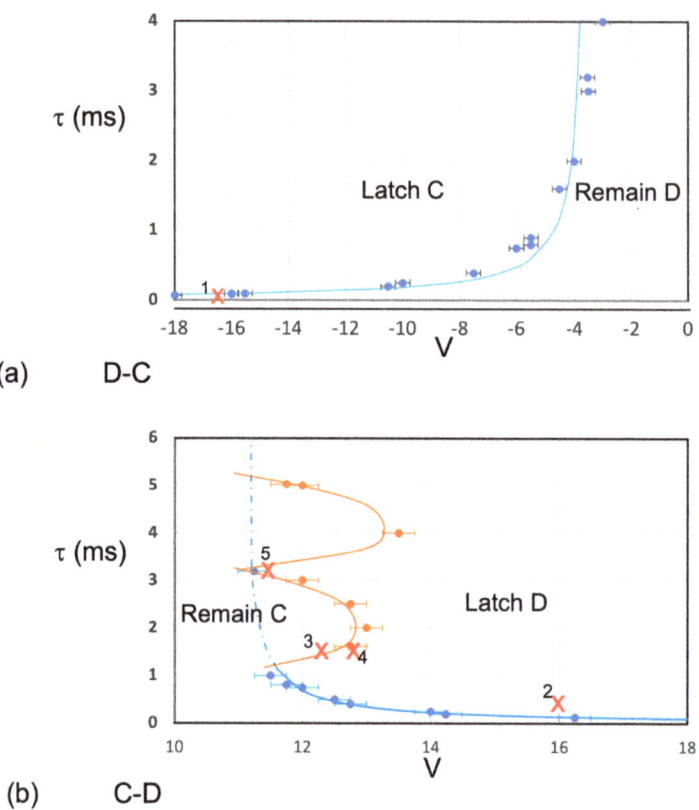

Figure 9. Modelling τV data for a 5 μm cell with strong anchoring, $W_\theta = 2.8 \times 10^{-4}$ N/m for (**a**) the D-C transition and (**b**) the C-D transition. The error bars indicate the difference from the lowest magnitude voltage tested which latched and the highest magnitude voltage that did not latch. Each data point represents halfway between these values calculated. The blue lines represent the best fits to Equation (5) with the parameters listed in Table 3. The fit for the C-D latching threshold includes the point at (11.2 V, 3.2 ms) but the fit is shown as a dotted line in this region, since there was no latching immediately above. The orange points at 1.6 ms, 2.5 ms, and 5 ms represent the anomalous latching, with the orange lines guides for the eye only. The five points studied in more detail are indicated by red crosses.

Figure 10 shows the typical trajectory of the defects when undergoing normal latching. For these and the subsequent figures, only the near-grating regions are shown; the director profile extends for a further 3 μm or so to the top surface. Where the order parameter S is shown, variations in the range 0.2 to 0.4 are emphasised, to allow visualization of the defect and surface distortion regions. For all cases, the liquid crystal undergoes the normal $\varepsilon_0 \Delta \varepsilon E^2$ response in the bulk of the liquid crystal when the electric field is applied.

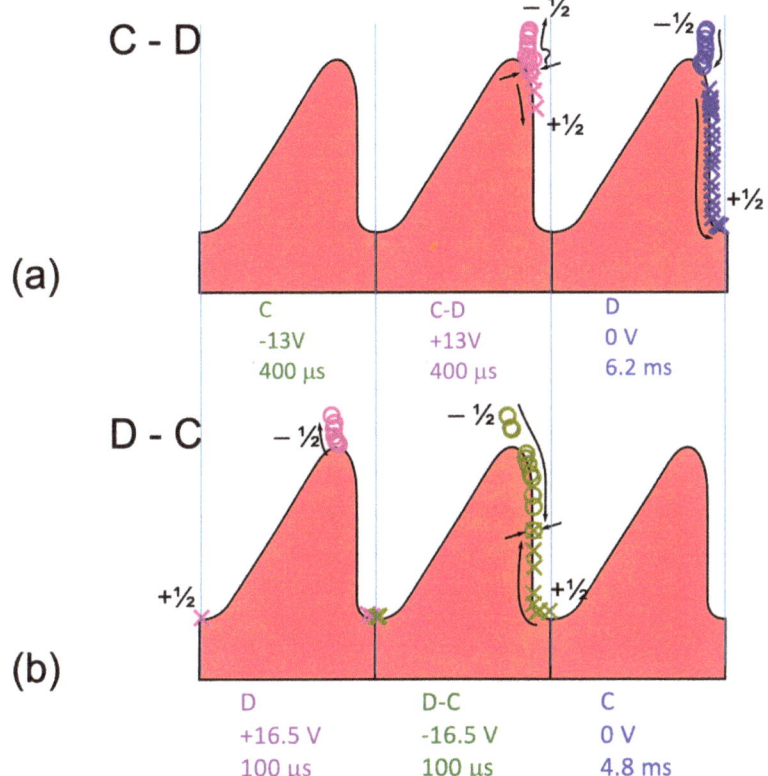

Figure 10. Pictorial summary of the defect dynamics in D-C latching (top) and C-D latching (bottom). Each diagram shows the defect trajectories (if any) during the (**a**) leading (DC balancing) pulse; (**b**) the trailing (latching pulse) pulse application; and C: the 0 V reorientation period. The pulse durations and amplitudes are indicated for each figure. Defect locations are indicated with circles ($-\frac{1}{2}$ defect) and crosses ($+\frac{1}{2}$ defect) and their direction of travel with arrows. The symbol colours indicate whether the field is positive (purple), negative (lime) or zero (blue). Pinched arrows indicate the location of defect annihilation. Note, in the D-C latching most of the defect movement occurs during the applied pulses, whereas for the C-D latching there is significant reorientation of the director, and hence defect positions, after the pulse has ended but before the defects settle onto the grating surface.

Since the material is strongly positive ($\Delta\varepsilon = +39$), the bulk director reorients towards the vertical condition regardless of whether the initial state is C or D, and the response time is expected to be between 1 ms (for ± 4 V) and 0.2 ms (for 10 V). This RMS response serves to both increase the local flexoelectric terms close to the surface, and to reorient the director away from the grating more towards the vertical condition.

The D-C simulation shown in Figure 10b and in Video S1 corresponds the conditions given by point 1 in Figure 9. The leading (positive) pulse pulls the $-\frac{1}{2}$ defect away from the grating peak to some distance within the bulk of the cell that depends on the impulse τV. This initial field has minimal effect on the $+\frac{1}{2}$ defect lying in the trough of the grating groove. This is consistent for all results, wherein the $+\frac{1}{2}$ defect moves towards the positive electrode and the $+\frac{1}{2}$ defect towards the negative electrode. On reversal of the field the trailing (negative) latching pulse propels this $+\frac{1}{2}$ defect up the grating sidewall, whilst the $-\frac{1}{2}$ defect returns to the grating surface and moved down the surface to meet it. The two defects annihilate, either during the latch pulse or, if within critical proximity, during the subsequent reorientation time. Should the voltage be insufficient for latching, the defects will resettle at their original positions on the grating following the reverse trajectories on the side walls of the grating.

The initial response, where the $-\frac{1}{2}$ defect moves into the bulk due to the leading pulse means that the trajectory for annihilation in the trailing pulse is lengthened. This is likely a cause for the breakdown of the simple 1D model and poor fits from Equation (5) since the resulting increase in latching time will be directly related to the magnitude of the field. Indeed, at low voltages ($\leq |10\text{ V}|$), the $-\frac{1}{2}$ defects do not leave the surface at all during the positive pulses. In this regime, the $+\frac{1}{2}$ defects also remain quite static at the bottom of the groove during the trailing negative pulse of the D-C latching sequence.

Similarly, the defect trajectories for C-D latching at point 2 of Figure 9 are shown in Figure 10b and Video S2. During the first (positive) pulse, no defects are observed; refraction of the electric field caused by the grating will cause a divergence at the point of inflexion on the side walls, but defects are prevented from forming by the surface curvature. Once the field is reversed in the trailing pulse, the field causes the nucleated defects to separate and move in the directions where the grating curvature stabilises each defect. In the instance shown, the $+\frac{1}{2}$ defect only moves part of the way towards the grating groove, but the $-\frac{1}{2}$ defect passes the top of the grating and begins to move into the bulk. On removal of the field, the $+\frac{1}{2}$ defect travels down the sidewall to the base of the groove and the $-\frac{1}{2}$ defect returns to the grating peak. Note, in this instance the D state forms during the relaxation process after the pulses have ended. If the pulse impulse is too small then either no defect nucleation occurs during the positive pulse or there is insufficient separation of the defects at the pulse end, leading to mutual annihilation; neither instance leads to latching.

The impulses at points 3 and 4 in Figure 9b are immediately below and above the latching threshold, respectively, in the range where the latching is anomalous and requires a significantly higher impulse than that expected from the reciprocal relationship of Equation (5). Figure 11 shows the position of the two defects at the end of the trailing pulse for both point 3 and point 4, together with the corresponding defect trajectories (also given in Video S3. The $-\frac{1}{2}$ defect follows similar trajectories in either case, moving from the initial nucleation point on the grating to deep within the bulk of the cell at the end of the pulse sequence. However, the $+\frac{1}{2}$ defect shows considerable differences in behaviour. For the lower voltage of point 3, the $+\frac{1}{2}$ defect detaches from the surface and enters the space between the grating grooves. After the pulse, the two opposing defects in the bulk of the liquid crystal move to mutually annihilate and the director profile relaxes to the C- state. Thus, despite a large impulse with the correct sign of trailing pulse to latch to the D state, latching does not occur and the C state is retained. Raising the voltage slightly to 13.0 V (point 4) then prevents the $+\frac{1}{2}$ defect from detaching from the surface and its movement is suppressed until the field is removed, when it again slides down the grating towards the groove where it remains, and the D state is latched.

The cause of such a dramatic change in behaviour is due to differences in the local director profile and defect orientations at the start of the trailing pulse. Reference [32] has shown quite substantial differences in defect trajectories on their paths to spontaneous annihilation due to their relative starting orientations. In particular, the $+\frac{1}{2}$ defect follows a path parallel to the local director. In this work, the situation is further complicated by the effect of the surface, the electric field, and its refraction. At the lower voltage, the director has reoriented less in the bulk of the sample and forms a path into the space between the grooves into which the $+\frac{1}{2}$ defect can move. With the higher field, the director is vertical closer to the defect, and the $+\frac{1}{2}$ defect movement into the bulk is impeded. As described in Section 1, defects on a strongly anchored surface will sit slightly further into the bulk than for weakly anchored, thereby are more likely to detach from the surface and move into the bulk of the liquid crystal.

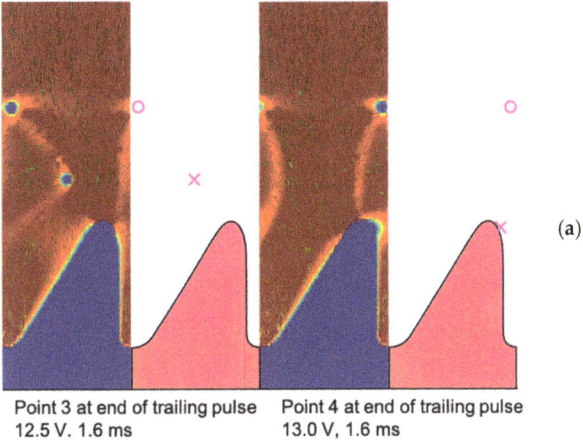

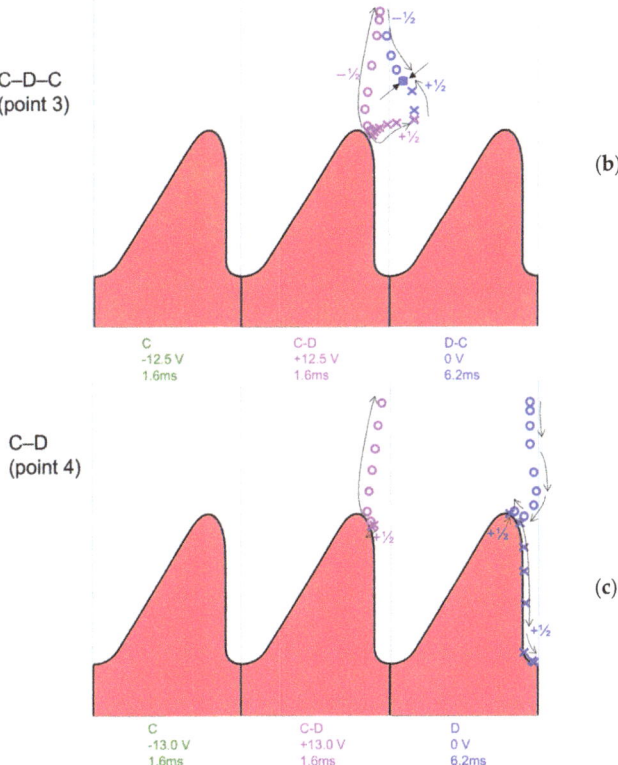

Figure 11. The anomalous latching mechanism either side of the threshold value. (**a**) Director and order profiles for the final frame of the trailing (positive) pulse. (**b**) Defect trajectories for point 3, below the latching threshold. (**c**) Defect trajectories for point 4 above the latching threshold.

Latching at point 5 in Figure 9 was also studied, as shown in Figure 12 and Video S4. Here, the 3.2 ms pulse is sufficiently long that the two defects move far into the bulk during the trailing (positive) pulse application. The director distortion around the grating surface also increases and supports the nucleation of a second pair of defects. As the original pair annihilate each other, this second generation behaves in the same fashion as the defects for shorter pulse widths, and thus are similar to the conventional latching mechanism. For voltages higher than the latching threshold at point 5, eventually this second-generation defect pair will enter the bulk together as the first pair did, which suppresses latching once more. Similarly, with even longer pulse widths, such as the data at 5 ms pulse widths in Figure 9, the second-generation pair also annihilates in the bulk. With successively longer pulse widths we see further generations of defects created in much the same way.

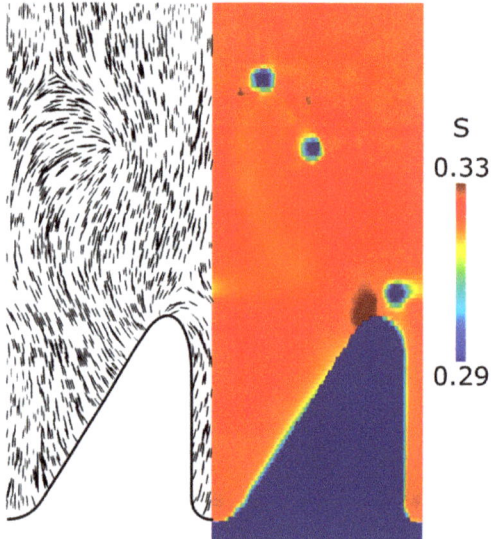

Figure 12. Modelling data of the continuous to defect latch highlighting the region near the grating during nucleation of a second latching defect pair. The average director orientations are shown on the left, and the simulated order parameter is given as a colourmap on the right. The scale for order parameter in the plot is shown to the right. The original defect pair are seen in the bulk near the top of the images and the second-generation pair are newly separated near the grating surface.

The simulation times for pulses longer than 5 ms were excessive, for which the maximum number of defect pairs was third generation (indicated by the upper orange loop in Figure 9b). However, up to seven generation have been found for very high anchoring energies of $W_\theta = 5 \times 10^{-4}$ N/m. The formation of higher-generation defects may also occur above the latching threshold, for the short pulse widths. It is clear from Figure 11 that the defects at point 5 are injected into the bulk of the cell and remain there until the field is removed, when they relax to the respective low energy states at the regions of highest surface curvature. However, at voltages above the threshold, further defect pairs can be formed at the top of the grating and propelled into the bulk. When the field is removed, the defects closest to the grating follow a similar path to those shown for point 4 in Figure 11, whereas the earlier defects move towards each other and mutually annihilate following similar paths as that of point 5 in Figure 11.

This behaviour of multiple defect pairs being propelled into the bulk of the sample will be strongly related to grating shape. For example, the gratings studied in reference [26] were more symmetrical and with less blaze than that used in the simulations here (see Figure 2c). In that case, further defect pairs were detected that were nucleated on the opposing sidewall of the grating, whereas here the multiple defect generation is being nucleated from the same point. It should also be recognised that, in the experimental studies of reference [45], reverse latching occurred at impulses above those for latching. That is, a negative-positive train would latch to D but increasing the voltage beyond a second threshold led to latching back to the C state. This transition was ascribed to the ionic content in the mixtures, causing a similar reverse latching to that in ferroelectric liquid crystals [46]. In this instance, ions diffuse to reduce the applied DC field, creating a reverse field. After the field is removed, the reverse field is retained briefly and before the ions have time to redistribute, which can be sufficient to cause latching from the desired state back to the opposite state. Instead of this explanation, it is possible that reverse latching might be due to forcing the defect pairs into the bulk where they annihilate (as found in this work). However, this would apply to C-D latching only, whereas reverse latching was observed for both C-D and D-C latching. Hence, the ionic viewpoint remains more likely because and was removed using commercially pure STN materials with typical room temperature conductivities of 10^{-11} Ω^{-1} cm^{-11} to reduce the ionic content [45]. However, the current work suggests that various types of behaviour are possible when including both multiple defects and ionic effects.

6. Conclusions

The ZBD device is an example of producing increased functionality through judicial use of topological defects in the nematic phase. In commercial devices, the shape and anchoring energy of the grating surface is designed to give bistable pretilts that can be addressed with wide operating windows using passive matrix addressing. That is achieved by having relatively weak anchoring, though sufficient to give bistability, and ensuring that the defects remain close to the surface throughout the electrical addressing. However, it has been shown here that there is a range of behaviour outside of these design constraints. The combination of grating shape and higher homeotropic anchoring has allowed defects to detach from the surface and migrate into the bulk of the nematic. Here, the defect trajectories are strongly influenced by the local director orientation. Multiple sets of defects have been induced, each interfering with each other and potentially leading to quite complex behaviour. Clearly, this behaviour is rich in possibilities. For example, defect pairs can be injected into the bulk of a liquid crystal sample, where they could interact with colloidal particles, micro-inclusions (such as droplets, bubbles, and microparticles), or the opposing interface.

Supplementary Materials: The following supporting information can be downloaded at: https://www.mdpi.com/article/10.3390/cryst12091291/s1. Video S1: Simulation of conventional D-C latching (point 1 in Figure 9). (+16.5 V, −16.5 V; τ = 0.1 ms); Video S2: Simulation of conventional C-D latching (point 2 in Figure 9). (−13.0 V, +13.0 V; τ = 0.4 ms); Video S3: Simulation of anomalous C-D latching immediately below (LHS) and above (RHS) the latching threshold where the C state after defect annihilation in the bulk (points 3 and 4, respectively, in Figure 9). (Point 3 = −12.5 V, +12.5 V; τ = 1.6 ms. Point 4 = −13.0 V, +13.0 V; τ = 1.6 ms); Video S4: Simulation of second-generation defects forming the final D state (point 5 in Figure 9). (−11.5 V, +11.5 V; τ = 3.2 ms).

Author Contributions: Conceptualization, J.C.J. and S.E.D.; methodology, S.A.J. and Z.R.G.; software, F.A.F.; validation, J.C.J.; formal analysis, S.A.J. and Z.R.G.; investigation, S.A.J. and Z.R.G.; resources, J.C.J.; data curation, S.A.J. and Z.R.G.; writing—original draft preparation, S.A.J. and J.C.J.; writing—review and editing, J.C.J.; visualization, J.C.J. and S.E.D.; supervision, J.C.J. and S.E.D.; project administration, J.C.J. and S.E.D.; funding acquisition, J.C.J. and S.E.D. All authors have read and agreed to the published version of the manuscript.

Funding: J.C.J. was funded through and advanced fellowship from the EPSRC, (EP/S029214/1). S.A.J. was funded by fully-funded studentship from Displaydata Ltd.; Z.R.G was jointly funded through the Soft Matter and Functional Interfaces CDT (SOFI) and Merck Ltd., Chilworth, Southampton.

Institutional Review Board Statement: Not applicable.

Informed Consent Statement: Not applicable.

Data Availability Statement: Data will be made available by request to the corresponding author.

Acknowledgments: Thanks are given to Displaydata Limited for fully funding a PGR studentship for S.A.J, to the Soft Matter and Functional Interfaces (SOFI) Centre of Doctoral Training and Merck Ltd. for funding Z.R.G, and to the EPSRC for an Advanced Manufacturing Fellowship for J.C.J. (EP/S029214/1). The authors also wish to thank Nikita Solodkov of the Soft Matter Physics group at Leeds University, Tim Spencer of Sheffield Hallam University, Guy Bryan-Brown of NVD Inc. and Steve Beldon of Displaydata Ltd. for useful discussions. Displaydata Ltd. is also thanked for provision of the proprietary grating embossing film and homeotropic photopolymers.

Conflicts of Interest: The authors declare no conflict of interest.

References

1. Mermin, N.D. The topological theory of defects in ordered media. *Rev. Mod. Phys.* **1979**, *51*, 591–648. [CrossRef]
2. Zurek, W.H. Cosmological experiments in condensed matter systems. *Phys. Rep.* **1996**, *276*, 177–221. [CrossRef]
3. Chuang, I.; Durrer, R.; Turok, N.; Yurke, B. Cosmology in the laboratory: Defect dynamics in liquid crystals. *Science* **1991**, *251*, 1336–1342. [CrossRef]
4. Bryan-Brown, G.P.; Brown, C.V.; Jones, J.C. Bistable Nematic Liquid Crystal Device. U.S. Patent No. 6,249,332, 16 October 1995.
5. Bryan-Brown, G.P.; Brown, C.V.; Jones, J.C.; Wood, E.L.; Sage, I.; Brett, P.; Rudin, J. Grating aligned bistable nematic device. *SID Int. Symp. Dig. Tech. Pap.* **1997**, *28*, 37.
6. Jones, J.C. Zenithal Bistable Displays: From concept to consumer. *J. SID.* **2008**, *16*, 143–154. [CrossRef]
7. Jones, J.C. Defects, flexoelectricity and RF Communications: The ZBD story. *Liq. Cryst.* **2017**, *44*, 2133–2160. [CrossRef]
8. Poulin, P.; Stark, H.; Lubensky, T.; Weitz, D. Novel colloidal interactions in anisotropic fluids. *Science* **1997**, *275*, 1770–1773. [CrossRef]
9. Lapointe, C.P.; Mason, T.G.; Smalyukh, I.I. Shape-controlled colloidal interactions in nematic liquid crystals. *Science* **2009**, *326*, 1083–1086. [CrossRef]
10. Muševič, I.; Škarabot, M.; Tkalec, U.; Ravnik, M.; Zumer, S. Two-dimensional nematic colloidal crystals self-assembled by topological defects. *Science* **2006**, *313*, 954–958. [CrossRef]
11. Muševič, I. *Liquid Crystal Colloids*; Springer: Cham, Switzerland, 2017.
12. Kitson, S.; Geisow, A. Controllable alignment of nematic liquid crystals around microscopic posts: Stabilisation of multiple states. *Appl. Phys. Lett.* **2002**, *80*, 3635–3637. [CrossRef]
13. Jones, J.C. Bistable Nematic Liquid Crystal Device. US Patent 7,371,362, 30 November 1999.
14. Lasak, S.; Davidson, A.; Brown, C.V.; Mottram, N.J. Sidewall control of static azimuthal bistable nematic alignment states. *J. Phys. D Appl. Phys.* **2009**, *42*, 085114.
15. Tai, J.-S.B.; Smalyukh, I.I. Three-dimensional crystals of adaptive knots. *Science* **2019**, *365*, 1449–1453. [CrossRef] [PubMed]
16. Pieranski, P.; Godinho, M.H. Flexo-electricity of the dowser texture. *Soft Matter* **2019**, *15*, 1469–1480. [CrossRef] [PubMed]
17. Čopar, S. Topology and geometry of nematic braids. *Phys. Rep.* **2014**, *538*, 1–37. [CrossRef]
18. Raynes, E.P. Improved contrast uniformity in twisted nematic liquid-crystal electro-optic display devices. *Electron. Lett.* **1974**, *10*, 141–142. [CrossRef]
19. Bigelow, J.E.; Kashnow, R.A. Observations of a bistable twisted nematic effect. *IEEE Trans. Electron Dev.* **1975**, *22*, 730–733. [CrossRef]
20. Nie, Z.; Day, S.E.; Fernández, F.A.; Willman, E.; James, R. Modelling of liquid crystals at the pixel edge. *SID Symp. Dig. Tech. Pap.* **2014**, *45*, 1382–1385. [CrossRef]
21. Jones, J.C. Bistable LCD. In *Handbook of Visual Display Technology*; Janglin, J.C., Cranton, W., Fihn, M., Eds.; Springer: Berlin/Heidelberg, Germany, 2012; pp. 1507–1543.
22. Thurston, R.N.; Cheng, J.; Boyd, G.D. Mechanically bistable liquid-crystal display structures. *IEEE Trans. Elec. Dev.* **1980**, *27*, 2069–2080. [CrossRef]
23. Bryan-Brown, G.P.; Jones, J.C. The Zenithal Bistable Display: A grating aligned bistable nematic liquid crystal device. In *E-Paper Displays*; Yang, B.R., Ed.; Wiley-SID Series in Display Technology; John Wiley and Sons: Hoboken, NJ, USA, 2022; Chapter 6; pp. 131–152.
24. Wood, E.L.; Bryan-Brown, G.P.; Brett, P.; Graham, A.; Jones, J.C.; Hughes, J.R. Zenithal Bistable Device (ZBD™) suitable for portable applications. In *SID Symposium Digest of Technical Papers*; Wiley: Hoboken, NJ, USA, 2000; Volume 31, pp. 124–127.
25. Jones, J.C. Novel geometries of the Zenithal Bistable Device. In *SID Symposium Digest of Technical Papers*; Wiley: Hoboken, NJ, USA, 2006; Volume 37, pp. 1626–1629.
26. Spencer, T.J.; Care, C.M.; Amos, R.M.; Jones, J.C. Zenithal bistable device: Comparison of modelling and experiment. *Phys. Rev. E* **2010**, *82*, 021702. [CrossRef]
27. Jones, S.A.; Bailey, J.; Walker, D.R.E.; Bryan-Brown, G.P.; Jones, J.C. Method for tunable homeotropic anchoring at microstructures in Liquid Crystal Devices. *Langmuir* **2018**, *34*, 10865–10873. [CrossRef]
28. Jones, J.C. Bistable nematic liquid crystals. In *The Handbook of Liquid Crystals*, 2nd ed.; Goodby, J.W., Collins, P.J., Kato, T., Tschierske, C., Gleeson, H.F., Raynes, P., Eds.; Wiley: Hoboken, NJ, USA, 2014; Chapter 4; Volume 8, pp. 87–145.

29. Porenta, T.; Ravnik, M.; Žumer, S. Effect of flexoelectricity and order electricity on defect cores in nematic liquid crystals. *Soft Matter* **2011**, *7*, 132–136. [CrossRef]
30. Kischka, C.; Parry-Jones, L.A.; Elston, S.J.; Raynes, E.P. Measurement of the flexoelectric coefficients e 1 and e 3 in nematic liquid crystals. *Mol. Cryst. Liq. Cryst.* **2008**, *480*, 103–110. [CrossRef]
31. Škarabot, M.; Mottram, N.J.; Kaur, S.; Imrie, C.T.; Forsyth, E.; Storey, J.M.D.; Mazur, R.; Piecek, W.; Komitov, L. Flexoelectric polarization in a nematic liquid crystal enhanced by dopants with different molecular shape polarities. *ACS Omega* **2022**, *7*, 9785–9795. [CrossRef] [PubMed]
32. Tang, X.; Selinger, J.V. Theory of defect motion in 2D passive and active nematic liquid crystals. *Soft Matter* **2019**, *15*, 587–601. [CrossRef]
33. Parry-Jones, L.A.; Elston, S.J. Flexoelectric switching in a zenithally bistable nematic device. *J. Appl. Phys.* **2005**, *97*, 093515. [CrossRef]
34. Parry-Jones, L.A.; Edwards, E.G.; Elston, S.J.; Brown, C.V. Zenithal bistability in a nematic liquid-crystal device with a monostable surface condition. *Appl. Phys. Lett.* **2003**, *82*, 1476–1478. [CrossRef]
35. Willman, E.; Fernández, F.A.; James, R.; Day, S.E. Modelling of weak anisotropic anchoring of nematic liquid crystals in the Landau-de Gennes theory. *IEEE Trans. Electron. Devices* **2007**, *54*, 2630–2637. [CrossRef]
36. James, R.; Willman, E.; Fernández, F.A.; Day, S.E. Finite element modeling of liquid crystal hydrodynamics with a variable degree of order. *IEEE Trans. Electron. Devices* **2006**, *53*, 1575–1582. [CrossRef]
37. Willman, E. Three-Dimensional Finite Element Modelling of Liquid Crystal Electro-Hydrodynamics. Ph.D. Thesis, University College London, London, UK, 2009.
38. Davidson, A.J.; Mottram, N.J. Flexoelectric switching in a bistable nematic device. *Phys. Rev. E* **2002**, *65*, 051710. [CrossRef]
39. Jones, J.C.; Amos, R.M. Relating display performance and grating structure of a zenithal bistable display. *Mol. Cryst. Liq. Cryst.* **2011**, *543*, 57–68, 823–834. [CrossRef]
40. Jones, S.A. Zenithal Bistable Display: Avenues for Improved Performance. Ph.D. Thesis, University of Leeds, Leeds, UK, 2019.
41. Qian, T.; Sheng, P. Generalized hydrodynamic equations for nematic liquid crystals. *Phys. Rev. E* **1998**, *58*, 7475–7485. [CrossRef]
42. Day, S.E.; Willman, E.J.; James, R.; Fernández, F.A. 67.4: Defect loops in the Zenithal Bistable Display. In *SID Symposium Digest of Technical Papers*; Wiley: Hoboken, NJ, USA, 2008; Volume 39, pp. 1034–1037.
43. Alexe-Ionescu, A. Flexoelectric polarization and second order elasticity for nematic liquid crystals. *Phys. Lett. A* **1993**, *180*, 456–460. [CrossRef]
44. Sonnet, A.M.; Virga, E.P. Dynamics of dissipative ordered fluids. *Phys. Rev. E* **2001**, *64*, 031705. [CrossRef] [PubMed]
45. Jones, J.C.; Beldon, S.; Brett, P.; Francis, M.; Goulding, M. Low voltage Zenithal Bistable Devices with wide operating windows. In *SID Symposium Digest of Technical Papers*; Wiley: Hoboken, NJ, USA, 2003; Volume 34, pp. 954–957.
46. Zhang, H.; Hirano, M.; Kobayashi, S.; Iimura, Y. Inverse switching phenomenon of surface-stabilized ferroelectric liquid crystals under a high voltage region. *Ferroelectrics* **1998**, *215*, 215–220. [CrossRef]

Article

Laser Written Stretchable Diffractive Optic Elements in Liquid Crystal Gels

Bohan Chen *, Zimo Zhao, Camron Nourshargh, Chao He, Patrick S. Salter, Martin J. Booth, Steve J. Elston and Stephen M. Morris *

Department of Engineering Science, University of Oxford, Parks Road, Oxford OX1 3PJ, UK
* Correspondence: bohan.chen2@eng.ox.ac.uk (B.C.); stephen.morris@eng.ox.ac.uk (S.M.M.)

Abstract: Direct laser writing (DLW) in liquid crystals (LCs) enables a range of new stimuli-responsive functionality to be realized. Here, a method of fabricating mechanically tunable diffraction gratings in stretchable LC gels is demonstrated using a combination of two-photon polymerization direct laser writing (TPP-DLW) and ultraviolet (UV) irradiation. Results are presented that demonstrate the fabrication of a diffraction grating that is written using TPP-DLW in the presence of an electric field in order to align and lock-in the LC director in a homeotropic configuration. The electric field is subsequently removed and the surrounding regions of the LC layer are then exposed to UV light to freeze-in a different alignment so as to ensure that there is a phase difference between the laser written and UV illuminated polymerized regions. It is found that there is a change in the period of the diffraction grating when observed on a polarizing optical microscope as well as a change in the far-field diffraction pattern when the film is stretched or contracted. These experimental results are then compared with the results from simulations. The paper concludes with a demonstration of tuning of the far-field diffraction pattern of a 2-dimensional diffraction grating.

Keywords: direct laser writing; diffraction gratings; stretchability; liquid crystals

---

Citation: Chen, B.; Zhao, Z.; Nourshargh, C.; He, C.; Salter, P.S.; Booth, M.J.; Elston, S.J.; Morris, S.M. Laser Written Stretchable Diffractive Optic Elements in Liquid Crystal Gels. Crystals 2022, 12, 1340. https://doi.org/10.3390/cryst12101340

Academic Editor: Ingo Dierking

Received: 11 August 2022
Accepted: 30 August 2022
Published: 22 September 2022

Publisher's Note: MDPI stays neutral with regard to jurisdictional claims in published maps and institutional affiliations.

Copyright: © 2022 by the authors. Licensee MDPI, Basel, Switzerland. This article is an open access article distributed under the terms and conditions of the Creative Commons Attribution (CC BY) license (https://creativecommons.org/licenses/by/4.0/).

1. Introduction

Diffraction gratings are optical components with precisely defined periodic structures that are designed to manipulate the spatial distribution of light [1]. For example, they can split a beam of light into a number of diffracted beams or direct a beam into one specific order. Diffraction gratings play an important role in a number of applications including monochromators, spectrometers, augmented reality (AR)/virtual reality (VR) displays, wavefront measurements, beam steering, structured light and optical encoders for high precision motion control [2–8]. The development of switchable and tunable diffraction gratings is also of considerable interest for potential future wearable technologies [9].

Tuning a diffraction grating often relies on applying some form of external stimuli to alter the optical properties of the material or the architecture of the grating. This external stimulus might be, for example, in the form of an electric field, a change in the temperature, or a change in the humidity conditions [10,11]. With the advent of wearable and conformable photonics technologies, there is a particular interest in using strain or mechanical deformation to tune photonic devices [9,12–17]. Towards this end, there are a range of materials that exhibit a reversible elastic response to mechanical strain [18–33], such as polydimethylsiloxane (PDMS), polyethylene terephthalate (PET), polyimide and Parylene-C, which would potentially enable the grating to be deformed by introducing external tensile strain leading to tuning of the optical diffraction pattern [16,34–39].

Two-photon direct laser writing (TPP-DLW) is a popular microfabrication technique as it enables small structures (of the order of microns and less) to be fabricated in a range of materials and has proved to be a particularly powerful fabrication tool for the development of a range of different metamaterial and nanophotonic platforms [40–42]. Liquid crystal

(LC) gels are a material that exhibit both elasticity and long range orientational order of the molecules (characterized by a unit vector referred to as the director). Using laser writing to structure and impart new functionality in polymerizable LC gels and elastomers has also attracted considerable interest in recent years [7,9,43–45]. For example, combining TPP-DLW with LC gels has enabled the development of a wide range of tunable stimuli-responsive components whose optical properties change when the material is subjected to an external stimulus, for example, LC microactuators, micro-soft robotic systems and LC sensors [46–49].

An advantage of using an LC gel compared with isotropic materials, such as photocurable polymer resins, is that these materials combine birefringence with the mechanical stretchability. Polymer-stabilized LCs also have the advantage of being able to lock-in and maintain any desired phase profile with the use of an external electric field and/or surface alignment [50]. By applying a voltage during fabrication, TPP-DLW can be used to lock-in different alignments of the director through photopolymerization so as to create a sculptured polymer network. After removing the voltage, and depending upon the relative lengthscales involved, the LCs in the non-polymerized regions tend to relax back to the initial alignment conditions leading to a spatially varying phase profile across the device [43]. Furthermore, compared with Si photolithography, TPP-DLW has the capability to build structures in three dimensions leading to more sophisticated architectures and multi-functional optical components [51–54]. Figure 1 shows examples of the functionality that can be unlocked using TPP-DLW in LC devices when an electric field is applied during fabrication. This includes the ability to hide features such as quick response codes [55], transport microparticles and generate microfluidics-inspired flow of boundaries between different topologically distinct states [56], fabricate tuneable diffractive optics elements in the form of diffraction gratings and computer-generated holograms [43,57], and the development of stretchable optical elements, which is the subject of this paper.

Figure 1. The range of functionality that our group has demonstrated using TPP-DLW in LC devices when an electric field is applied during fabrication [43,55–57]. This includes the development of switchable QR codes [55], microparticle transport and microfluidics inspired flow of defect walls [56], and diffractive optic elements in the form of one-dimensional diffraction gratings, Dammann gratings, and computer-generated holograms [43,57]. The focus of this paper is to demonstrate how this technique can be used to develop mechanically tuneable diffractive optic elements.

In this paper, we present a strategy to fabricate a mechanically tunable phase grating using a combination of laser writing and ultraviolet (UV) illumination to trigger polymerization in a photopolymerizable LC gel. After describing the laser writing process and demonstrating how the laser writing conditions influence the morphology of the polymerized structures, the fabrication process employed to manufacture stretchable diffraction gratings in LC gels is demonstrated. Results are then presented that show tuning of the period of the diffraction grating and it is shown how the diffraction pattern in the far-field is changed when the LC gel is extended or contracted. The degree of tuning is discussed and, where possible, results from simulations are presented to compare with our experimental findings. To conclude, a mechanically tunable 2-dimensional phase grating is demonstrated, and the pros and cons of the approach presented herein are discussed.

2. Materials and Methods

2.1. Laser Writing Conditions

In both this work and our previous work on TPP-DLW laser writing in LCs, we have used a custom-built system which is shown in Supplementary Materials, Figure S1. The laser source was a Titanium: Sapphire femtosecond laser (Spectra-Physics Tsunami) emitting at $\lambda = 780$ nm that was excited by a diode-pumped solid-state CW laser (Spectra-Physics Millennia V) emitting at $\lambda = 532$ nm. The laser pulse width was 100 fs by mode-locking at a repetition of 80 MHz.

Before fabricating the gratings into the LC gels, we considered the impact of the laser writing conditions (i.e., power, scan speed, and fabrication voltage) on the resulting polymerized structures. Results from these characterizations are presented in Figure 2 (different fabrication voltages and laser intensity) and Figure 3 (different laser powers and scanning speeds for different orientations of the laser polarization relative to the rubbing direction of the LC device). The LC mixture that was used consisted of three elements: 69 wt.% of the nematic LC mixture, E7 (Synthon Chemicals Ltd.), 30 wt.% of the reactive mesogen RM257 (1,4-Bis-[4-(3-acryloyloxypropyloxy) benzoyloxy]-2-methylbenzene (Synthon Chemicals Ltd.)), and 1 wt.% of the Photoinitiator Irgacure 819 (Ciba-Geigy). The components were then added to an empty vial and left in an oven at 70–75 °C overnight to mix the components via thermal diffusion. In both characterization tests, the structures were fabricated in a glass cell with anti-parallel rubbed polyimide alignment layers and indium tin oxide (ITO) electrodes on the inner surfaces of both substrates (Instec-LC5.0). The cell gap was measured to be 4.9 μm from the interference fringes recorded on a UV-Vis spectrometer (Agilent 8454). The ITO transparent electrodes enable a uniform electric field to be applied to the sample during laser microfabrication and the rubbed polyimide alignment layers ensure that, in the absence of an external electric field, the LC director would be homogeneously (planar) aligned in the plane of the device with a pre-tilt angle of approximately 4°. The filled glass cells were then loaded into the TPP-DLW system for fabrication.

In the laser writing process, a high energy ultrafast pulse train was focused into the bulk of LC sample through an objective lens with a 0.45 NA so that two-photon absorption only occurred within a relatively small volume (voxel size: 1 μm in the lateral dimension and 7 μm in the axial direction) in order to cross-link the reactive mesogens via a free-radical polymerization reaction that freezes-in the LC alignment permanently. By moving the translation stage with respect to the laser focus, a 3-dimensional polymer network structure can be constructed in the LC film [32,33]. Diffusion of the polymer network means that the resulting polymer structures are often slightly larger than the voxel sizes quoted above. Our experience is that if there is overlap between the focal volume and the substrate, the polymer walls are strongly tethered to the substrate.

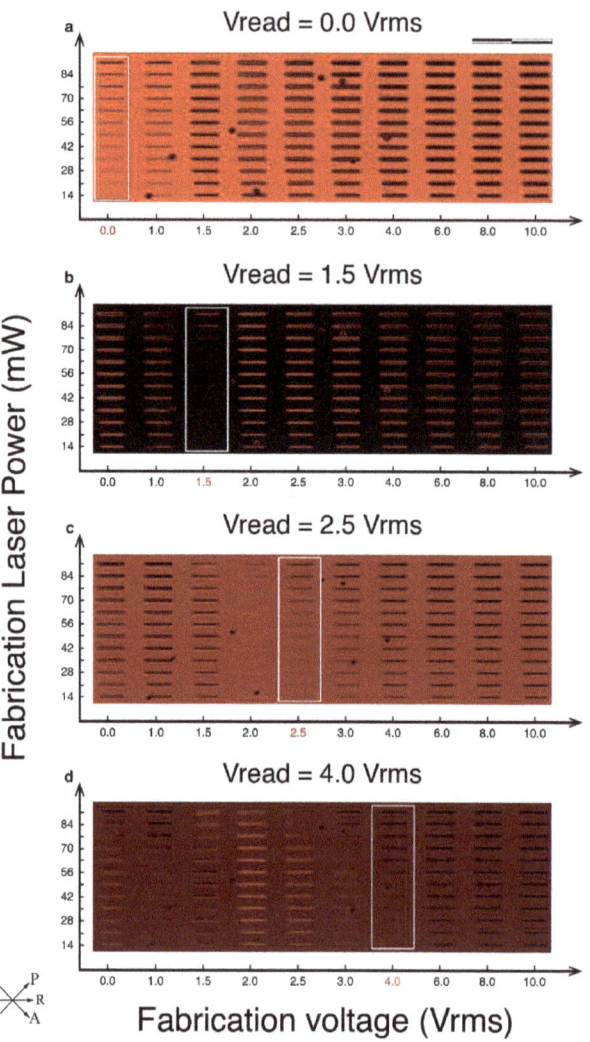

Figure 2. Laser written polymer walls in a polymerizable LC mixture for different fabrication laser powers and voltages applied to the LC during fabrication. The scale bar in the figure is 200 μm. The laser polarization was set to be perpendicular to the rubbing direction of the LC device. Images were recorded on a polarizing optical microscope when the rubbing direction was aligned at 45° to the transmission axes of the polarizer and analyzer pair, and with the aid of a narrow bandpass filter to remove all wavelengths beyond 640 nm–660 nm to avoid any further polymerization post-fabrication.

For the laser written polymerized structures shown in Figure 2, the voltage applied during the fabrication was varied (left to right on each subfigure) as was the laser power while the polarization was kept at an orientation that was perpendicular to the rubbing direction. Applying a voltage during fabrication causes a rotation out-of-plane of the LC director towards a homeotropic alignment at high voltages. Writing in the presence of a fabrication voltage causes the LC director alignment to be preserved at the moment of exposure to the laser writer. If, for example, the laser writing takes place with a large voltage applied, this results in a homeotropic alignment of the nematic LC being retained. The sample was translated at 100 μm/second during the fabrication process. Polarizing optical microscope (POM) images of the structure were then taken with the device between a pair of crossed polarizers and the rubbing direction of the cell oriented at 45° to the transmission axes of the polarizers. Between each of the subfigures, the voltage applied during imaging (the read voltage) was varied. In the case when the read voltage is $V = 0$, there is a clear increase in the thickness of the lines as the writing voltage was increased. This is due to the elastic distortion of the LC. Away from the polymerized region, the director orientation is dictated by the alignment layers; within the polymerized region, the director orientation is determined by the writing voltage. The distance over which the director profile relaxes from the polymerized to the surface aligned orientation is dictated by the elastic constant of the LC and the orientation in the polymerized region. Hence, while changing the writing voltage does not impact the width of the polymerized lines, the lines appear wider due to the additional relaxation distance. Considering again the case where the read voltage was zero, we see that changing the writing voltage changes the intensity in the centre of the lines; this is a result of the retardance being fixed to a given value in the polymerized region. Increasing the writing power increased the contrast when the read voltage matched the write voltage. It also caused a slight increase in the line thickness when the read voltage did not match the write voltage.

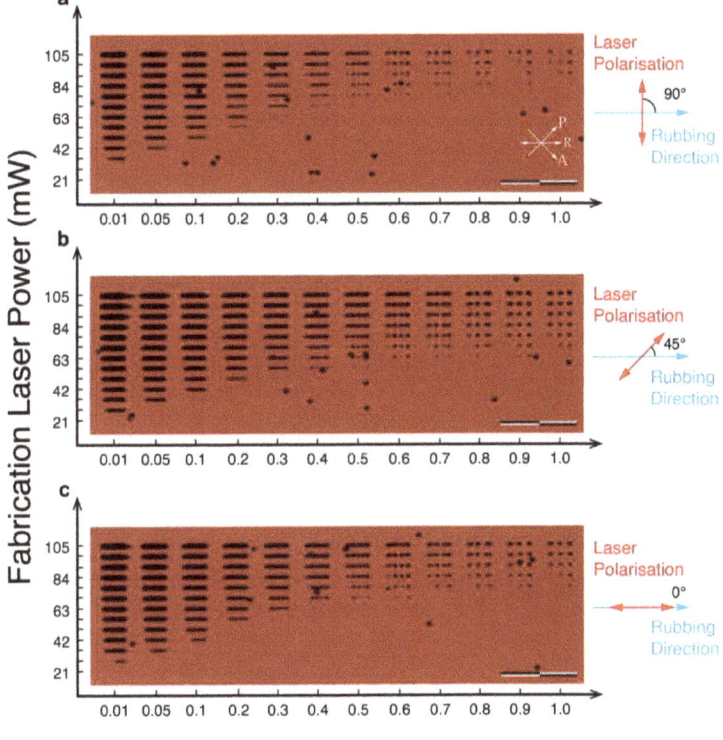

Figure 3. Laser written polymer walls in a polymerizable LC mixture for different fabrication laser powers and fabrication stage speeds when no voltage was applied to the LC during fabrication. Images were recorded on a polarizing optical microscope with an illumination source at 650 nm when the polarization of the writing laser was aligned at 90° (top), 45° (middle), or 0° (bottom) relative to the rubbing direction of the LC device. For these images, the rubbing direction was aligned at 45° to the transmission axes of the polarizer and analyzer pair and the scale bar in each image represents a length of 200 μm.

In Figure 3, the voltage applied during the writing process was fixed at 30 Vrms to ensure the LC was homeotropically aligned. Within each subfigure, the writing speed was then varied, and the laser power increased. The polarization of the writing laser relative to the rubbing direction was varied between each of the subfigures. The images were all taken with the LC glass cell between crossed polarizers, with a narrow bandpass filter at 650 nm after the illumination source to avoid any unwanted polymerization post-fabrication and with no voltage applied during inspection. Theoretically, if the laser is polarized perpendicular to the director, it only encounters the ordinary component of the refractive index. Otherwise, the laser should encounter a combination of the ordinary and extraordinary indices, which can cause a serious focal splitting aberration at a high numerical aperture [58]. However, according to the comparison in Figure 3, the polymer walls fabricated with different writing polarizations look qualitatively similar, primarily because the device is switched at a high voltage so it is effectively homeotropic and the aberration should be the same for all incident orientations of the laser linear polarizations. Furthermore, as the device here is quite thin (4.9 µm), the birefringent medium is contained purely within the laser focal intensity distribution, reducing the effect of the aberration. There is a noticeable increase in the linewidth when the laser power was increased, and when the writing speed was reduced, as both cause a greater exposure of the laser to the LC, which results in more polymerization. As the writing speed increases, we first see a reduction in the linewidth and then a breakup of the lines into discrete dots. These dots are smaller and more circular when the polarization is perpendicular to the rubbing direction which is again a result of the optical aberrations.

In practice, we tend to fabricate polymer structures at the maximum voltage that can safely be applied to the cell, as this always results in almost no in-plane retardance and hence shows some contrast at any voltage below this. We use a write speed of 0.1 mm/s as this is slow enough to allow smooth lines even at low powers, while not being so slow that the resolution suffers. Laser powers are typically chosen to be around 50 mW as this ensures that the polymer structures are fabricated consistently while again not causing a significant loss in resolution. These parameters are used herein for the fabrication of the diffractive optic elements in the stretchable LC gels. In the following, we describe the strategy employed in this work to fabricate phase gratings in LC gels using TPP-DLW and describe the steps taken to form a transparent stretchable free-standing film that preserves the phase grating once the LC film has been delaminated from glass substrates. Finally, the two experimental methods used in this work to characterize and analyze the stretchable gratings in the LC gel are discussed. Specifically, these two methods are: (1) direct observation of the grating on a POM and (2) observation of the diffraction pattern in the far-field.

2.2. Design and Fabrication of the Phase Grating

The configuration of our one-dimensional diffraction grating is illustrated in Figure 4. In this case, the low refractive index regions are defined by polymer walls of nematic LC that are homeotropically aligned and that have been frozen-in using the laser writing process in the presence of an electric field. The high refractive index regions, on the other hand, are defined by nematic LC regions between the homeotropic polymer walls that are then locked-in by subjecting the sample to UV illumination. To ensure that there is a phase difference between the high and low refractive index regions, the transmission as a function of voltage for the LC mixture was recorded (see Supplementary Figure S2) where it was found that the transmission reaches saturation at around 60 Vrms. Therefore, 100 Vrms was deemed more than sufficient to align the LC director homeotropically. After locking-in the low refractive index regions with the laser written localized polymer walls at 100 Vrms, the remaining LC-polymer mixture was then polymerized with UV light without a voltage to ensure that the polymer walls have a phase difference with the non-laser written regions.

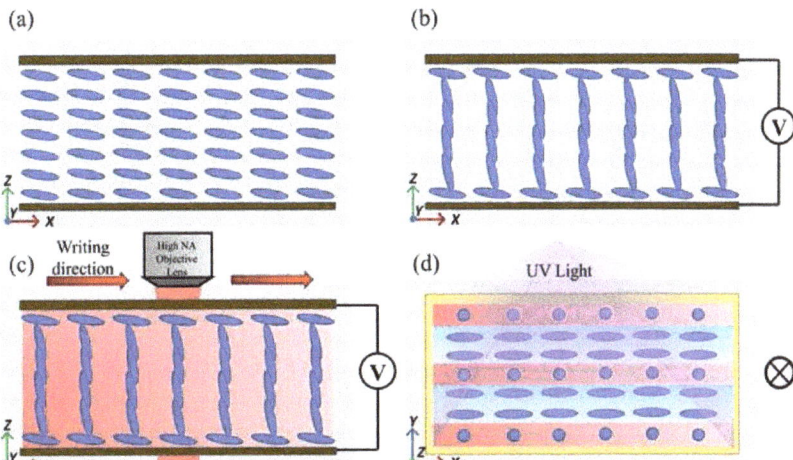

Figure 4. The one-dimensional phase grating fabricated using a combination of laser writing and ultraviolet (UV) illumination. (**a**) The nematic LC was homogeneously aligned such that the director lay in the plane of the glass cell with a pre-tilt at the substrate surfaces of 4°; (**b**) The illustration of the LC director switching to a homeotropic alignment with a voltage amplitude of 100 Vrms; (**c**) The laser writer scanned across the LC sample to form thin walls of polymer network along the *x*-direction that locked-in the homeotropic state; (**d**) After laser fabrication, the voltage was then removed and the sample was exposed to UV illumination to polymerize the remaining LC so as to freeze-in the alignment of the director between the polymer walls after the electric field had been removed. (**a**–**c**) show a side view of the LC between glass substrates while (**d**) shows a top view after UV polymerization where the pink regions correspond to the polymer walls for which the LC director was homeotropically aligned whereas the blue regions correspond to the bulk regions that have been polymerized using UV illumination and that consist of an LC director that was gradually relaxed from homeotropic alignment. The gratings were polarization dependent, which means that a diffraction pattern was only observed when the polarization of the incident light source was aligned parallel to the rubbing direction.

For the stretchable films, a 20 µm-thick glass cell (Instec-LC20.0) was used because the diffusion in the axial direction is sufficient to tether the polymer walls to the glass substrate. Here, the airgap represents not just the thickness of the LC layer between the glass substrates but also represents an approximate thickness for the free-standing film after the photopolymerization and delamination processes. The other properties of the glass cell (e.g., ITO layers, alignment layers etc.) were the same as that used in the previous section. The fabrication process for the stretchable phase gratings is illustrated in Figure 5. After capillary filling of the LC mixture (Figure 5a(ii)), the glass cells were then placed on a hotplate at 70 °C for several minutes to ensure that the LC mixture filled the cell uniformly. Finally, wires were attached to the glass cells using indium shot so that an electric field could be applied to the LC layer.

For this work, the TPP-DLW system was used to write a phase grating consisting of 65 parallel polymer walls with a 12 µm spacing between each polymer wall. Here, the laser written region was defined to be within the bulk of the LC layer, which then results in a polymer network that is denser in the center of the layer, but weaker close to the glass substrates. This leads to polymer structures that are tethered to the substrates and ensures that they do not drift or move even when the LC layer is subjected to an applied voltage. The fabrication accuracy in the *x-y* plane is less than 100 nm. Each polymer wall was written in the presence of a relatively high voltage (100 Vrms) with a power of 48 mW (Figure 5a(iii)) to ensure that a homeotropic alignment of the nematic LC had been obtained.

The reason why these parameters were chosen was that 65 polymer walls that are spaced at 12 μm intervals results in a grating with a total width of 768 μm, which is slightly larger than the laser diode spot diameter used to generate and inspect the diffraction patterns. A power of 48 mW was used for the laser writing process as this had been demonstrated to result in high fidelity polymer structures. A full list of the fabrication parameters used in this study is provided in Table S1 in the Supplementary Information.

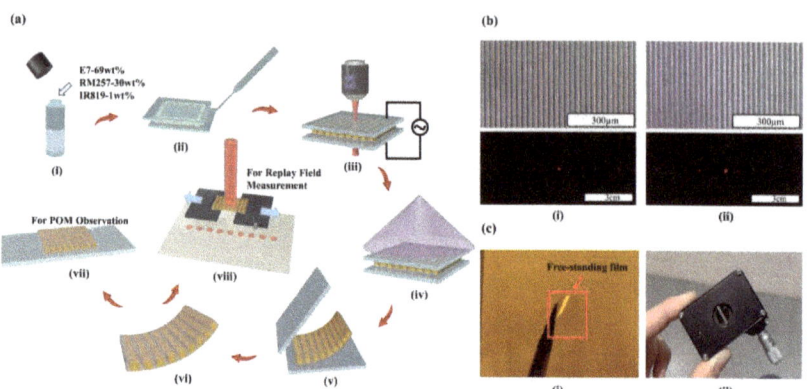

Figure 5. A stretchable one-dimensional diffraction grating in a liquid crystal (LC) gel. (**a**) Illustration of the fabrication and measurement process used to construct and characterize the mechanically stretchable grating: (i) mixture preparation, (ii) capillary filling of glass cell, (iii) fabrication of a 1-D diffraction grating with a voltage applied to the LC cell, (iv) bulk photo-polymerization of the film using ultraviolet illumination, (v) delamination of the cell to form a free-standing LC film that is approximately 20 μm thick, (vi) extraction of the flexible film, (vii) polarising optical microscope (POM) observation of the film, (viii) far (Replay) field observation of the diffraction pattern. (**b**) The POM (top) and far (replay)-field (bottom) images of the phase grating before (i) and after (ii) delamination. (**c**) Photographs of the LC free-standing film: (i) after delamination and (ii) mounted on an adjustable slit for stretchability and reversibility measurements.

After the formation of the phase grating, the samples were then illuminated with UV light to trigger polymerization in the remaining unreacted regions of the sample (Figure 5a(iv)). The samples were photopolymerized with UV light (peak wavelength of λ = 365 nm) with a power density of 10 mWcm^{-2} for 10 min. Once the film had been fully cross-linked, a scalpel was used to detach one of the four corners of the sample (Figure 5a(v)). After removing the top glass substrate of the sample, the free-standing film was then extracted from the bottom glass layer with a razor blade and tweezers. An example of a free-standing film is presented in Figure 5a(vi). Figure 5c(i) shows a photograph of the LC free-standing film captured with a digital single lens reflex (SLR) camera. The films were first attached to a glass substrate for an initial observation on a POM before being removed and then transferred to a mechanically adjustable slit for further characterization.

2.3. Characterization of the Stretchable Phase Gratings

To observe the phase gratings directly, the patterned free-standing film was positioned on a glass substrate (Figure 5a(vii)) and placed on a POM. An image of the phase grating can be seen in Figure 5b. Figure 5b(i),(ii) show the grating before and after the delimitation, respectively. It can be seen that the overall width of each grating shrinks a little bit from the initial value (768 μm), which indicates that the film contracts during the delamination process and therefore might be slightly thicker than 20 μm. The reason for this is that the detachment of the film decreases the internal pressure within the LC gel forcing it to shrink and become thicker to compensate for this pressure decrease.

In order to determine the mechanical tunability of the stretchable phase grating after bulk UV curing and delamination, the patterned film was placed across an adjustable mechanical slit consisting of two metallic plates whose separation could be altered using a micrometer (Thorlabs), as shown in Figure 5c(ii). By adjusting the micrometer setting, the separation between the metallic plates could be varied resulting in either an extension or relaxation of the LC film. The slit was initially set to have a separation of 1.30 mm, which was then increased in step sizes of 0.05 mm until the film was eventually broken so that the full tunable range could be determined [59,60]. To test the reversibility in terms of successive extensions and contractions of the film, the free-standing films were stretched to an extent that would not lead to the films being broken. By comparing the tunable characteristics of the film before and after relaxation, the reversibility could be determined.

To complement the direct observations of the grating on a POM, measurements of the diffraction patterns observed in the far-field were also recorded in replay imaging system, which is shown in Supplementary Figures S3 and S4. In this case, the separation of the metallic plates on which the film was placed was varied and a laser diode operating at $\lambda = 635$ nm was incident on the diffraction grating. A white screen placed at 23 cm from the sample in the far-field allowed us to observe the diffraction pattern and a full-color Thorlabs CCD camera (DCC1240C) was used to record images of the diffraction pattern in order to determine the relative intensities in each diffraction order. By measuring the spatial separation between the zeroth and 1st order spots, r, the grating period, d, can be estimated from

$$d = \frac{a\lambda}{r} \quad (1)$$

where a represents the distance between the diffraction grating and the screen in the replay field. Here, we have used the standard grating equation, $d \sin \theta_m = m\lambda$, ($m$ is an integer representing the diffraction order of interest) and employed the small angle approximation, $\sin \theta \approx \tan \theta = \frac{r}{a}$). By tracking the change in the separation r as the film is either extended or contracted, we can determine the change in the grating period and compare this with our observations on the microscope.

3. Results and Discussion

3.1. Demonstration of a Laser-Written Diffraction Grating in a Free-Standing LC Gel

As mentioned previously, the grating contains 65 polymer walls spaced at 12 µm, which results in a total grating width of 768 µm after the laser writing process. However, a film formulation of 30 wt.%-RM257 would typically shrink to around 94% of the original size after delamination, as can be seen in Figure 6a, which shows that the grating is now 723 µm in length. From the enlarged image, the dark stripes show the laser written polymer walls which have locked-in a homeotropic alignment of the LC director. The remaining non-laser written regions, on the other hand, have been exposed to UV illumination for bulk polymerization. Figure 6b represents the corresponding diffraction pattern of the phase grating in the far field. Results for the intensities in the different diffraction orders are shown in Figure 6c.

Simulations were conducted to predict the intensity profile of the different orders in the diffraction pattern and to compare the results from simulations with the results from the experiments [55,61–63]. To model the phase grating, the tilt angle of the LC director was determined from simulations for both the laser written regions that have frozen-in a homeotropic alignment due to the application of a high voltage (100 Vrms) during fabrication and the regions in between the laser written walls that have been stabilized by UV illumination with no voltage applied (0 Vrms). For the simulations, we have assumed that the thickness of the film was 20 µm, the initial pre-tilt angle to be 4^0, the grating period to be 12 µm, and the birefringence of the LC mixture to be $\Delta n = 0.17$ ($n_e = 1.71$, $n_o = 1.54$). We also applied the one-constant approximation for the elastic constants (i.e., $K_{11} = K_{22} = K_{33} = K = 11.6$ pN). By determining the director orientation in both the laser written and UV aligned regions this enables the periodic optical phase difference after the grating to be simulated so that the diffraction pattern in the far-field can be determined using a fast Fourier transform (FFT) as shown in Figure 6d.

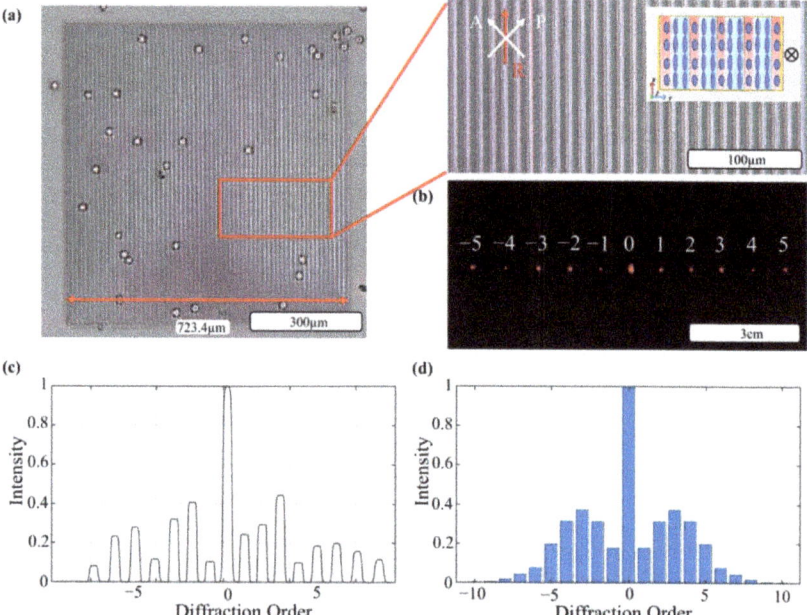

Figure 6. Characteristics of a laser-written diffraction grating in a free-standing LC film. (**a**) Polarizing optical microscope (POM) and (**b**) replay field images after delamination of the film. The white arrows labelled P and A refer to the orientations of the polarizer and analyzer, respectively, and the red arrow labelled R represents the rubbing direction of the alignment layers in the glass cell. (**c**) Experimental results of the intensity profile for the different diffraction orders. The results have been normalized to the value of the intensity of the zeroth order. (**d**) Simulated normalized intensity distribution for the free-standing film in the replay field after delamination.

For the phase grating in the free-standing film, the intensity profile predicted by simulation does agree, to some extent, qualitatively with the experimental intensity distribution recorded by the CCD camera as shown in Figure 6c. However, the experimental intensity distribution shows some degree of asymmetry that is not replicated in the simulations. One potential reason for this is that the LC director is biased in one direction because of the pre-tilt at the substrate surfaces. This biasing of the LC director could result in the asymmetry observed in the intensities recorded in the far-field diffraction pattern. In the experiments, the sample is relatively thick (20 µm); however, for the simulations, the sample is considered to be thin in accordance with Raman-Nath diffraction, which assumes that the propagation path is straight across the bulk of the film and any variation in direction is ignored. Supplementary Figure S5a shows the director tilt as a function of both the position throughout the bulk and the position within each grating period, while (b) represents the retardance distribution induced by the laser written grating with a period of 12 µm without and with a voltage of 100 Vrms applied, respectively. It confirms the LC phase grating can be fabricated by the TPP-DLW system not only from the perspective of simulations but also from that of polarization measurements. The images in Figure S5b were captured using a Mueller matrix microscope [64–66].

3.2. Tuning the Grating Period

Figure 7a shows POM images of the two extreme cases in terms of stretching: the initial unperturbed grating and the grating when the film was stretched. The film was stretched from a length of 724 ± 3 μm to 825 ± 3 μm, resulting in a change in grating period from 11.31 ± 0.05 μm to 12.89 ± 0.05 μm, which indicates that the film can be stretched by 114% of its initial length. Figure 7b shows the change in the diffraction pattern in the far-field as the film is stretched resulting in a decrease in the spacing between successive orders. From these images, the intensities of each diffraction order were tracked by the CCD camera and are shown in the plot presented in Figure 7c as the film was stretched. During the stretching process, the separation between the zeroth order and either of the first orders decreases from 13.0 mm to 11.4 mm, which corresponds to a change in the grating period from 11.3 ± 0.2 μm to 12.9 ± 0.2 μm. This result is in very good agreement with the results obtained from the POM images for the change in grating period for the same extension of the film. It was found that as the film was extended, the intensity in the zeroth order decreased dramatically while the intensity in the first order increased.

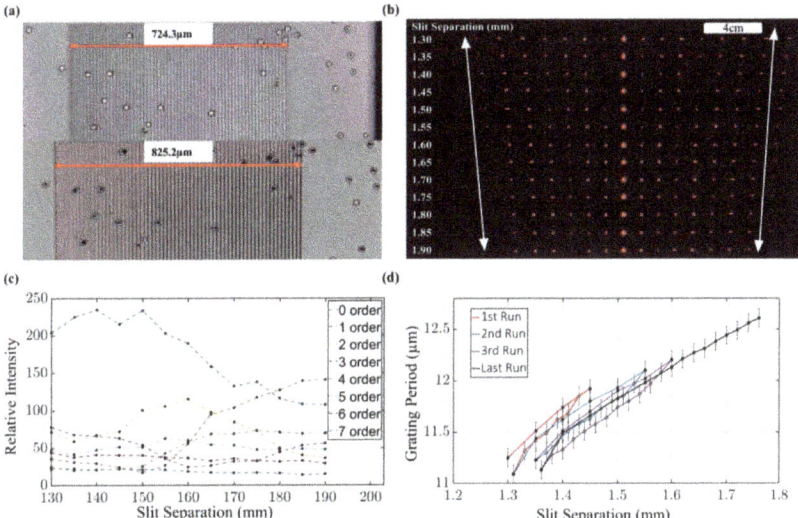

Figure 7. Mechanically tuning the 1D diffraction grating. (**a**) Polarization optical microscope (POM) images which show the grating in the undistorted state and the grating after the film has been extended by the maximum amount. (**b**) Replay field images which show the change in the diffraction pattern observed in the far-field as the grating is extended and contracted. (**c**) Intensity of each diffraction order as a function of slit separation. The dashed lines between the measured values are to guide the eye only and the estimated error in these relative intensity measurements is ±4%. (**d**) Grating period as a function of the separation between the metal plates (the slit) on which the samples were placed and which triggered either an extension or contraction of the film. The values for the grating period were obtained from POM images. Solid squares represent an extension of the film while open circles indicate a contraction of the film.

The film was stretched by varying the opening of the slit in step sizes of 0.05 mm upon increasing the separation and step sizes of 0.02 mm upon decreasing the separation of the slit. This process was repeated three times until the film was stretched to the point that it was broken. Figure 7d presents results of the grating period determined from the POM images as a function of the separation of the slit on the adjustable mount (which in turn causes an extension of the film). The results show that the grating period can be returned to its original size before being stretched again, demonstrating the elastic behavior and reversibility of the free-standing film. Specifically, the grating period returns to the same value (between 11.10 μm and 11.30 μm). As shown in Supplementary Figure S6, the results for the stretchability and reversibility obtained from the far-field measurements appear to be in good agreement with the results obtained from the POM images.

The dependence of the grating period on the slit separation follows a roughly linear dependence. In Figure 7d, the gradients for each of these lines of best fit appear to be approximately the same, which means the film can contract back to its previous size for the same slit separation as before. Therefore, it indicates that the film did not slide too much during the stretching and contraction processes. According to the experimental data, the film only slid on the slit by 0.05 mm during the whole process. The replay field images obtained for the stretching and contraction process can be found in Supplementary Figure S6. Supplementary Figure S7 shows two extremes in terms of stretching.

3.3. Comparison of the Laser Written and UV Polymerization Regions

In order to investigate the difference in the mechanical stretchability between the laser written diffraction grating and the surrounding UV-polymerized regions that had not been subjected to laser writing, the locations of the 20-micron spacer beads trapped in the film were tracked during the stretching process. It was noticed that whenever the film was contracted, the gratings appeared to shrink more than the regions of the polymerized LC that did not consist of laser written gratings. This indicated that the polymer network formed in the laser-written regions appeared to be more rigid than those in the non-written UV-illuminated regions. By tracking the positions of the spacer beads at various locations in the film during the stretching process, the different 'stretchability' for the different regions could be evaluated.

Figure 8 presents the results obtained when six spacer beads randomly distributed through the film were monitored during the stretching process. A–F represent the six spacer beads (Figure 8a) that were tracked with image processing while the film was subjected to a mechanical strain to elongate the film in a direction parallel to the grating period. Since the gradient of the straight line fit in Figure 8b is smaller than unity, under the same force, the strain of AB is larger than that of EF, which verifies the assumption that the pure polymer film is 'softer' than the high-voltage laser written region. Therefore, the stiffness of polymer network fabricated with the laser writing at a high voltage is a bit larger than that observed for UV bulk curing, which indicates that the elasticity of the film in the different regions is not the same. In Figure 8c, the distance between C and D remains the same while the film was subjected to a mechanical stress. This means that the film does not show any deformation along an axis perpendicular to the direction in which the film is being extended. Therefore, the stretchable LC film only elongates in the direction parallel to the applied force. This would allow one to deform the sample in two dimensions separately, providing a potential route to the development of sophisticated 2D-gratings that exhibit different tuning characteristics if elongated along different directions. In order to ensure that the volume remain unchanged, the film thickness must reduce.

3.4. A Stretchable 2D Diffraction Grating

To conclude, we also demonstrate the feasibility of mechanically tuning more complex 2-dimensional diffraction gratings. Dammann gratings play a key role as optical array generators as they can uniformly distribute the optical energy among the designed diffraction orders. Conventional Dammann gratings are binary-phase gratings that generate equal-intensity spot arrays. Generally speaking, they cannot be tuned continuously. However, the mechano-optical approach presented here could potentially lead to analogue tuning of the grating. As an example, the laser written pattern in Figure 9a was designed to have a specific phase profile such that it leads to a diffraction pattern consisting of a 2-d array of 6×6 spots in the form of a rhombus shape. Based upon the stretchability achieved demonstrated herein, the initial angles of the rhombus in Figure 9a were set to be $84°$ and $96°$. When the film was subjected to a mechanical stress, the rhombus shape in Figure 9a then gradually transformed into a square with angles of $90°$ and $90°$, respectively, at maximum strain. Similarly, the diffraction pattern altered from a diamond shape to a square array during stretching (Figure 9b shows the diffraction pattern before and after stretching). It can be seen that the separation between each laser spot in the horizontal direction contracts resulting in the array of spots forming a square arrangement. The repeatability of this specific device was not investigated in detail. However, we believe that as the formulation and the fabrication process are the same as that used for the film presented in Figure 7d (with the exception that this one is a more sophisticated grating) we would expect this 2-D stretchable grating film to display a similar level of repeatability in terms of an extension and contraction of the film. Mechanically tuning complex 2D diffraction elements is of potential importance for tunable AR/VR devices, tunable multiple imaging processes, mechanical beam steering and next generation optical sensors.

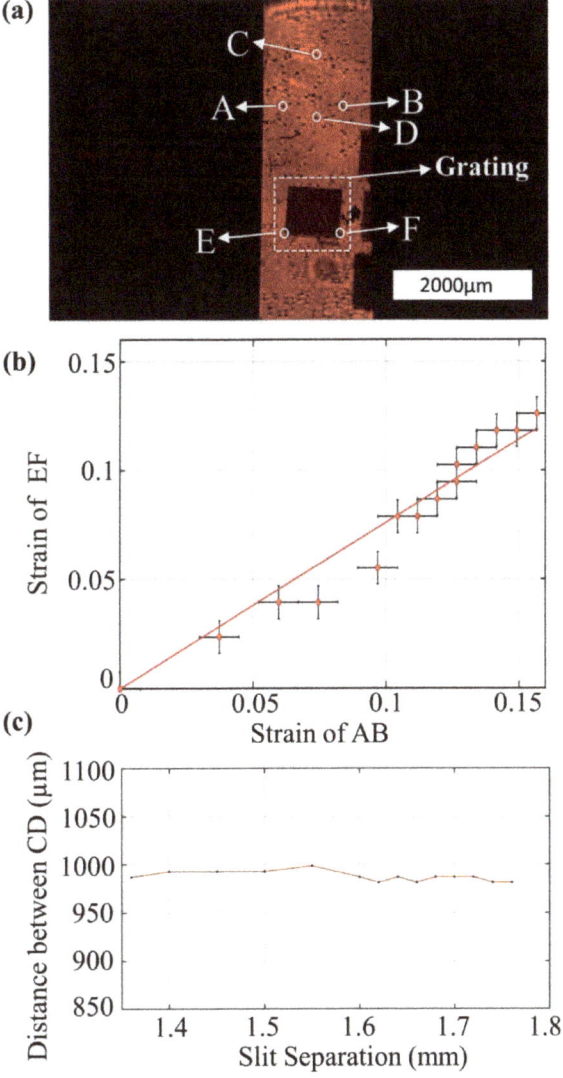

Figure 8. Comparison of the stiffness of the polymer network in either the laser written or UV polymerized regions in the stretchable films. (**a**) Photograph of the stretchable film when illuminated

with a 635 nm laser diode and captured with the CCD camera. The six spacer beads are labelled A–F. By tracking the separation between A & B and E & F as the film is stretched, the relationship between the strain for the region occupied by the (laser written) grating and that of the surrounding (UV polymerized) film can be determined. (**b**) The strain for the region between E and F as a function of the strain experienced for the region between A and B. The red line represents a line of best fit. (**c**) Separation between spacer beads C and D as a function of the slit separation. The red line is to guide the eye.

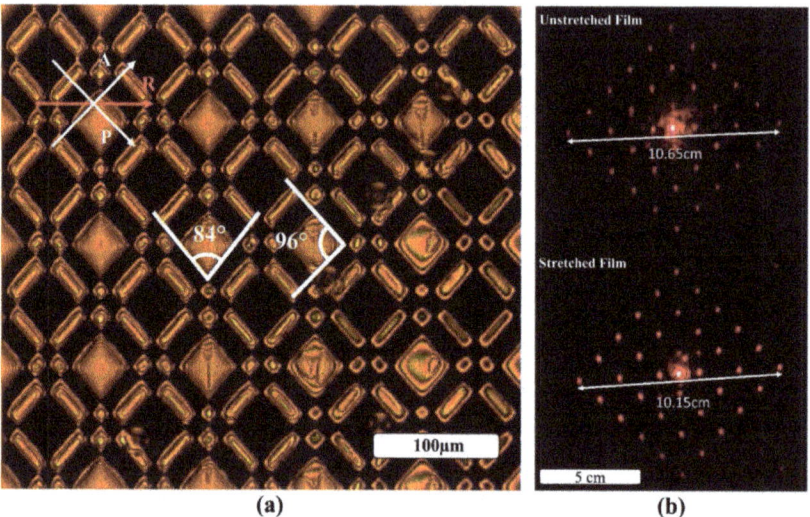

Figure 9. Demonstration of mechanical stretching of a free-standing laser written film with a 2D diffraction pattern in the far-field. (**a**) a POM image of a film (20 µm) with a laser written 6 × 6 2D Dammann grating in a rhombus shape. The dark regions correspond to the homeotropic aligned director locked-in using TPP-DLW at a voltage of 100 Vrms, while the bright regions correspond to the domains of the LC gel photopolymierzed using UV exposure without an electric field; (**b**) Far-field images of the change in the diffraction pattern in the replay field before stretching and when the film was stretched.

4. Conclusions

In summary, we have proposed and demonstrated tunable one- and two-dimensional phase gratings written in thin (20 µm-thick) LC gels that have been patterned with a combination of TPP-DLW and bulk UV illumination. These phase gratings can be mechanically tuned by subjecting the gels to an external tensile strain along the grating period. Results are also presented that show the mechanical tuning of the far-field diffraction pattern of a 2D Dammann grating which gradually changes from a rhombus shape to a square shape. These results confirm that the mechanical tunability could be applied to other tunable diffractive optical elements; for example, continuously tunable holograms. It was shown that a patterned free-standing LC gel consisting of a mixture of 30 wt.% RM257 and the nematic LC E7 can be stretched by 14% of its initial size leading to a change in the grating period of approximately 2 microns. This tuning of the grating period was verified both by observing changes in the diffraction grating directly on a POM and by tracking the change in the separation of the diffraction orders in the far-field. In addition, results were obtained that indicated there was a difference in the mechanical strength between the laser written and non-written regions by tracking the movement of the spacer beads that were trapped in the LC gel. The results indicated that the UV-illuminated polymerized regions were softer than the laser-written regions within the gratings, and that the film did not deform along an axis orthogonal to the direction in which the films were stretched. These results demonstrate the important potential of mechano-optical diffractive optic elements that have been fabricated in LC gels using both laser writing and UV to form polymer networks. Improvements in the mixture formulation will ensure that larger tuning ranges can be achieved in the future.

Supplementary Materials: The following supporting information can be downloaded at: https://www.mdpi.com/article/10.3390/cryst12101340/s1, Figure S1. Schematic of the optical layout of the two-photon polymerisation direct laser writing system. The main subsystems are highlighted by dashed rectangular frames. The red lines refer to the path of excitation for the two-photon absorption (TPA) process at λ = 780 nm as the beam from the Ti:Sapphire laser propagates to the LC samples. The yellow path represents the illumination of the sample and image capture with the CCD camera. Figure S2. (a) Schematic of the experimental setup used to record the transmission of monochromatic light through the polymerizable LC when sandwiched between crossed polarisers. ND: neutral density filter; PD: photodiode. The light from the laser first travels through a polarizer used to provide linear polarised light before the beam is attenuated by a variable neutral density (ND) filter (Thorlabs NDC-50C-4). The beam then propagates to the LC sample and through an analyzer that is crossed with respect to the first polariser before the intensity is then recorded by a photodiode. The LC sample was positioned at 45° respect to each one of the polarizers. (b) Normalized transmission as a function of voltage for the LC mixture. The transmission decreases to zero when the sample is subjected to a voltage amplitude of 100 Vrms corresponding to a homeotropic alignment of the LC director. Figure S3. Schematic of the optical setup for characterizing the diffractive optical elements. The red solid line in the optical path represents the emission from the laser diode (λ = 635 nm), while the purple dashed line indicates the illumination path from the LED light source with a central wavelength of λ = 660 nm. Figure S4. (a) An example photograph showing the diffraction gratings in the LC when observed using the system presented in Figure S3 with the LED source turned on. (b) An example photograph of the sample when the LED was switched off and the laser diode (used to observe the diffraction pattern) was turned on. (c) An example photograph showing the LC sample when both the LED and the laser diode were switched on. This enables the position of the laser beam relative to the laser-written diffraction grating to be determined. (d) A photograph of the experiment whereby the far-field diffraction pattern can be seen on a white screen. Figure S5. (a) The tilt in the LC director as a function of position within the cross section of a grating period. This shows that the LC director is tilted at $\pi/2$ within the laser written regions but relaxes to a non-homeotropic alignment within the regions intended to be polymerized using UV illumination. (b) The retardance distribution of a laser written grating in an LC sample before UV polymerization and delamination. The image on the left represents the phase retardance induced by the grating with a period of 12 µm without an applied voltage while the image on the right shows the retardance when a voltage of 100 Vrms is applied to the LC. According to the comparison of the retardance distribution without and with a voltage, it's notable that the laser writing process locked-in a periodic phase profile which functions as a phase grating when no voltage is applied but that the grating vanishes when a large voltage was applied. Figure S6. Photographs of the polarizing optical microscope (a) and the replay field (b) for all the extremes in terms of stretching and contracting the film for the three separate cycles. In (a), the fourth order spots are linked by solid white lines to highlight the tuning and reversibility of the film. (b) represents the polarizing optical microscope images which are correspond to relevant far field images. In both methods, it can be observed that the diffraction grating is stretched further after each contraction. Figure S7. Comparison between the initial diffraction pattern and the final extreme in terms of stretching. Table S1. Fabrication parameters in the laser written gratings test.

Author Contributions: Conceptualization, B.C., S.M.M., S.J.E., M.J.B. and P.S.S.; methodology, B.C., Z.Z., C.N., C.H., S.J.E., P.S.S., M.J.B. and S.M.M.; validation, B.C., Z.Z., C.N.,C.H. and S.J.E.; formal analysis, B.C., Z.Z., C.N.,C.H. and S.J.E.; investigation, B.C., Z.Z., C.N., C.H., S.J.E., P.S.S., M.J.B. and S.M.M.; resources, P.S.S., S.J.E., M.J.B. and S.M.M.; data curation, B.C., Z.Z., C.N.,C.H. and S.J.E.; writing—original draft preparation, B.C., Z.Z., C.N., C.H. and S.M.M.; writing—review and editing, B.C., Z.Z., C.N., C.H., S.J.E., P.S.S., M.J.B. and S.M.M.; supervision, S.M.M., S.J.E., M.J.B., C.H. and P.S.S.; project administration, S.M.M. and S.J.E.; funding acquisition, S.M.M., P.S.S., S.J.E. and M.J.B. All authors have read and agreed to the published version of the manuscript.

Funding: This research was part-funded by EPSRC in the UK: grant EP/R004803/01 (EPSRC Fellowship Patrick Salter), grant EP/R511742/1 for an Impact Acceleration project, and grant EP/T517811/1 (a CASE Conversion studentship award with Merck Ltd. for Camron Nourshargh). This research was also supported by the John Fell Fund (Oxford University Press) and The Royal Society (UK) who provided resources for the laser writing facility.

Data Availability Statement: The data presented in this study are openly available in Oxford University Research Archive at 10.5287/bodleian:E92K24a4P, reference number 508697.

Conflicts of Interest: The authors declare no conflict of interest.

References

1. Bonod, N.; Neauport, J. Diffraction Gratings: From Principles to Applications in High-Intensity Lasers. *Adv. Opt. Photonics* **2016**, *8*, 156. [CrossRef]
2. Goncharsky, A.; Goncharsky, A.; Durlevich, S. Diffractive Optical Element for Creating Visual 3D Images. *Opt. Express* **2016**, *24*, 9140. [CrossRef]
3. Levola, T. Diffractive Optics for Virtual Reality Displays. *J. Soc. Inf. Disp.* **2006**, *14*, 467. [CrossRef]
4. Xiong, J.; Yin, K.; Li, K.; Wu, S.-T. Holographic Optical Elements for Augmented Reality: Principles, Present Status, and Future Perspectives. *Adv. Photonics Res.* **2021**, *2*, 2000049. [CrossRef]
5. Loewen, E.G. Diffraction Gratings for Spectroscopy. *J. Phys. E* **1970**, *3*, 201. [CrossRef]
6. He, Z.; Gou, F.; Chen, R.; Yin, K.; Zhan, T.; Wu, S.-T. Liquid Crystal Beam Steering Devices: Principles, Recent Advances, and Future Developments. *Crystals* **2019**, *9*, 292. [CrossRef]
7. Lightman, S.; Bin-Nun, M.; Bar, G.; Hurvitz, G.; Gvishi, R. Structuring Light Using Solgel Hybrid 3D-Printed Optics Prepared by Two-Photon Polymerization. *Appl. Opt.* **2022**, *61*, 1434. [CrossRef]
8. Blanchard, P.M.; Fisher, D.J.; Woods, S.C.; Greenaway, A.H. Phase-Diversity Wave-Front Sensing with a Distorted Diffraction Grating. *Appl. Opt.* **2000**, *39*, 6649. [CrossRef]
9. He, J.; Kovach, A.; Wang, Y.; Wang, W.; Wu, W.; Armani, A.M. Stretchable Optical Diffraction Grating from Poly(Acrylic Acid)/Polyethylene Oxide Stereocomplex. *Opt. Lett.* **2021**, *46*, 5493. [CrossRef]
10. Münchinger, A.; Hahn, V.; Beutel, D.; Woska, S.; Monti, J.; Rockstuhl, C.; Blasco, E.; Wegener, M. Multi-Photon 4D Printing of Complex Liquid Crystalline Microstructures by in Situ Alignment Using Electric Fields. *Adv. Mater. Technol.* **2022**, *7*, 2100944. [CrossRef]
11. Del Pozo, M.; Delaney, C.; Bastiaansen, C.W.M.; Diamond, D.; Schenning, A.P.H.J.; Florea, L. Direct Laser Writing of Four-Dimensional Structural Color Microactuators Using a Photonic Photoresist. *ACS Nano* **2020**, *14*, 9832–9839. [CrossRef] [PubMed]
12. Wang, Y.; Li, Z.; Xiao, J. Stretchable Thin Film Materials: Fabrication, Application, and Mechanics. *J. Electron. Packag.* **2016**, *138*, 020801. [CrossRef]
13. He, Y.; Gao, L.; Bai, Y.; Zhu, H.; Sun, G.; Zhu, L.; Xu, H. Stretchable Optical Fibre Sensor for Soft Surgical Robot Shape Reconstruction. *Opt. Appl.* **2021**, *51*, 589–604. [CrossRef]
14. Tham, N.C.Y.; Sahoo, P.K.; Kim, Y.-J.; Murukeshan, V.M. Ultrafast Volume Holography for Stretchable Photonic Structures. *Opt. Express* **2019**, *27*, 12196. [CrossRef] [PubMed]
15. Xu, M.-J.; Huang, Y.-S.; Ni, Z.-J.; Xu, B.-L.; Shen, Y.-H.; Guo, M.-Q.; Zhang, D.-W. Two-Dimensional Stretchable Blazed Wavelength-Tunable Grating Based on PDMS. *Appl. Opt.* **2020**, *59*, 9614. [CrossRef]
16. Yin, K.; Lee, Y.-H.; He, Z.; Wu, S.-T. Stretchable, Flexible, Rollable, and Adherable Polarization Volume Grating Film. *Opt. Express* **2019**, *27*, 5814. [CrossRef]
17. Simonov, A.N.; Akhzar-Mehr, O.; Vdovin, G. Light Scanner Based on a Viscoelastic Stretchable Grating. *Opt. Lett.* **2005**, *30*, 949. [CrossRef]
18. Oh, J.Y.; Rondeau-Gagné, S.; Chiu, Y.-C.; Chortos, A.; Lissel, F.; Wang, G.-J.N.; Schroeder, B.C.; Kurosawa, T.; Lopez, J.; Katsumata, T.; et al. Intrinsically Stretchable and Healable Semiconducting Polymer for Organic Transistors. *Nature* **2016**, *539*, 411–415. [CrossRef]
19. Tee, B.C.K.; Ouyang, J. Soft Electronically Functional Polymeric Composite Materials for a Flexible and Stretchable Digital Future. *Adv. Mater.* **2018**, *30*, 1802560. [CrossRef]
20. Kim, K.; Park, Y.; Hyun, B.G.; Choi, M.; Park, J. Recent Advances in Transparent Electronics with Stretchable Forms. *Adv. Mater.* **2019**, *31*, 1804690. [CrossRef]
21. Wang, Y.; Liu, X.; Li, S.; Li, T.; Song, Y.; Li, Z.; Zhang, W.; Sun, J. Transparent, Healable Elastomers with High Mechanical Strength and Elasticity Derived from Hydrogen-Bonded Polymer Complexes. *ACS Appl. Mater. Interfaces* **2017**, *9*, 29120–29129. [CrossRef] [PubMed]
22. Brown, D.H.; Das, V.; Allen, R.W.K.; Styring, P. Monodomain Liquid Crystal Elastomers and Elastomeric Gels: Improved Thermomechanical Responses and Phase Behaviour by Addition of Low Molecular Weight LCs. *Mater. Chem. Phys.* **2007**, *104*, 488–496. [CrossRef]
23. Biggins, J.S.; Warner, M.; Bhattacharya, K. Elasticity of Polydomain Liquid Crystal Elastomers. *J. Mech. Phys. Solids* **2012**, *60*, 573–590. [CrossRef]

24. Khosla, S.; Lal, S.; Tripathi, S.K.; Sood, N.; Singh, D. Optical Properties of Liquid Crystal Elastomers. *Am. Inst. Phys.* **2011**, *1393*, 303–304.
25. Shahinpoor, M. Electrically Activated Artificial Muscles Made with Liquid Crystal Elastomers. In Proceedings of the Smart Structures and Materials 2000: Electroactive Polymer Actuators and Devices (EAPAD), New Port Beach, CA, USA, 6–8 March 2000; p. 187.
26. Yu, Y.; Nakano, M.; Ikeda, T. Liquid-Crystalline Elastomers with Photomechanical Properties. In Proceedings of the Optical Science and Technology, the SPIE 49th Annual Meeting, Denver, CO, USA, 2–6 August 2004; p. 1.
27. DeSimone, A.; Gidoni, P.; Noselli, G. Liquid Crystal Elastomer Strips as Soft Crawlers. *J. Mech. Phys. Solids* **2015**, *84*, 254–272. [CrossRef]
28. Xie, P.; Zhang, R. Liquid Crystal Elastomers, Networks and Gels: Advanced Smart Materials. *J. Mater. Chem.* **2005**, *15*, 2529. [CrossRef]
29. Urayama, K. Selected Issues in Liquid Crystal Elastomers and Gels. *Macromolecules* **2007**, *40*, 2277–2288. [CrossRef]
30. Li, M.-H.; Keller, P. Artificial Muscles Based on Liquid Crystal Elastomers. *Philos. Trans. R. Soc. A Math. Phys. Eng. Sci.* **2006**, *364*, 2763–2777. [CrossRef]
31. Wei, R.B.; Zhang, H.; He, Y.; Wang, X.; Keller, P. Photoluminescent Nematic Liquid Crystalline Elastomer Actuators. *Liq. Cryst.* **2014**, *41*, 1821–1830. [CrossRef]
32. Geiger, S.; Michon, J.; Liu, S.; Qin, J.; Ni, J.; Hu, J.; Gu, T.; Lu, N. Flexible and Stretchable Photonics: The Next Stretch of Opportunities. *ACS Photonics* **2020**, *7*, 2618–2635. [CrossRef]
33. Chen, W.; Liu, W.; Jiang, Y.; Zhang, M.; Song, N.; Greybush, N.J.; Guo, J.; Estep, A.K.; Turner, K.T.; Agarwal, R.; et al. Ultrasensitive, Mechanically Responsive Optical Metasurfaces via Strain Amplification. *ACS Nano* **2018**, *12*, 10683–10692. [CrossRef]
34. Simonov, A.N.; Grabarnik, S.; Vdovin, G. Stretchable Diffraction Gratings for Spectrometry. *Opt. Express* **2007**, *15*, 9784. [CrossRef] [PubMed]
35. Yin, D.; Feng, J.; Ma, R.; Zhang, X.-L.; Liu, Y.-F.; Yang, T.; Sun, H.-B. Stability Improved Stretchable Metallic Gratings with Tunable Grating Period in Submicron Scale. *J. Light. Technol.* **2015**, *33*, 3327–3331. [CrossRef]
36. Yin, K.; Lee, Y.; He, Z.; Wu, S. Stretchable, Flexible, and Adherable Polarization Volume Grating Film for Waveguide-based Augmented Reality Displays. *J. Soc. Inf. Disp.* **2019**, *27*, 232–237. [CrossRef]
37. Mahpeykar, S.M.; Xiong, Q.; Wei, J.; Meng, L.; Russell, B.K.; Hermansen, P.; Singhal, A.V.; Wang, X. Stretchable Hexagonal Diffraction Gratings as Optical Diffusers for In Situ Tunable Broadband Photon Management. *Adv. Opt. Mater.* **2016**, *4*, 1106–1114. [CrossRef]
38. Xu, L.; Liu, N.; Ge, J.; Wang, X.; Fok, M.P. Stretchable Fiber-Bragg-Grating-Based Sensor. *Opt. Lett.* **2018**, *43*, 2503. [CrossRef]
39. Ghisleri, C.; Potenza, M.A.C.; Ravagnan, L.; Bellacicca, A.; Milani, P. A Simple Scanning Spectrometer Based on a Stretchable Elastomeric Reflective Grating. *Appl. Phys. Lett.* **2014**, *104*, 061910. [CrossRef]
40. Kowerdziej, R.; Ferraro, A.; Zografopoulos, D.C.; Caputo, R. Soft-Matter-Based Hybrid and Active Metamaterials. *Adv. Opt. Mater.* **2022**, *28*, 2200750. [CrossRef]
41. Sakellari, I.; Yin, X.; Nesterov, M.L.; Terzaki, K.; Xomalis, A.; Farsari, M. 3D Chiral Plasmonic Metamaterials Fabricated by Direct Laser Writing: The Twisted Omega Particle. *Adv. Opt. Mater.* **2017**, *5*, 1700200. [CrossRef]
42. Chi, T.; Somers, P.; Wilcox, D.A.; Schuman, A.J.; Johnson, J.E.; Liang, Z.; Pan, L.; Xu, X.; Boudouris, B.W. Substituted Thioxanthone-Based Photoinitiators for Efficient Two-Photon Direct Laser Writing Polymerization with Two-Color Resolution. *ACS Appl. Polym. Mater.* **2021**, *3*, 1426–1435. [CrossRef]
43. Sandford O'Neill, J.; Salter, P.; Zhao, Z.; Chen, B.; Daginawalla, H.; Booth, M.J.; Elston, S.J.; Morris, S.M. 3D Switchable Diffractive Optical Elements Fabricated with Two-Photon Polymerization. *Adv. Opt. Mater.* **2022**, *10*, 2102446. [CrossRef]
44. Li, G.; Lee, D.; Jeong, Y.; Cho, J.; Lee, B. Holographic Display for See-through Augmented Reality Using Mirror-Lens Holographic Optical Element. *Opt. Lett.* **2016**, *41*, 2486. [CrossRef]
45. Katz, S.; Kaplan, N.; Grossinger, I. Using Diffractive Optical Elements: DOEs for Beam Shaping-Fundamentals and Applications. *Laser Tech. J.* **2018**, *13*, 83–86.
46. Del Pozo, M.; Delaney, C.; Pilz da Cunha, M.; Debije, M.G.; Florea, L.; Schenning, A.P.H.J. Temperature-Responsive 4D Liquid Crystal Microactuators Fabricated by Direct Laser Writing by Two-Photon Polymerization. *Small Struct.* **2022**, *3*, 2100158. [CrossRef]
47. Del Pozo, M.; Sol, J.A.H.P.; Schenning, A.P.H.J.; Debije, M.G. 4D Printing of Liquid Crystals: What's Right for Me? *Adv. Mater.* **2022**, *34*, 2104390. [CrossRef]
48. McCracken, J.M.; Tondiglia, V.P.; Auguste, A.D.; Godman, N.P.; Donovan, B.R.; Bagnall, B.N.; Fowler, H.E.; Baxter, C.M.; Matavulj, V.; Berrigan, J.D.; et al. Microstructured Photopolymerization of Liquid Crystalline Elastomers in Oxygen-Rich Environments. *Adv. Funct. Mater.* **2019**, *29*, 1903761. [CrossRef]
49. Ku, K.; Hisano, K.; Kimura, S.; Shigeyama, T.; Akamatsu, N.; Shishido, A.; Tsutsumi, O. Environmentally Stable Chiral-Nematic Liquid-Crystal Elastomers with Mechano-Optical Properties. *Appl. Sci.* **2021**, *11*, 5037. [CrossRef]
50. Kizhakidathazhath, R.; Higuchi, H.; Okumura, Y.; Kikuchi, H. Effect of Polymer Backbone Flexibility on Blue Phase Liquid Crystal Stabilization. *J. Mol. Liq.* **2018**, *262*, 175–179. [CrossRef]
51. Vaezi, M.; Seitz, H.; Yang, S. A Review on 3D Micro-Additive Manufacturing Technologies. *Int. J. Adv. Manuf. Technol.* **2013**, *67*, 1721–1754. [CrossRef]

52. Dierking, I. Polymer Network-Stabilized Liquid Crystals. *Adv. Mater.* **2000**, *12*, 167–181. [CrossRef]
53. Baldacchini, T. *Three-Dimensional Microfabrication Using Two-Photon Polymerization: Fundamentals, Technology, and Applications*; William Andrew: Norwich, NY, USA, 2015; ISBN 9780323354059.
54. Do, M.T.; Li, Q.; Nguyen, T.T.N.; Benisty, H.; Ledoux-Rak, I.; Lai, N.D. High Aspect Ratio Submicrometer Two-Dimensional Structures Fabricated by One-Photon Absorption Direct Laser Writing. *Microsyst. Technol.* **2014**, *20*, 2097–2102. [CrossRef]
55. Tartan, C.C.; Sandford O'Neill, J.J.; Salter, P.S.; Aplinc, J.; Booth, M.J.; Ravnik, M.; Morris, S.M.; Elston, S.J. Read on Demand Images in Laser-Written Polymerizable Liquid Crystal Devices. *Adv. Opt. Mater.* **2018**, *6*, 1800515. [CrossRef]
56. Sandford O'Neill, J.J.; Salter, P.S.; Booth, M.J.; Elston, S.J.; Morris, S.M. Electrically-Tunable Positioning of Topological Defects in Liquid Crystals. *Nat. Commun.* **2020**, *11*, 2203. [CrossRef]
57. Zhao, Z.; Chen, B.; Salter, P.S.; Booth, M.J.; O'Brien, D.; Elston, S.J.; Morris, S.M. Multi-Element Polychromatic 2-Dimensional Liquid Crystal Dammann Gratings. *Adv. Mater. Technol.* **2022**. [CrossRef]
58. Woska, S.; Münchinger, A.; Beutel, D.; Blasco, E.; Hessenauer, J.; Karayel, O.; Rietz, P.; Pfleging, S.; Oberle, R.; Rockstuhl, C.; et al. Tunable Photonic Devices by 3D Laser Printing of Liquid Crystal Elastomers. *Opt. Mater. Express* **2020**, *10*, 2928. [CrossRef]
59. Wood, S.M.; Castles, F.; Elston, S.J.; Morris, S.M. Wavelength-Tuneable Laser Emission from Stretchable Chiral Nematic Liquid Crystal Gels via in Situ Photopolymerization. *RSC Adv.* **2016**, *6*, 31919–31924. [CrossRef]
60. Castles, F.; Morris, S.M.; Hung, J.M.C.; Qasim, M.M.; Wright, A.D.; Nosheen, S.; Choi, S.S.; Outram, B.I.; Elston, S.J.; Burgess, C.; et al. Stretchable Liquid-Crystal Blue-Phase Gels. *Nat. Mater.* **2014**, *13*, 817–821. [CrossRef]
61. Deshmukh, R.R.; Jain, A.K. Effect of Anti-Parallel and Twisted Alignment Techniques on Various Propertiesof Polymer Stabilised Liquid Crystal (PSLC) Films. *Liq. Cryst.* **2016**, *43*, 436–447. [CrossRef]
62. Harvey, J.E.; Pfisterer, R.N. Understanding Diffraction Grating Behavior: Including Conical Diffraction and Rayleigh Anomalies from Transmission Gratings. *Opt. Eng.* **2019**, *58*, 1. [CrossRef]
63. Harvey, J.E.; Pfisterer, R.N. Understanding Diffraction Grating Behavior, Part II: Parametric Diffraction Efficiency of Sinusoidal Reflection (Holographic) Gratings. *Opt. Eng.* **2020**, *59*, 1. [CrossRef]
64. Wang, Y.; He, H.; Chang, J.; He, C.; Liu, S.; Li, M.; Zeng, N.; Wu, J.; Ma, H. Mueller Matrix Microscope: A Quantitative Tool to Facilitate Detections and Fibrosis Scorings of Liver Cirrhosis and Cancer Tissues. *J. Biomed. Opt.* **2016**, *21*, 71112. [CrossRef]
65. He, C.; He, H.; Chang, J.; Chen, B.; Ma, H.; Booth, M.J. Polarisation Optics for Biomedical and Clinical Applications: A Review. *Light Sci. Appl.* **2021**, *10*, 194. [CrossRef] [PubMed]
66. He, C.; Chang, J.; Salter, P.S.; Shen, Y.; Dai, B.; Li, P.; Jin, Y.; Thodika, S.C.; Li, M.; Tariq, A.; et al. Revealing Complex Optical Phenomena through Vectorial Metrics. *Adv. Photonics* **2022**, *4*, 026001. [CrossRef]

MDPI
St. Alban-Anlage 66
4052 Basel
Switzerland
Tel. +41 61 683 77 34
Fax +41 61 302 89 18
www.mdpi.com

Crystals Editorial Office
E-mail: crystals@mdpi.com
www.mdpi.com/journal/crystals